D0007397

Approaches to Gene Mapping in Complex Human Diseases

Approaches to Gene Mapping in Complex Human Diseases

Edited by

Jonathan L. Haines

Department of Molecular Physiology and Biophysics
Program in Human Genetics
Vanderbilt University School of Medicine
Nashville, TN

Margaret A. Pericak-Vance

Section of Medical Genetics
Center for Human Genetics
Duke University Medical Center
Durham, NC

A John Wiley & Sons, Inc., Publication

New York • Chichester • Weinheim • Brisbane • Singapore • Toronto

This text is printed on acid-free paper. ♾

Published simultaneously in Canada.

Library of Congress Cataloging in Publication Data:

Approaches to gene mapping in complex human diseases / edited by
Jonathan L. Haines and Margaret A. Pericak-Vance.
p. cm.
Includes bibliographical references and index.
ISBN 0-471-17195-6 (cloth : alk. paper)
1. Genetic disorders. 2. Human gene mapping. I. Haines,
Jonathan L., 1957– . II. Pericak-Vance, Margaret ann.
[DNLM: 1. Hereditary Diseases—genetics. 2. Chromosome Mapping,
3. Genetic Markers. 4. Genotype. QZ 50 A653 1998]
RB155.5.A67 1998
616'.042—dc21
DNLM/DLC
for Library of Congress 97-42666

Printed in the United States of America

10 9 8 7 6 5 4 3 2

Contents

Foreword

The promise of the Human Genome Initiative that rapid automated genotyping would become available for genetic mapping has been fulfilled. Hundreds of simple Mendelian traits have now been mapped, and in the majority of cases the responsible gene has been cloned.

From a historical perspective, the evolution of mapping techniques began with Morton in 1955 with his now classic paper "Sequential Tests for the Detection of Linkage." The method was amenable only to sibships, though with much tedious calculation, large families could be analyzed. The next milestone occurred in 1971, when Elston and Stewart published their efficient algorithm to determine the likelihood of a pedigree, which became the basis of the well-known computer program LIPED written by Ott.

The weak link was the number of markers available, in the order of 30–40 blood groups and serum proteins, which were tedious to genotype and in general were not very polymorphic. The 1980 publication by Botstein and his colleagues, which advocated the use of restriction fragment length polymorphisms (RFLPs) as markers for the construction of genetic linkage maps ushered in a new era. The pace of human gene mapping increased exponentially. Though RFLPs were a major advance in genotyping, they were overshadowed by the discovery by Weber and May in 1989 of microsatellites, which are both abundant and highly polymorphic. These findings have culminated in the mapping and gene cloning of the great majority of the major Mendelian disorders in man.

Geneticists have realized for many years that many of the common disorders that affect man have a major genetic component. Examples are manic depression and schizophrenia, adult diabetes, osteoporosis, and hypertension. With the advent of dense maps, gene mappers felt that they could, in theory at least, unravel the genes responsible for a major component of these disorders. Many hurdles needed to be surmounted, however. The use of large multigeneration families in early linkage studies of bipolar disorder and schizophrenia produced what were later found to be spurious linkages. It soon became clear that new approaches were necessary if mapping genes involved in complex disorders was to be successful. A number of theoretical population geneticists soon rallied to the task leading to a number of innovative new approaches of mapping genes for complex human diseases. New or modified approaches began appearing in numerous scientific journals.

These approaches have been assimilated into this book. The editors have assembled experts in the field to provide scientists with a comprehensive guide to human disease gene mapping in complex diseases in one volume. Most publications on gene mapping in complex disorders tend to emphasize the analytic techniques. This volume on the other hand covers all areas, not just the statistical methodology. These include the overall design, the clinical phenotype and subphenotypes, family and sib pair ascertainment, computer software, data analysis, and interpretation. The decision not to include the detailed theoretical background for the linkage analysis will be seen as a boon to most of us, who are mainly interested in the application of the techniques. This volume will, hopefully, allow researchers to avoid the many pitfalls involved in complex linkage analysis. All major areas are covered and should be studied before a researcher embarks on complex linkage analysis. All of us should be grateful to the contributors for providing in a straightforward readable fashion all the key elements involved in finding genes for common/complex diseases.

I take great pleasure in writing a foreword to this volume. As one who has been involved in human gene mapping from its early beginnings, it is especially gratifying to see two of my former students produce a comprehensive and up-to-date volume in this rapidly advancing field. They allow me and, hopefully, many others to understand the entire process involved in the mapping of complex human disease genes.

P. Michael Conneally
Distinguished Professor
Department of Molecular and Medical Genetics
Indiana University Medical School
Indianapolis, IN

Preface

We have written this book to capture the state of the art in the emerging field of gene mapping in common and genetically complex traits. While we cannot hope to provide a completely comprehensive review of all the background, methods, and designs, we have tried to discuss the most useful and most common approaches to this most difficult of problems. It is amazing to note that just 25 years ago at the first Human Gene Mapping Conference in 1973 (New Haven Conference, 1973), fewer than 100 genes of any type had been mapped to any chromosome and several chromosomes had no mapped genes at all. It was only 15 years ago that the first disease gene was mapped using recombinant DNA methods (Gusella et al., 1983). Thus virtually all the methods and approaches we describe have been developed over the past 15 years, and most of these in the past 5 years. Hundreds of Mendelian disease genes have now been mapped, leading to major breakthroughs in our understanding of these diseases. While we have learned much from these endeavors, mapping the genes for more common traits has presented an even greater challenge. Only recently has the torrent of new resources arising from the Human Genome Initiative and the subsequent development of new statistical methods made it possible to map these genes. The emerging interest in mapping such genes is driven by several factors:

1. Understanding genetic forms of common disorders can provide insight into the pathogenesis of the environmental etiologies of these same diseases.
2. Controlling for known genetic susceptibility will improve our ability to identify and characterize additional genes, other risk factors, and gene–gene and gene–environment interactions.
3. Applying successful therapeutic intervention may depend on the underlying genotype of the susceptible individual.

This book has grown out of a 4-day course on mapping genetically complex diseases that we have been teaching since 1994. Our goal is to provide a comprehensive overview of common and genetically complex disease gene mapping, as well as a guide to the often difficult design aspects, to allow interested readers (physician-scientists, students, and other researchers) to understand the entire mapping process without getting too heavily into the statistical and laboratory details. We hope readers will gain a healthy respect for the fact that there are as many ways to study common and genetically complex diseases as there are genes underlying them.

It would not have been possible to put together this book without substantial help from many individuals, not the least of whom are all our coauthors. We thank Cheryl Pizzano and Annie Bernard for their help in typing the manuscripts, and Jason Stajich, Casey Crotty, Kathryn Jones, and Stephen Brown for their help with the illustrations. We also thank John Rogus, Daniel Weeks, Suzanne Leal, and Susan Albright for commenting on and reviewing some of the chapters. We especially thank our editor, Ann Boyle, for her immense patience with the delays, changes, and problems that accompanied the realization of this project.

Jonathan L. Haines
Margaret A. Pericak-Vance

REFERENCES:

Gusella JF, Wexler NS, Conneally PM, Naylor SL, Anderson MA, Tanzi RE, Watkins PC, Ottina K, Wallace MR, Sakaguchi AY, et al. (1983) A polymorphic DNA marker genetically linked to Huntington's disease. *Nature* 306:234–238.

New Haven Conference (1973): First International Workshop on Human Gene Mapping. Birth Defects Original Article Series X: 3, 1974. New York: The National Foundation.

Contributors

ARTHUR S. AYLSWORTH, Division of Genetics and Metabolism, Department of Pediatrics and the Neuroscience Research Center, The School of Medicine
The University of North Carolina at Chapel Hill, Chapel Hill, North Carolina

COLETTE BLACH, Section of Medical Genetics, Center for Human Genetics, Duke University Medical Center, Durham, North Carolina

PAMELA E. COHEN, Pacific Southwest Regional Genetics Network, California Department of Health Services, Berkeley, California

L. ADRIENNE CUPPLES, Department of Epidemiology and Biostatistics, Boston University School of Medicine, Boston, Massachusetts

LINDSAY A. FARRER, Departments of Medicine, Epidemiology, and Biostatistics, Boston University School of Medicine, Boston, Massachusetts, and Department of Neurology, Harvard University School of Medicine, Boston, Massachusetts

DAVID E. GOLDGAR, Unit of Genetic Epidemiology, International Agency for Research on Cancer, Lyon, France

JONATHAN L. HAINES, Department of Molecular Physiology and Biophysics, Program in Human Genetics, Vanderbilt University School of Medicine, Nashville, Tennessee

CAROL HAYNES, Section of Medical Genetics, Center for Human Genetics
Duke University Medical Center, Durham, North Carolina

DOUGLAS A. MARCHUK, Department of Genetics, Duke University Medical Center, Durham, North Carolina

DEBORAH A. MEYERS, Center for the Genetics of Asthma and Complex Diseases, University of Maryland School of Medicine, Baltimore, Maryland

KAMEL BEN OTHMANE, Division of Neurology, Department of Medicine, Duke University Medical Center and Health System, Durham, North Carolina

MARGARET A. PERICAK-VANCE, Section of Medical Genetics, Center for Human Genetics, Duke University Medical Center, Durham, North Carolina

JOELLEN M. SCHILDKRAUT, Community Family Medicine, Duke University Comprehensive Cancer Center, Durham, North Carolina

MARCY C. SPEER, Section of Medical Genetics, Center for Human Genetics, Duke University Medical Center, Durham, North Carolina

JEFFERY M. VANCE, Division of Neurology, Department of Medicine, Duke University Medical Center and Health System, Durham, North Carolina

CHANTELLE WOLPERT, Division of Neurology, Center for Human Genetics, Duke University Medical Center, Durham, North Carolina

JIANFENG XU, Center for the Genetics of Asthma and Complex Diseases, University of Maryland School of Medicine, Baltimore, Maryland

Approaches to Gene Mapping in Complex Human Diseases

1

Overview of Mapping Common and Genetically Complex Human Disease Genes

Jonathan L. Haines
Department of Molecular Physiology and Biophysics
Program in Human Genetics
Vanderbilt University School of Medicine
Nashville, Tennessee

Margaret A. Pericak-Vance
Section of Medical Genetics
Center for Human Genetics
Duke University Medical Center
Durham, North Carolina

INTRODUCTION

We are nothing but a ragbag of disappeared ancestors
—Mark Twain

Disease gene mapping in humans has a long history, predating even the identification of DNA as the genetic molecule (Watson and Crick, 1953), and the determina-

Approaches to Gene Mapping in Complex Human Diseases, Edited by Jonathan L. Haines and Margaret A. Pericak-Vance. ISBN 0-471-17195-6

tion of the number of human chromosomes (Ford and Hamerton, 1956; Tjio and Levan, 1956). In fact, as early as the 1930s some simple statistical methods had been developed (Bernstein, 1931; Fisher, 1935). However, these methods were severely limited in their application. Not only were genetic markers lacking (the ABO blood type was one of the few that had been described), but these methods were restricted to small nuclear pedigrees, perhaps including grandparents. Any calculations had to be done by hand of course, making analysis very laborious.

There were two hurdles to be overcome before human gene mapping could be performed routinely. First, appropriate statistical methods were lacking, as were ways of automating the laborious calculations of the statistics. Second, sufficient genetic markers to cover the entire human genome had to be developed. Morton (1955), building on the work of Haldane and Smith (1947) and Wald (1947), described the use of maximum likelihood approaches in a sequential test to test for linkage between two loci. He used the term "lod score" (for logarithm of the odds of linkage) for his test statistic. This score is the basis of most linkage analyses being performed today, and it represents a milestone in human gene mapping. However, the complex calculations had to be done by hand, severely limiting the use of this approach. In 1971, Elston and Stewart described a general approach for calculating the likelihood of any non-consanguineous pedigree. This was extended by Lange and Elston (1975) to include pedigrees of arbitrary complexity. Soon thereafter the first general-purpose computer program for linkage in humans, LIPED (Ott, 1976), was described. Thus the first of the two major hurdles had been overcome.

By the mid-1970s there were 40–50 red cell antigen and serum protein polymorphisms available as genetic markers. A few markers could be arranged into some initial linkage groups, but these markers covered only approximately 5–15% of the human genome. In addition to this limited coverage, genotyping these polymorphisms was labor-intensive, time-consuming, and often quite technically demanding. This remaining hurdle was crossed with the description of restriction fragment length polymorphisms (RFLPs) by Botstein et al. (1980). Not only are these markers easier to genotype in a standard manner, they occur with great frequency throughout the genome, opening up the remaining 85–95% of the human genome for the first time.

With these tools in place, the field of human disease gene mapping blossomed. The first successful linkage using RFLPs was reported by Gusella et al. (1983), mapping the Huntington disease gene to chromosome 4p. This was the beginning of the approach to disease gene identification often termed "positional cloning." Table 1.1 presents a list of the first few disease genes to be identified through positional cloning. It is noteworthy that all these diseases are inherited in a simple manner: autosomal dominant, autosomal recessive, or X-linked (i.e., Mendelian inheritance).

This book is designed to introduce the reader to the problems inherent in mapping genes for common and genetically complex human traits, and to approaches to solving those problems. We define a common and genetically complex trait or disease as one that is not inherited in an easily determined manner. The inheritance pattern for traits such as Huntington disease (autosomal dominant), cystic fibrosis (autosomal recessive), and Duchenne muscular dystrophy (X-linked recessive) can be

TABLE 1.1 Some Early Successes in Positional Cloning

Disease	Year Linked	Ref.	Year Cloned	Ref.
Chronic granulomatous disease	1986	Baehner et al.	1986	Royer-Pokora et al.
Duchenne muscular dystrophy	1982	Murray et al.	1987	Koenig et al.
Cystic fibrosis	1985	Tsui et al.	1989	Rommens et al.
Nephrogenic diabetes insipidus	1987	Knoers et al.	1989	Bichet et al.
Neurofibromatosis type I	1987	Barker et al.	1990.	Wallace et al.
Hyperkalemic periodic paralysis	1990	Fontaine et al.	1990	Fontaine et al.
Retinitis pigmentosa type 4	1989	McWilliam et al.	1990	Dryja et al.
Alport syndrome	1984	Menlove et al.	1990	Barker et al.
Early-onset Alzheimer disease	1987	St. George-Hyslop et al.	1991	Goate et al.
Charcot–Marie–Tooth type 1A	1989	Vance et al.	1991	Lupski et al.
Marfan syndrome	1990	Kainulainen et al.	1991	Dietz et al.
Retinitis pigmentosa type 7	1991	Farrar et al.	1991	Farrar et al.
Kennedy disease	1986	Fischbeck et al.	1991	La Spada et al.
Fragile X syndrom	1983	Filippi, et al.	1991	Yu et al.
Familial polyposis coli	1987	Bodmer et al.	1991	Kinzler et al.
Myotonic dystrophy	1954, 1971, 1983	Mohr, Renwick, et al. Eiberg et al.	1992	Harley et al.
Charcot–Marie–Tooth 1B	1980	Bird et al.	1993	Hyazaka et al.
X-Linked Charcot-Marie-Tooth Diease	1985	Gal et al.	1993	Bergoffen et al.
Huntington disease	1983	Gusella et al.	1993	HDCRG

easily and correctly explained by the action of mutation(s) in a single major gene. This action (usually referred to as the *genetic model* underlying the trait) is described in terms of disease allele frequency, the relationship between genotype and phenotype, and any mutation rate. Although confounding factors such as genetic heterogeneity, variable penetrance, and phenocopies may exist, it is generally possible, with a known genetic model in hand, to determine the best and most efficient approach to mapping the responsible gene. However, the inheritance patterns for traits such as Alzheimer disease, multiple sclerosis, insulin-dependent diabetes, and hypertension (with the exception of some rare subtypes) do not fit any simple genetic explanation, making it far more difficult to identify the best approach to mapping the unknown underlying effect. In addition to the confounding factors involved in single gene disorders such as genetic heterogeneity and phenocopies, gene–gene and gene–environment interactions must be considered when a complex trait is dissected.

With the attention now being paid to genetically complex traits, it is important to change our thinking about what, exactly, we are trying to find. No longer are we simply searching for the one or few rare mutations in a single gene that cause a rare and devastating disease. We are now searching for alterations in more than one gene that alone or in concert (and we do not know which) either increase or decrease the risk of developing a trait. Such alterations may not be rare at all, but could be common polymorphisms. For example, the risk for late-onset Alzheimer disease is increased up to 20-fold for individuals carrying two copies of the apolipoprotein E-4 (*APOE-4*) allele, while the risk may be cut in half for those carrying one copy of the *APOE-2* allele. The frequency of the *APOE-4* allele in the general population is about 16%, while that for the *APOE-2* allele is 7% (Corder et al., 1993, 1994; Saunders et al., 1993).

THE COMPONENTS OF A MAPPING STUDY

Each genetically complex trait has its own peculiarities that require special attention. However, a guiding paradigm can be applied to most situations. Over the past 18 years, the general paradigm that has emerged for application to single gene disorders has been termed *positional cloning.* While this term has sometimes been used to refer only to the physical mapping and actual cloning of the disease-causing mutation, we use it here in the more general sense, that is, to include the complete process from initial study design through mutational analysis.

The classical positional cloning approach consists of identifying families, collecting blood samples, genotyping markers, and performing analyses for initial localization, fine genetic mapping, physical mapping, and mutation identification. This process is essentially linear, although there are some steps that are highly recursive in nature, particularly in the genotyping, initial localization, and fine mapping steps. This property has made it easier to apply positional cloning as a general approach to many different Mendelian traits.

In contrast to the classical approach, the current approaches to mapping common and genetically complex traits are not linear, and many steps are works in progress, subject to further defining, refining, or replacement. Figure 1.1 is a diagrammatic interpretation of the steps as we see them today. Each of these steps has its own peculiarities and key factors that must be considered. Particularly for complex traits, the answers and paths that define the best approach will be trait specific. This fact is perhaps underappreciated, and it contrasts highly with the classical positional cloning approach. Thus the amount of confusion surrounding the issue of how to approach complex traits is not entirely surprising.

This section discusses each of the steps in Figure 1.1. The purpose is not a detailed description of each step, but a review of the major points of consideration and a guide to the chapters that cover these points in more detail.

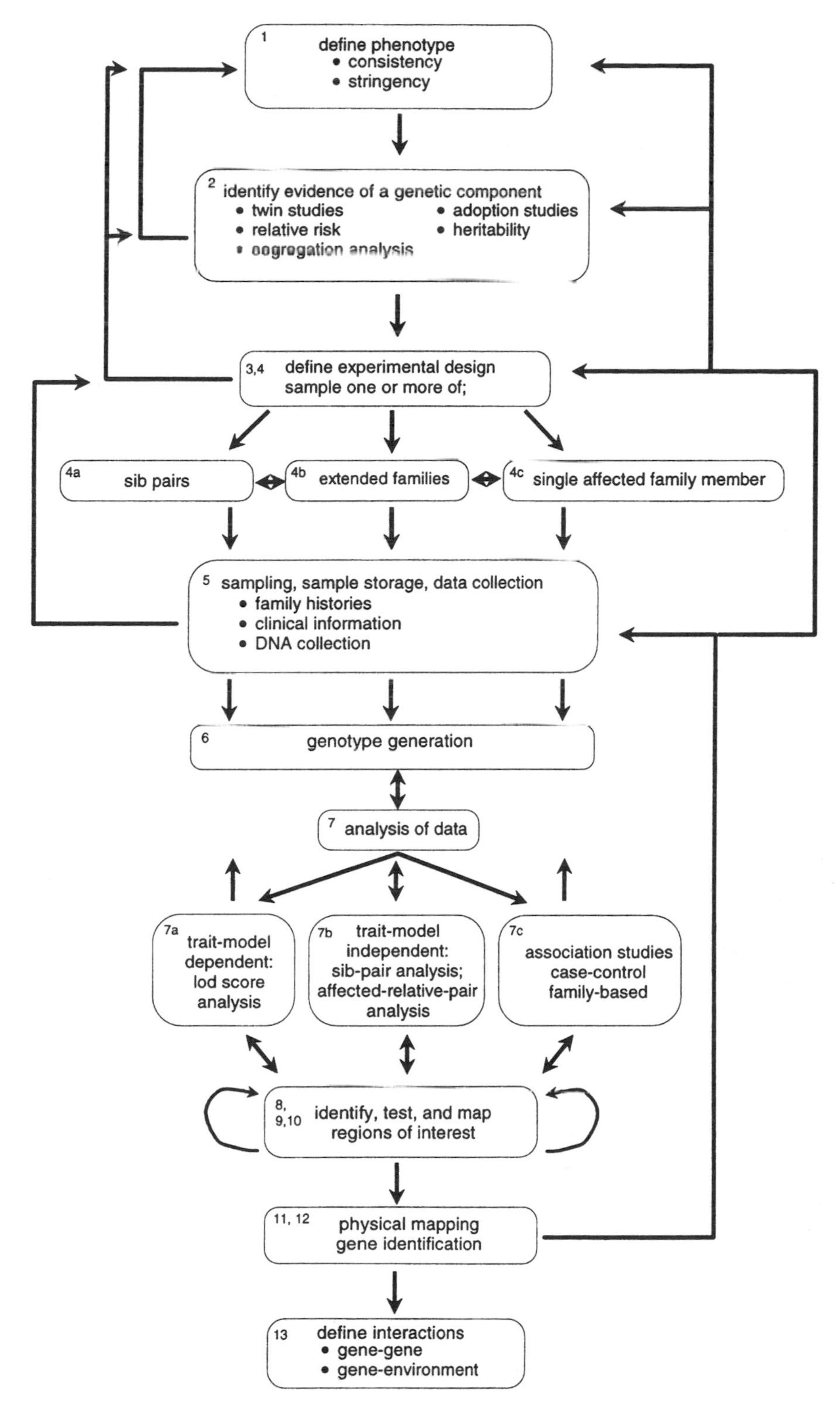

Figure 1.1 *The intricate mapping approach needed for common and genetically complex traits.*

1. Define Phenotype

The first step in any mapping process is to know what is to be mapped. This may sound trivial, but many factors must be considered. These can be broken down into two categories: determining the clinical parameters and determining what clinical data to collect.

Determining the Clinical Parameters It is not enough to define a trait as "Huntington disease" or "diabetes." In the former case there can be wide variation in the symptoms, with some only psychological or very mild motor disturbances, and the age at examination plays a critical role. In the latter case there are known subtypes [insulin-dependent (IDDM) vs. non-insulin-dependent (NIDDM)], as well as age effects. In addition, blood glucose levels (a quantitative trait) are strongly associated with the diabetes (a qualitative trait) and could perhaps be used as a surrogate measure. One critical role of the clinician in the study design is to assess the various diagnostic procedures and tools and determine which ones best define a consistent phenotype. In addition, mapping genetically complex diseases usually requires large data sets, to supply enough power to unravel genetic effects. For this reason family data collection often extends across multiple centers. It is critical when this occurs that consensus diagnostic criteria be established a priori and used by all centers. For example, the establishment of a consensus diagnostic scheme played an important role in a successful complex disease linkage study in late-onset familial Alzheimer disease (AD) (McKhann et al., 1984).

Determining the Clinical Data to Collect The phenotype assignment must be done in a rigorously consistent fashion. Even a few misassigned phenotypes may have major negative implications in the analysis, possibly leading to both false positive and false negative results. For example, given the recombination fraction (θ) of 0.01 and an estimated misclassification rate of 10%, twice as much data must be sampled to obtain the same power of analysis as would be needed if no misclassification existed (Ott, 1992). Thus which data will be used to assign the trait status must be determined. Must clinical records be obtained and reviewed for consistency on every patient? Is the report of a patients relative sufficient? Is a note in a medical record sufficient? Or is direct examination for the study required? The goal is to minimize any source of error and uncertainty. This is discussed in more detail in Chapter 3.

2. Determine That the Trait Has a Genetic Component

It is critical that as much as possible be known about the genetic basis of a trait for linkage studies before the expensive and time-consuming process of data collection and genotyping is begun. That a trait "runs in families" is insufficient evidence, since this phenomenon can occur for several reasons, including common environmental exposure and biased ascertainment, as well as a true genetic disposition. There are numerous lines of evidence that can be examined, including family studies, segregation analysis, twin studies, adoption studies, heritability studies, and population-based risks to relatives of probands. For most traits being contemplated, some such data already exist in the literature, and a thorough and considered review of this literature may provide most of the necessary information, and point out any missing data. The data may not only indicate the strength of the genetic effect on the trait, but also give some indication of the underlying genetic model. For example, there may be obvious evidence of a single "major" gene, such as in Huntington disease, or multiple genes interacting in complex ways, such as in multiple sclerosis (Sadovnick et al., 1996). This is discussed in more detail in Chapter 5.

3. Develop Experimental Design

Steps 1 and 2 are not independent. Review of the available data may indicate that a trait as originally defined has little, or even no, evidence of a genetic component. However, there may be strong evidence that a particular subform of the trait is strongly genetic. For example, there had for many years been debate about the role of genetics in Alzheimer disease. However, it became increasingly clear that a subform of AD with onset before age 65 existed and is, although rare, strongly influenced by the action (within families) of a single gene. By restricting further ascertainment and analysis to families with such early-onset AD, three different single genes have now been identified (Goate et al., 1991; Levy-Lehad et al., 1995; Rogaev et al., 1995; Sherrington et al., 1995).

The exact approach to the mapping process should be outlined as completely as possible before the project gets under way. With the clinical phenotype in hand, it is possible to determine the best strategy for steps 4–7 below as described in Chapter 11. Underlying this entire process is the need for keeping track of the immense amount of data being collected. Approaches to databasing are discussed in Chapter 10.

4. Ascertain Families

Family ascertainment is perhaps the longest and most labor-intensive step in the entire process. Determination of the type(s) of family to collect (e.g., sib pairs, cousin pairs, extended families) is based on the knowledge of phenotype and any evidence for the genetic model. For example, if a single major gene is suspected, large extended families are likely to prove most efficient. For a more complex genetic model, small families, such as sib pairs, may prove most efficient, whereas if a trait phenotype could be defined in numerous ways, a restrictive set of criteria may limit the available family material to a single type (most likely sib pair).

It is imperative that the ascertainment of families proceed with careful consideration of the wishes of the participating families, and that their rights of participation/nonparticipation and confidentiality be protected. There are many sources of potential ascertainment, such as support groups, hospital clinics, and private referrals. These issues are discussed in detail in Chapter 6.

It is also important to have some sense of the sample size required to identify the genes being sought. For single gene disorders, formal power studies using standard simulation programs are possible because the underlying genetic model can be assigned with reasonable confidence. For genetically complex traits, however, the underlying genetic model is usually unknown; thus simulation studies can only give general guidelines to the potential detection of genes. Chapter 7 discusses sample power.

5. Collect Biological Samples

In most cases, collecting biological samples is done simultaneously with family ascertainment (Chapter 6). This usually consists of obtaining 10–40 mL of blood, from which DNA is extracted. However, other options, such as initiating cell lines, using buccal brushes, or extracting DNA from pathological tissue specimens, may be preferable under some circumstances. These methods are discussed in Chapter 8.

6. Genotype Samples

The specific approaches toward genotype generation are discussed in Chapter 9. It is important to note that genotype generation should not move forward independently of analysis, since many laboratory errors, such as sample mix-ups or contamination, may be detected only in the analytical process. Data generation also takes very different tacks depending on the goal. For an initial genomic screen (Chapters 9 and 11), high-throughput, highly automated genotyping using a standard set of markers is desirable. However, for follow-up or saturation genotyping, a less automated approach may prove more efficient for more intensive efforts.

7. Analyze the Genotypic Data

The analysis of the resulting clinical and genotypic data is itself complex, and refinements of current and development of new methods are constantly becoming available. The type(s) of analyses chosen depend on the type of trait, the type of families to collect, the likely underlying genetic model, and the potential power of the dataset. The various methods of analysis include genetic model-dependent (lod score) (Chapter 12), sib pair analysis (Chapter 13), affected relative pair analysis (Chapter 14), and association studies (Chapter 15). These are not mutually exclusive approaches, and the best overall design may include multiple techniques. In addition, the choice of method may depend on the stage of the mapping process. For example, sib pair analysis may be used in the initial genomic screening, but the detailed analysis of specific genomic regions may use association studies. Association studies, in particular, are being further developed and extended, in preparation for the plethora of new single nucleotide polymorphisms that are likely to appear in the near future as the microchip genotyping technology matures (Chapters 9 and 15).

8. Identify Genomic Regions of Interest

The initial genomic screen is only the first step in analyzing the genotypic data to locate the underlying genes. There has been much discussion of what constitutes a "significant" result from a genomic screen, with different opinions about the appropriate P values to be used. This issue is discussed in more detail in Chapter 11. However, the goal from this stage of analysis is not a confirmed gene localization (a very happy event, should it occur!), but a small subset of genomic regions that *might* harbor such genes. These regions must then be explored in much more detail, perhaps using analytical techniques different from those used for the initial genomic screen.

9. Retest Genomic Regions of Interest

How best to follow up specific genomic regions is an area of much discussion and methodological development. The primary goals of this process—to eliminate any false positive results and to confirm any true positives—can be achieved via two approaches. The first is to maximize the amount of genetic (e.g., segregation) information in the families. This is usually done by genotyping additional markers within the proposed region(s) so that each person is fully informative for that region. If this region is truly involved in the etiology of the trait, then the support for this region should be increased. The second is to test this specific hypothesis in a second independent data set. Independence does not necessarily require collection by a different group of researchers (although this is certainly possible); rather, it means using families and data not used for the initial exploration. It should be noted that this is often called a "replication" data set, although the goal is *not* replication, but rather the testing of the specific hypothesis that this region harbors a susceptibility gene. The methods used in this step are often the same, although potentially more detailed than those used for the initial genomic screen. This is discussed in Chapter 11.

10. Fine-Map Genomic Region(s) to Identify the Critical Region

Once one or a few regions have been confirmed as harboring susceptibility genes, the goal is to narrow that region as much as possible. For single gene disorders (e.g., autosomal dominant, recessive, or X-linked traits), identification of the critical recombinants defining the region is possible (Chapter 12). For common and genetically complex traits, however, the potential complexity of the underlying genetic model prevents the interpretation of any single recombinant individual in the same manner. The most common methods for fine-mapping a region are:

(a) To examine the region for any known genes that might have a biological function related to the trait. If such a gene exists, it should be tested using the methods described in Chapter 17.

(b) To perform saturation genotyping of the region and analyze the data not only for linkage, but also for allelic association. Any evidence of an allelic association could suggest either a direct action of the polymorphism in question (e.g., a functional polymorphism within a gene), or linkage disequilibrium to such a functional difference. This is discussed in more detail in Chapter 15.

11. Identify Physical Maps

Once a region has been sufficiently narrowed, the switch may be made to the wet laboratory aspects of gene identification. The term "sufficiently narrowed" has changed its meaning somewhat over time. Early in the history of positional cloning, it often meant a region of no more than 100,000–300,000 base pairs of DNA. In general, this is quite a bit less than 1 centiMorgan on a genetic map—ambitious for single gene disorders, well nigh impossible for genetically complex traits. Currently, regions spanning 1–2 million base pairs of DNA are considered reasonable. As the results of the Human Genome Initiative continue to accumulate, the size of the region considered "reasonable" will continue to expand.

The first step in physical map identification is to determine whether or not such maps already exist or can be built quickly using existing data from public databases (Chapter 16). If not, these maps may have to be constructed (Chapter 17).

12. Gene Identification

When the physical maps are available, the primary goal is to identify any genes that reside within the target region. At this point it is likely that many of these genes will not have been identified in any manner, some will be identified only as uncharacterized "ESTs" (Chapter 17), while a very few will have known function. Again, the first place to look is in the public databases (Chapter 16). Various techniques for gene transcript identification are discussed in Chapter 17.

Even with the forward charge of the Human Genome Initiative, only small pieces of most genes will have been identified, with little or no characterization (of proposed function, protein structure, or even gene intron/exon structure) available. Pending the compilation of such data, it is incumbent upon the researcher to generate it.

With genes in hand, it is necessary to examine them for variations that are specifically associated with the trait. For single gene disorders, this has been a laborious but straightforward process of identifying an extremely rare mutation that usually severely disrupts the normal function of the gene. However, for common and genetically complex traits, the variation may not be rare (in fact, it may be a common polymorphism), and there may be no apparent deleterious function on the gene in question. If this is the case, it becomes difficult to "know" that the right variation has been pinpointed. Clues can be taken from the strength of any allelic association (Chapter 15) found, but the ultimate proof comes with testing the function of the gene in biological systems.

While most positional cloning efforts have claimed success based on finding rare mutations in a gene this is, strictly, not sufficient evidence. More conclusive is evidence arising from biological systems (e.g., cultured cells, animal models, or human trials) that the trait can be either induced by introduction of the mutation or ameliorated by blocking the action of the abnormal gene. In genetically complex traits, where the responsible variation may be a common polymorphism, it is even more critical that such evidence be found before success is declared.

Tests in biological systems can be of several types. Perhaps the most common is to test the action of the gene in mice. With transgenic mice the proposed mutation is introduced into the germ line of the mouse, and the resulting offspring are examined for evidence of the abnormal phenotype. With knockout mice the action of the gene in question is eliminated and the offspring are examined for evidence of an abnormal phenotype. Similar experiments may be performed in cultured cells, where the control of the mutation or knockout process is generally easier. The cellular phenotype may be difficult to discern, however, and its relationship to the overt trait phenotype in humans may be remote.

13. Define Interactions

Another critical component in the dissection of a genetically complex disease is an understanding of potential interactions between the genes that underlie the trait, and between genes and other risk factors (usually environmental). This is perhaps the least well developed of any of the steps, as it is only now becoming possible to identify and examine more than one gene (and/or risk factor) for complex diseases. This step also requires integration of the techniques used in both genetics and epidemiology, a process that is just now developing. Current approaches to examining interactions are discussed in Chapter 18.

KEYS TO A SUCCESSFUL STUDY

The feasibility of using the linkage approach for the dissection of genetically complex disease depends on a number of different variables. These include the strength of the genetic component, the number of genes involved and the magnitude of their individual effects, and the consistency of diagnosis, together with the frequency of phenocopies and the amount, type, and power of the available family material. Although every component of a mapping study is important, there are two overriding keys to success in mapping common and genetically complex human traits. These points, often overlooked in the rush to initiate studies, must be carefully considered if the study is not later to grind to a halt while these matters are belatedly addressed.

Foster Interaction of the Necessary Expertise

To appropriately carry out any gene mapping study, one must use techniques from three different fields of expertise. These fields are clinical, molecular/wet laboratory, and genetic epidemiology. The first provides the necessary diagnostic and patient ascertainment skills needed to define the phenotype and help collect samples. The second provides the genotyping and cloning skills necessary to help locate and identify the gene(s) of interest. The third provides the statistical and analytical framework for the proper design of the study, and the analysis of the generated data.

During the 15-year history of mapping and cloning genes primarily for single gene disorders, the limited number of possible underlying genetic modes of action has allowed the development of a generally applicable paradigm that can be implemented by researchers having expertise in only one of the three fields. Appropriate and perhaps only occasional consultation with experts in the other fields may be necessary. When common and genetically complex traits are to be mapped, however, no generally applicable paradigm is possible because the underlying genetic modes of action are likely to be different for each trait. Thus experts in each of these fields must be intimately involved in all aspects of the study. Even with this expertise in place, it is essential that the study not be divided into separate parts with little interaction between the various researchers. For example, genetic epidemiologists should be involved in the discussion of the clinical phenotype to determine the effect of potential changes to the definition of the phenotype on the power to detect any underlying genes.

Develop a Careful Study Design

It may seem self-evident that a careful study design is necessary for a successful study. However, it is not enough to decide on a general design of "collect sib pairs, genotype, analyze using sib pair statistical methods." As is explained in the following chapters, each step in the process requires substantial thought, and the decisions made for one step will have implications for the other. Much as engineers and architects have to carefully ferret out unintended side effects of any change in their designs, lest a catastrophic failure ensue, geneticists must consider carefully all aspects of the experimental design lest they doom themselves to making inappropriate conclusions based on inadequately obtained results.

REFERENCES

Baehner RL, Kunkle M, Monaco AP, Haines JL, Conneally PM, Palmer C, Heerema N, Orkin SH (1986): DNA linkage analysis of X-chromosome-linked chronic granulomatous disease. *Proc Natl Acad Scie USA* 83:3398–3401.

Barker D, Wright E, Nguyen K, Cannon L, Fain P, Goldgar D, Bishop DT, Carey J, Baty B, Kivlin J (1987): Gene for von Recklinghausen neurofibromatosis is in the pericentromeric region of chromosome 17. *Science* 236:1100–1102.

Barker DF, Hostikka SL, Zhou J, Chow LT, Oliphant AR, Gerken SC, Gregory MC, Skolnick MH, Atkin CL, Tryggvason K (1990): Identification of mutations in the *COL4A5* collagen gene in Alport syndrome. *Science* 248:1224–1227.

Bergoffen J, Scherer SS, Wang S, Scott MO, Bone LJ, Paul DL, Chen K, Lensch MW, Chance PF, Fishbeck KH (1993): Connexin mutations in X-linked Charcot–Marie–Tooth disease. *Science* 262:2039–2042.

Bichet DG, Razi M, Arthus M-F, Lonergan M, Tittley P, Smiley RK, Rock G, Hirsch DJ (1989): Epinephrine and dDAVP administration in patients with congenital nephrogenic diabetes insipidus: Evidence for pre-cycle AMP V(2) receptor defective mechanism. *Kidney Int* 36:859–866.

Bodmer WF, Bailey CF, Bodmer J, Bussey HJ, Ellis A, Gorman P, Lucibello FC, Murday VA, Rider SH, Scambler P (1987): Localization of the gene for familial adenomatous polyposis on chromosome 5. *Nature* 328:614–616.

Botstein D, White RL, Skolnick MH, Davies RW. (1980): Construction of a genetic linkage map in man using restriction fragment length polymorphisms. *Am J Hum Genet* 32:314–331.

Bird, TD, Ott J, Giblett ER (1980): Linkage of Charcot-Marie-Tooth Neuropathy to the Duffy Locus on chromosome 1. *Am J Hum Genet* 32:99A.

Corder EH, Saunders AM, Strittmatter WJ, Schmechel DE, Gaskell PC, Small GW, Roses AD, Haines JL, Pericak-Vance MA. Apolipoprotein E4 gene dose and the risk of Alzheimer disease in late onset families. *Science.* 1993; 261:921–923.

Corder EH, Saunders AM, Risch NJ, Strittmatter WJ, Schmechel DE, Gaskell PC, Rimmler JB, Locke PA, Conneally PM, Schmader KE, Small GW, Roses AD, Haines JL, Pericak-Vance MA. Apolipoprotein E type 2 allele decreases the risk of late onset Alzheimer disease. *Nat Genet.* 1994; 17:180–184.

Dietz HC, Cutting GR, Pyeritz RE, Maslen CL, Sakai LY, Corson GM, Puffenberger EG, Hamosh A, Nathakumar EJ, Curristin SM (1991): Marfan syndrome caused by a recurrent de novo missense mutation in the fibrillin gene. *Nature* 352:337–339.

Dryja TP, McGee TL, Hahn LB, Cowley GS, Olsson JE, Reichel E, Sandberg MA, Berson L (1990): Mutations with the rhodopsin gene in patients with autosomal dominant retinitis pigmentosa. *New Engl J Med* 323:1302–1307.

Eiberg H, Mohr J, Nielson LS, Simonsen N (1983): Genetics and linkage relationships of the C3 polymorphism: Discovery of C3-se linkage and assignment of LES-C3-DM-Se-PEPD-Lu synteny to chromosome 19. *Clin Genet* 24:159–170.

Elston RC, Stewart J. (1971): A general model for the analysis of pedigree data. *Hum Hered* 21:523–42.

Farrar GJ, Kenna P, Jordan SA, Kumar-Singh R, Humphries MM, Sharp EM, Sheilds DM, Humphries P (1991) A three-base-pair deletion in the peripherin-*RDS* gene in one form of retinitis pigmentosa. *Nature* 354:478–480.

Farrer GJ, Jordan SA, Kenna P, Humphries MM, Kumar-Singh R, McWilliam P, Allamand V, Sharp S, Humphries P (1992): Autosomal dominant retinitis pigmentosa: Localization of a disease gene (*RP6*) to the short arm of chromosome 6. *Genomics* 11:870–874.

Filippi G, Rinaldi A, Archidiacono N, Rocchi M, Balazs I, Siniscalco M (1993): Brief report: Linkage between G6PD and fragile-X syndrome. *Am J Hum Genet* 15:113–119.

Fischbeck KH, Ionasecu V, Ritter AW, Ionasecu R, Davis K, Ball S, Bosch P, Burns T, Hausmanowa-Petrusewicz I, Borkowska J (1986): Localization of the gene for the X-linked spinal muscular atrophy. *Neurology* 36:1595–1598.

Fontaine B, Khurana TS, Hoffman EP, Bruns GA, Haines JL,Trofatter JA, Hanson MP, Rich J, McFarlane H, Yasek DM (1990): Hyperkalemic periodic paralysis and the adult muscle sodium channel alpha-subunit gene. *Science* 250:1000–1002.

Gal A, Mucke J, Theile H, Wieacker PF, Ropers HH, Weinker TF (1985): X-linked dominant Charcot–Marie–Tooth disease: Suggestion of linkage with a cloned DNA sequence from the proximal Xq. *Hum Genet* 70:38–42.

Goate A, Chartier-Harlin MC, Mullan M, Brown J, Crawford F, Fidani L, Giuffra L, Haynes A, Irving N, James L (1991): Segregation of a missense mutation in the amyloid precursor protein gene with familial Alzheimer's disease. *Nature* 349:704–706.

Groden J, Thiliveris A, Samowitz W, Carlson M, Gelbert L, Albertsen H, Joslyn G, Stevens J, Spiro L, Robertson M (1991): Identification and characterization of the familial adenomatous polyposis coli gene. *Cell* 66:589–600.

Gusella JF, Wexler NS, Conneally PM, Naylor SL, Anderson MA, Tanzi RE, Watkins PC, Ottina K, Wallace MR, Sakaguchi AY (1983): A polymorphic DNA marker genetically linked to Huntington's disease. *Nature* 306:234–238.

Harley HG, Brooks JD, Rundle SA, Crow S, Reardon W, Buckler AJ, Harper PS, Housman DE, Shaw DJ (1992): Expansion of an unstable DNA region and phenotypic variation in myotonic dystrophy. *Science* 355:546–546.

Huntington's Disease Collaborative Research Group 1993: A novel gene containing a trinucleotide repeat that is expanded and unstable on Huntington's disease chromosomes. *Cell* 72:971–983.

Kainulainen K, Pulkkinen L, Savolainen A, Kaitila I, Pelton L (1990): Location on chromosome 15 of the gene defect causing Marfan syndrome. *New Engl J Med* 323:935–939.

Kinzler KW, Nilbert MC, SU LK, Vogelstein B, Bryan TM, Levy DB, Smith KJ, Preisinger AC, Hedge P, McKechnie D (1991): Identification of *FAP* locus genes from chromosome 5q21. *Science* 253:661–665.

Knoers N, van den Heyden H, van Oost B, Monnenes L, Willems J, Ropers H (1987): Tight linkage between nephrogenic diabetes insipidus and DXS52. *Cytogenet Cell Genet* 46:640.

Knowlton RG, Cohen-Haguenauer O, Cong NV, Frezal J, Brown VA, Barker D, Braman JW, Schumm JW, Tsui LC, Buchwald. A polymorphic DNA marker linked to cystic fibrosis is located on chromosome 7. *Nature* 318:380–382.

Koenig M, Hoffman EP, Bertelson CJ, Monaco AP, Feener C, Kunkel LM (1987): Complete cloning of the Duchenne muscular dystrophy (DMD) cDNA and preliminary genomic organization of DMD gene in normal and affected individuals. *Cell* 50:509–517.

Lange K, Elston RC. (1975): Extensions to pedigree analysis. I. Likelihood calculations for simple and complex pedigrees. *Hum Hered.* 25:95–105.

La Spada LA, Wilson EM, Lubahn DB, Harding AE, Fishbeck KH. Androgen receptor gene mutations in X-linked spinal and bulbar muscular atrophy. *Nature* 352:77–79.

Leppert M, Dobbs M, Scrambler P, O'Connell, Nakamura Y, Stauffer D, Woodward S, Burt R, Huges J, Gardner E (1987): The gene for familial polyposis coli maps to the long of chromosome 5. *Science* 238:1411–1413.

Levy-Lahad E, Wasco W, Poorkaj P, Romano DM, Oshima J, Pettingell WH, Yu CE, Jondro PD, Schmidt SD, Wang K, Crowley AC, Fu Y-H, Guenette SY, Galas D, Nemens E, Wijsman EM, Bird TD, Schellenberg GD and Tanzi RE (1995) Candidate gene for the chromosome 1 familial Alzheimer's disease locus. *Science* 269:973–977.

Lupski JR, de Oca-Luna RM, Slaughaupt S, Pentao L, Guzzetta V, Trask BJ, Saucedo-Cardenas O, Barker DF, Killian JM, Garcia CA (1991)-DNA duplication associated with Charcot–Marie–Tooth disease type 1A. *Cell* 66:219–232.

McKhann G, Drachman, D, Folstein M (1984): Clinical diagnosis of Alzheimer's disease: Report of the NINCDS-ADRDA Work Group under the auspices of the Department of Health and Human Services Task Force on Alzheimer's disease. *Neurology* 34:939–944.

McWilliam P, Farrer GJ, Kenna P, Bradley DG, Humphries MM, Sharp EM, McConnell DJ, Lawler M, Sheils D, Ryan C, Stevens K, Daiger SP, Humphries P (1989): Autosomal dominant retinitis pigmentosa (ADRP): localization of ADRP gene to the long arm of chromosome 3. *Genomics* 5:619–622.

Menlove L, Aldridge J, Schwartz C, Atkin C, Hasstedt S, Kunkle L, Bruns G, Latt S, Skolnick M (1984): Linkage between Alport syndrome–like hereditary nephritis and X-linked RFLPS. *Am J Hum Genet* 36:146S.

Mohr J (1954): *A Study of Linkage in Man*. Copenhagen: Munksgaard.

Morton NE. (1955): Sequential test for the detection of linkage. *Am J Hum Genet* 7: 277–318.

Mucke GA, Theile J, Wieacker PF, Ropers HH, Weinker TF (1985): X-Linked dominant Charcot–Marie–Tooth disease: Suggestion of link with a cloned DNA sequence from the proximal Xq. *Hum Genet* 70:38–42.

Murray JM, Davis KE, Harper PS, Meredith L, Mueller CR, Williamson R (1982): Linkage relationship of a cloned DNA sequence on the short arm of the X chromosome to Duchenne muscular dystrophy. *Nature* 300:69–71.

Ott, J (1976): A computer program for linkage analysis of general human pedigrees. *Am J Hum Genet* 28:528–529.

Ott J (1992): Strategies for characterizing highly polymorphic markers in human gene mapping. *Am J Hum Genet* 41:283–290.

Renwick JH, Bolling DR (1971): An analysis procedure illustrated on a triple linkage of use for prenatal diagnosis of myotonic dystrophy. *J Med Genet* 8:399–406.

Rogaev E, Sherrington R, Rogaeva EA, Levesque G, Ikeda M, Liang G, Chi H, Lin C, Holman K, Tsuda T, Mar L, Sorbi S, Nacmias B, Placentini S, Amaducci L, Chumakov I, Cohen D, Lannfelt L, Fraser PE, Rommens JM and St. George-Hyslop PH (1995): Familial Alzheimer's disease in kindreds with missense mutations in a gene on chromosome 1 related to the Alzheimer's disease type 3 gene. *Nature* 376, 775–778.

Rommens JM, Iannuzzi MC, Kerem B, Drumm ML, Melmer G, Dean M, Rozmahel R, Cole JL, Kennedy D, Hidaka N (1989): Identification of the cystic fibrosis gene: Chromosome walking and jumping. *Science* 245:1059–1065.

Royer-Pokora, Kunkle LM, Monaco AP, Goff SC, Newburger PE, Baehner RL, Cole FS, Curnutte JT, Orkin SH (1986): Cloning the gene for an inherited disorder—chronic granulomatous disease—on the basis of its chromosomal location. *Nature* 322:32–48.

Sadovnick AD, Ebers GC, Dyment DA, Risch NJ (1996): Evidence for genetic basis in multiple sclerosis. The Canadian Collaborative Study Group. *Lancet* 347:1728–1730.

Saunders AM, Strittmatter WJ, Breitner JC, Schmechel D, St. George-Hyslop PH, Pericak-Vance MA, Joo SH, Rosi BL, Gusella JF, Crapper-MacLachlan DR, Growden J, Alberts MJ, Hulette C, Crain B, Goldgaber D and Roses AD (1993): Association of apolipoprotein E allele 4 with late-onset familial and sporadic Alzheimer's disease. *Neurology* 43, 1467–1472.

Seizinger BR, Rouleau GA, Ozelius LJ, Lane AH, Faryniarz AG, Chao MV, Huson S, Korf BR, Parry DM, Pericak-Vance MA (1987): Genetic linkage of von Rechlinghausen neurofibromatosis to the nerve growth factor receptor gene. *Cell* 49:589–594.

Sherrington R, Rogaev E, Liang Y, Rogaeva EA, Levesque G, Ikeda M, Chi H, Lin C, Li G, Holman K, Tsuda T, Mar L, Foncin J-F, Bruni AC, Montesi MP, Sorbi S, Rainero I, Pinessi L, Nee L, Chumakov I, Pollen D, Brookes A, Sanseau P, Polinsky RJ, Wasco W, DaSilva HAR, Haines JL, Pericak-Vance MA, Tanzi RE, Roses AD, Fraser PE, Rommens JM and St. George-Hyslop PH (1995): Cloning of a novel gene bearing missense mutations in early familial Alzheimer's disease. *Nature* 375, 754–760.

St George-Hyslop PH, Tanzi RE, Polinsky RJ, Haines JL, Nee L, Watkins PC, Myers RH, Feldman RG, Pollen D, Drachman D, Growdon J, Bruni A, Foncin JF, Salmon D, Frommelt P, Amaducci L, Sorbi S, Piacentini S, Stewart GD, Hobbs WJ, Conneally PM, Gusella JF (1987): The genetic defect causing familial Alzheimer's disease maps on chromosome 21. *Science* 235:885–890.

Tsui LC, Buchwald M, Barker JC, Knowlton R, Schumm JW, Eiberg H, Mohr J, Kennedy D, Plavsic N (1995): Cystic fibrosis locus defined by a genetically linked polymorphic DNA marker. *Science* 230:1054–1057.

Vance JM, Nicholson GA, Yamaoka LH, Stajich J, Stewart CS, Speer MC, Hung WY, Roses AD, Barker D, Pericak-Vance MA (1989): Linkage of Charcot–Marie–Tooth neuropathy type 1a to chromosome 17. *Exp Neurol* 104:186–189.

Wainwright BJ, Scambler PJ, Schmidtke J, Watson EA, Law HY, Farrall M, Cooke HJ, Eidberg H, Williamson R (1985): Localization of cystic fibrosis locus to human chromosome 7cen-q22. *Nature* 318:384–385.

Wallace MR, Marchuk DA, Anderson LB, Letcher R, Odeh HM, Saulino AM, Fountain JW, Brereton A, Nicholson J, Mitchell AL (1990): Type 1 neurofibromatosis gene: Identification of a large transcript disrupted in three NF1 patients. *Science* 250:1749.

White R, Woodward S, Leppert M, O'Connell, Hoff M, Herbst J, Lalouel JM, Dean M, Vande Woude G (1985): A closely linked genetic marker for cystic fibrosis. *Nature* 318: 382–384.

Xu GF, O'Connell P, Viskochil D, Cawthon R, Robertson M, Culver M, Dunn D, Stevens J, Gestland R, White (1990): The neurofibromatosis type 1 gene encodes a protein related to GAP. *Cell* 62:599–608.

Yu S, Pritchard M, Kremer E, Lynch M, Nancarow J, Baker E, Holman K, Mulley JC, Warren ST, Schlessinger D (1991): Fragile X genotype characterized by an unstable region of DNA. *Science* 252:1179–1181.

2

Basic Concepts in Genetics

Marcy C. Speer

Section of Medical Genetics
Center for Human Genetics
Duke University Medical Center
Durham, North Carolina

This chapter explores the underpinnings for observational and experimental genetics. Concepts ranging from laws of Mendelian inheritance through molecular and chromosomal aspects of DNA structure and function are defined; their ultimate utilization in linkage mapping of simple Mendelian disease and common and genetically complex disease is presented. The chapter concludes with clinical examples of the various types of DNA mutation and their implications for human disease.

INTRODUCTION

For centuries, the hereditary basis of human disease has fascinated both scientists and the general public. The Talmud gives behavioral proscriptions regarding circumcision in sons born after a male sibling who died of a bleeding disorder, suggesting that the ancient Hebrews knew of hemophilia; nursing students in Britain in the 1600s tracked the recurrence of spina bifida in families; and questions as to whether Abraham Lincoln and certain celebrity sports figures had Marfan syndrome sometimes arises in casual dinner conversation.

In many respects, the study of the genetic factors in disease today remains, as it

Approaches to Gene Mapping in Complex Human Diseases, Edited by Jonathan L. Haines and Margaret A. Pericak-Vance. ISBN 0-471-17195-6

has for centuries, dependent on careful description of human pedigree data in which patterns of transmission from parent to offspring are characterized. For example, Gregor Mendel provided the groundwork for the study of human genetics by carefully constructing quantified observations of the frequency of variable characteristics in the pea plant. The importance of detailed pedigree analysis was exemplified recently in the delineation of patterns of transmission of the fragile X syndrome, the most common genetic cause of mental retardation. Careful documentation of pedigrees from families with more than one person with fragile X syndrome led to description of the aptly named Sherman paradox (Sherman et al., 1985) in which different recurrence risks for relatives of various types were described. From this, the complicated workings of unstable DNA harbored in expanding trinucleotide repeats were later elucidated (e.g., Verkerk et al., 1991; Burke et al., 1994).

HISTORICAL CONTRIBUTIONS

Segregation and Linkage Analysis

In 1865 Gregor Mendel, an Austrian monk (Fig. 2.1), published his findings on the inheritance in the pea plant of a series of traits, including seed texture (round or wrinkled), seed color (yellow or green), and plant height (tall or short). He described three properties of heritable factors that explained his quantified observations of these scorable (discontinuous or qualitative) traits. The first property was *unit inheritance*, which is now considered the basis for defining the gene. In essence, he hypothesized that a *factor* was transmitted from parent to offspring in an unchanged form. Such a factor produced an observable trait. This idea represented a radical departure from the scientific thinking at the time, which suggested that parental characteristics were *blended* in the offspring.

The remaining two properties Mendel described are known as his first and second laws. Mendel's first law of the segregation of hereditary characters describes the behavior of factors controlling observable traits such as flower color or plant height as a single unit. He proposed that these factors were transmitted independently and with equal frequency to germ cells (egg and sperm). In experiments for a variety of traits, Mendel crossbred the offspring (the F1 generation) of two phenotypically different, pure-breeding parental strains with one another. The offspring of these matings (the F2 generation) expressed the grandparental traits in a 3:1 ratio (Fig. 2.2). Serendipitously, some of the traits Mendel had chosen to study were simple dominant traits, such that the presence of one factor was sufficient to express the trait, with the other trait being recessive (expressed only in the absence of the dominant factor); later work showed that the factor defining a characteristic need not be dominant to the other. For instance, if each factor contributes to the trait equally, as in codominant systems, then three different classes of offspring from the same cross described above are possible: the two parental traits and a third intermediate trait. These classes occur in proportions parental to intermediate to parental of 1:2:1.

Figure 2.1. *Mendel's garden at the old monastery at Brno. On the right is the door to the Mendel museum, which contains exhibits celebrating his life and research. It is located in what used to be the monastery refectory (dining hall). Mendel's apartment window(s) overlooked the garden. The garden is planted in ornamental red and white flowers, with the first two (farthest from the camera) labeled P (parental generation), the next single red one is labeled F1, the next row of four (three red and one white) is labeled F2, and the next nine labeled F3. At the far left, under a tree, is a large statue of Mendel that once stood in the town square just outside the garden gates. The square is called Mendelovaplatz. (Photo, taken in 1986, courtesy of Arthur S. Aylsworth, M.D.)*

Mendel's second law, that of independent assortment, extended his observations from the transmission of a single trait from parent to offspring to the interaction of two traits. Mendel's second law predicts that factors controlling different traits will be transmitted (will *segregate*) to offspring independently from one another. For instance, seed texture will segregate to offspring independently of seed color. In one experiment Mendel crossed pure-breeding round, green seed plants to pure-breeding wrinkled, yellow seed plants. Since round is dominant to wrinkled and green is dominant to yellow, the resulting seeds (F1 generation) yielded entirely round, green seed plants. These F1 plants were then crossed to one another. These experiments confirmed the predictions from his theory of independent assortment: the seed (offspring) types in the F2 generation were 9 round/green: 3 round/yellow: 3 wrinkled/green: 1 wrinkled/yellow (Fig. 2.3). Note that in the absence of dominance at the trait loci, the expected proportions of F2 seed classes would be in a 1:1:1:1 ratio. Any observed departure from these expected ratios using identical parental crossing strategies suggests that two traits fail to segregate independently and may

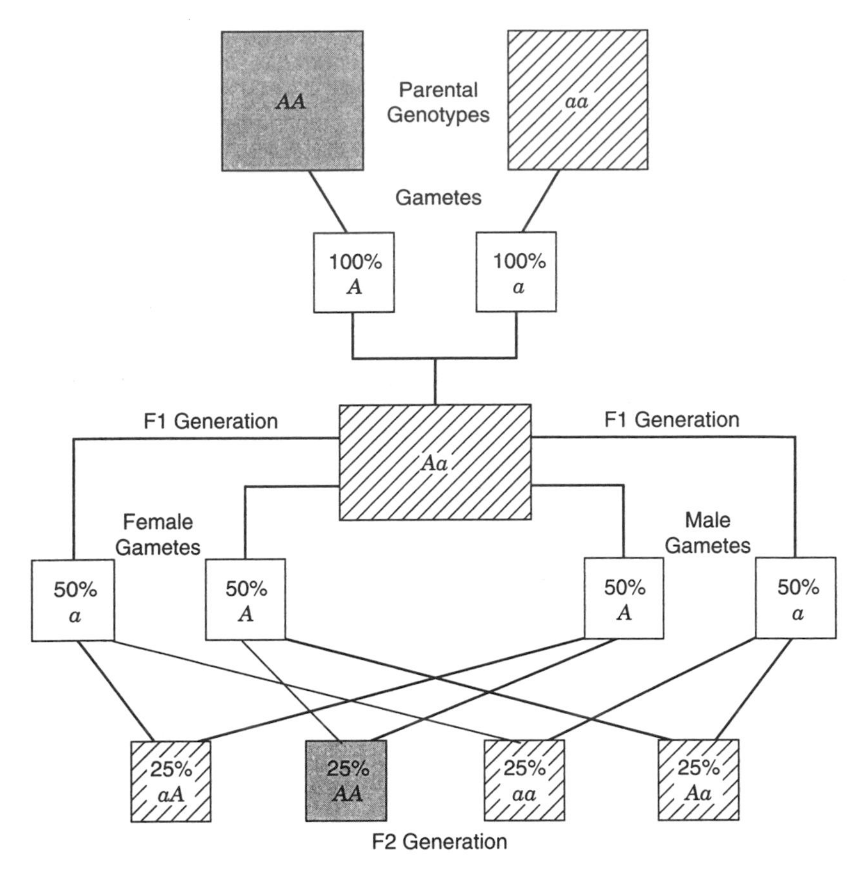

Figure 2.2. *Principles of Mendel's first law of segregation of heritable characters for a dominant trait.*

be physically *linked*. We will describe later how the failure of two traits to segregate independently can be exploited to find genes and diagnose genetic disorders.

Mendel's observations remained largely obscure until the early 1900s when they were independently rediscovered by plant geneticists and by Sir Archibald Garrod, who was studying the human hereditary disorder alkaptonuria (Garrod, 1902). Garrod's work provided the basis of our understanding of alleles and of genetic linkage. Although Mendel's experiments have been the subject of considerable debate with respect to scientific integrity, they remain one of the most important contributions of critical descriptive science in the history of genetics.

Hardy–Weinberg Equilibrium

Another historical landmark in genetics occurred in the early 1900s as evolutionary biologists attempted to explain why the frequency of a dominant trait or disease in

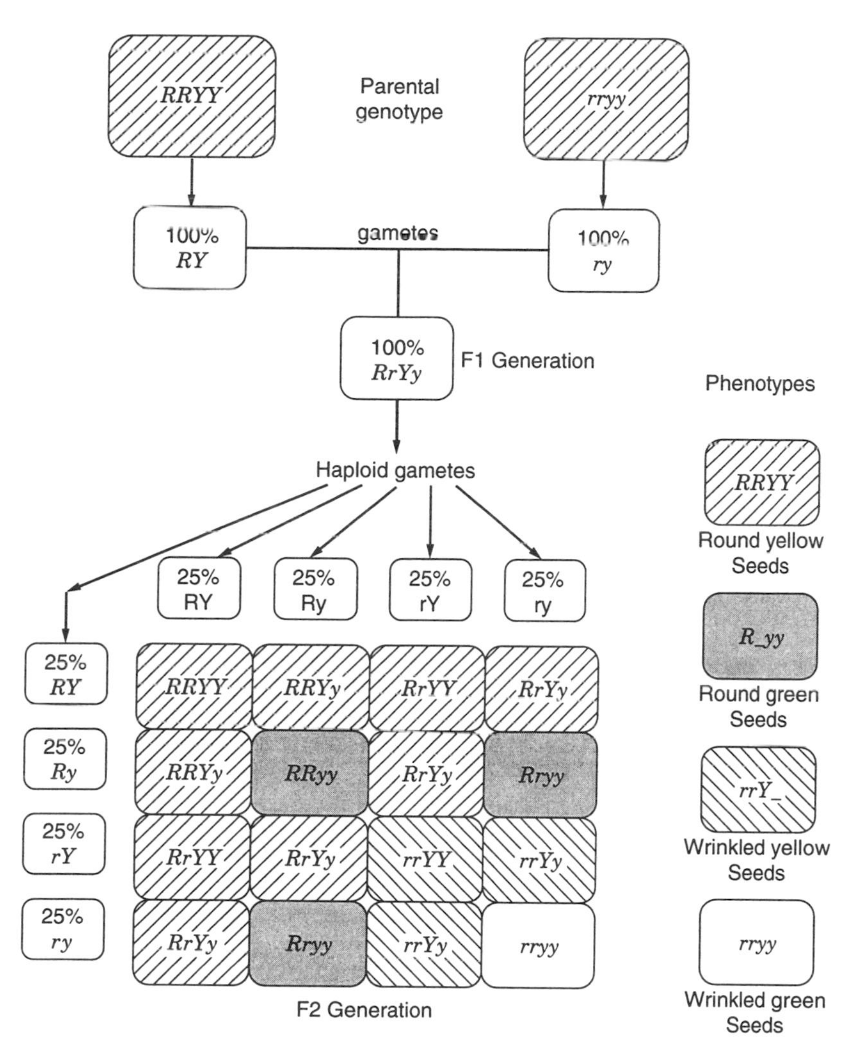

Figure 2.3. *Principles of Mendel's second law of independent assortment with a dominant trait.*

the population did not increase until, over many generations, everyone in the population was affected. The answer to this question was provided independently by Hardy (1908) and Weinberg (1908), who predicted the behavior of alleles in a population using the binomial theorem. Their proof, now called the Hardy–Weinberg theorem, shows that in a large population, trait (genotypic) frequencies for autosomal traits will achieve and remain in a state of equilibrium after one generation of random mating. Several evolutionary forces can alter equilibrium frequencies, including selection for or against a phenotype, migration into or out of a population, new mutation, and genetic drift. For a sex-linked trait, the attainment of equilibrium will require more than one generation.

Specifically, in a two-allele autosomal system with alleles A and a (having frequencies p and q, respectively) $p + q = 1$. The Hardy–Weinberg theorem predicts that the frequencies of genotypes AA, Aa, and aa are p^2, $2pq$, and q^2. Various manipulations of these algebraic formulas potentially allow many useful calculations, such as carrier frequencies of diseases, disease prevalence, and gross estimates of penetrance. Some example applications of the Hardy–Weinberg theorem are shown in Box 2.1.

DNA, GENES, AND CHROMOSOMES

The Structure of DNA

When Mendel described his genetic factor, he did not know what the underlying biological factor was. It would not be until 90 years later that the actual genetic molecule was identified. Mendel's fundamental unit of inheritance is termed the gene, which contains the information for synthesizing proteins necessary for human development, cellular and organ structure, and biological function. DNA, or deoxyribonucleic acid, is the molecule that comprises the gene and encodes information for synthesizing both proteins and RNA (ribonucleic acid). DNA is present in the nucleus of virtually every cell in the body. It is made up of three components: a sugar, a phosphate, and a base. In DNA, the sugar is deoxyribose, while in RNA the sugar is ribose. The four bases in DNA are the pyrimidines adenine and guanine, and the purines cytosine and thymine. A DNA sequence is often described as an ordered list of bases, each represented by the first letter of its name (e.g., ACTGAAACTTGATT). A *nucleoside* is a molecule made of a base and a sugar; a *nucleotide* is made by adding a phosphate to a nucleoside. A single strand of DNA is a *polynucleotide*, consisting of nucleotides bonded together.

A single strand of DNA is, however, unstable. The double-helical nature of DNA, which confers stability to the molecule, was hypothesized in 1953 by J. D. Watson and F. H. C. Crick. Their cohesive theory of the structure of DNA accounted for some of the previously identified properties of DNA. A fascinating account of the internecine struggles in science surrounding this discovery was provided later by Watson (1968).

Specifically, Watson and Crick postulated that DNA is a double-stranded structure and that the two strands of DNA are arranged in an antiparallel orientation. In the central portion of the molecule, a hydrogen bond links a base with its complement such that a purine always bonds with a pyrimidine (e.g., adenine always bonds with thymine and guanine always bonds with cytosine). The conformation of the resultant molecule is the double helix, which undergoes several levels of compacting to fit within the cell (Fig. 2.4).

The sequence of DNA bases represents a code for synthesizing proteins. The fundamental unit of this genetic code is termed a *codon*, which consists of three nu-

BOX 2.1. USEFUL APPLICATIONS OF THE HARDY–WEINBERG THEORY

Example 1. Cystic fibrosis (CF), an autosomal recessive disease, has an incidence of 1/400. What is the frequency CF carriers in the general population?

(a) The frequency of the CF allele (q) is calculated as $\sqrt{1/400} = 1/20$
(b) The frequency of CF carriers is calculated as $2pq = 2\ (19/20)\ (1/20) = 0.095$

Example 2. The frequency of the allele (q) for an autosomal dominant disorder is 1/100. What is the frequency of the disease itself in the population?

(a) Since the frequency of the disease allele is 1/100, the frequency of the normal allele: $(p) = 1 - 1/100 = 99/100$.
(b) Since the disease is dominant, both heterozygous carriers and homozygous individuals are affected with the disease:

$$q^2 + 2pq = (1/100)^2 + 2\ (99/100)\ (1/100) = 0.0199$$

Example 3. An autosomal dominant disorder with incomplete penetrance (f) has a population prevalence of 16/1000. If the allele frequency for the normal allele p is 0.99, what is the estimated penetrance of the disease allele?

(a) Since $p = 0.99$, then $q = 0.01$.
(b) As in Example 2, both heterozygous and homozygous gene carriers are affected (assuming no reduced penetrance). Therefore,

$$f(q^2) + f(2pq) = 0.016$$

$$f(q^2 + 2pq) = 0.016$$

$$f(0.0199) = 0.016$$

$$f = 0.804$$

cleotides. Since there are four different nucleotides (one made with each of the four bases), and a codon is made of three nucleotides, there are $4^3 = 64$ different codons. However, these 64 codons specify only 20 different amino acids, which are the building blocks of proteins. Thus, the genetic code is *degenerate*: different codons may code for the same amino acid. Some codons act as punctuation. For instance, one specific codon in a string of DNA signals the molecular code "interpreter" to start, and then the reading of the DNA strand proceeds in three base pair chunks;

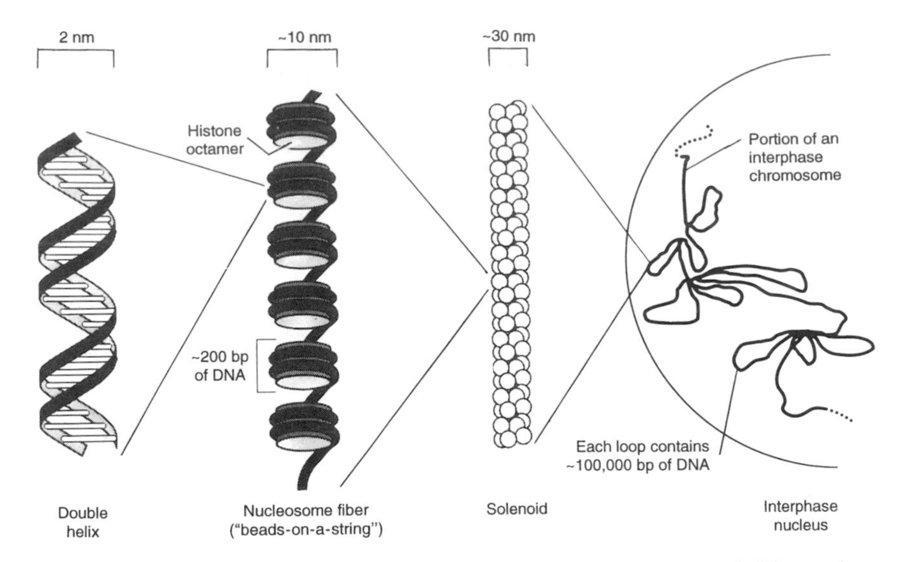

Figure 2.4. *The DNA double helix is packaged and condensed in several different forms. (Reprinted by permission from Thompson et al., eds.,* Genetics in Medicine, *5th ed. W.B. Saunders Company, Philadelphia, 1991.)*

several other specific codons signal the reading process to cease. These reading signals are called *start* and *stop* codons, respectively.

Not all the DNA in a cell actually codes for a protein product; in fact, a significant proportion of the DNA sequence does not lead to protein formation. The human genome has been estimated to contain approximately 3 billion base pairs of DNA and about 50,000–100,000 genes. Within a gene, *exons* are the portions utilized (transcribed) to make proteins. *Introns* are the sequences between exons and are not transcribed. The size and number of introns and exons varies dramatically between genes (Fig. 2.5). Given an average gene size of 30,000 base pairs, genes

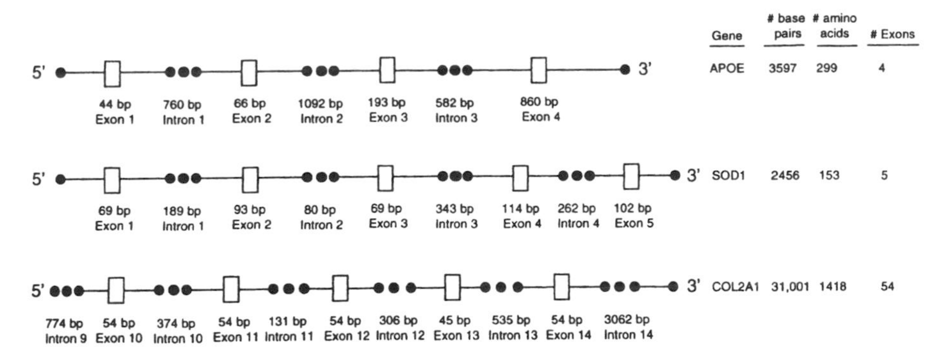

Figure 2.5. *Intron and exon sizes vary between genes.*

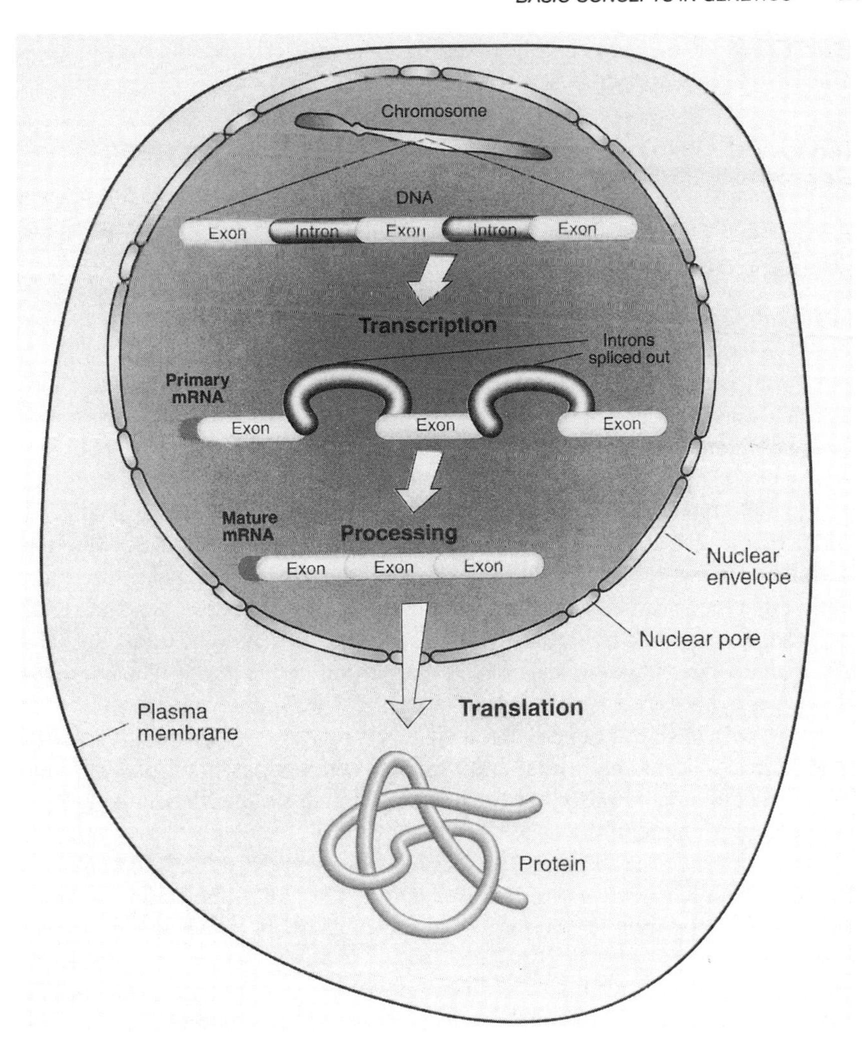

Figure 2.6. The central dogma of genetics: DNA → RNA → protein. (Reprinted by permission from Jorde et al., eds., Medical Genetics, *C.W. Mosby, St. Louis, MO, 1995).*

thus account for a maximum of 1% of the DNA sequence in humans.

The central dogma of genetics is that the utilization of DNA is unidirectional such that DNA → RNA → protein (Fig. 2.6). Specifically genes are encoded in the DNA. Then, in the nucleus, messenger RNA (mRNA) is *transcribed* (produced) from the DNA. mRNA subsequently undergoes a series of post-transcriptional modifications: the introns are spliced out, a *cap* is added at the 5′ end of the molecule, and a string of adenylate residues (poly-A tail) is added to the 3′ end. The mRNA is then transported out of the nucleus into the cytoplasm, where it is *translated* into protein by means of cellular machinery called the ribosomes. Many excellent resources describe the very complicated process of transcription and translation (e.g., Strachan and Read, 1996).

Genes and Alleles

The physical site or location of a gene is called its *locus*. At any particular gene site, or locus, there can exist different forms of the gene, called *alleles*. Except on the sex chromosomes of males, an individual has two alleles at each locus. These alleles are analogous to the *factors* identified in the 1800s by Mendel. *Homozygosity* is defined as the presence of two alleles that are indistinguishable from one another. In *heterozygotes*, the two alleles can be distinguished from another. Males with a normal chromosome complement are *hemizygous* for all X chromosome loci, since they have only one copy of the X chromosome.

The difference between two alleles may be as subtle as a single base pair change, such as the thymine-to-alanine substitution that alters the B chain of hemoglobin A from its wild type to its hemoglobin sickle cell state. Some base pair changes have no deleterious effect on the function of the gene; nevertheless, these functionally neutral changes in the DNA still represent different forms of a gene. Alternatively, allelic differences can be as extensive as large, multicodon deletions such as those observed in Duchenne muscular dystrophy. Any locus having two or more alleles, each with a frequency of at least 1% in the general population, is considered to be *polymorphic* (i.e., having many forms).

Differences in alleles can be scored via laboratory testing. The ability to score allele differences accurately within families, between families, and between laboratories is critically important for linkage analysis in both simple Mendelian and genetically complex common disorders. Allele scoring strategies may be as simple as + or – for the presence or absence of a deletion or point mutation, or as complicated as assessing the allele size in base pairs of DNA. The latter application is common when highly polymorphic, microsatellite repeat markers (Chapter 9) are used in linkage analysis.

A measure frequently utilized to quantitate the extent of polymorphism of a gene or marker system is the *heterozygosity* value *H*, which is calculated as follows:

$$H = 1.0 - \sum_{i=1}^{n} p_i^2$$

where *n* is the number of alleles at the locus and *p* is the frequency of the *i*th allele at the locus. For example, the heterozygosity value of a three-allele marker with frequencies of allele 1 = 0.25, allele 2 = 0.30, and allele 3 = 0.45 is calculated as: 1 – [(0.25 × 0.25) + (0.30 × 0.30) + (0.45 × 0.45)] = 0.645. This measure can be interpreted as the probability that an individual randomly selected from the general population will be heterozygous at the locus. Polymorphic loci or genetic markers (Chapter 9) are critically important to linkage analysis because they allow each individual in a family a high probability of being heterozygous for the locus. The investigator may then be able to deduce the parental origin of each allele and identify recombinant and nonrecombinant gametes.

Genes and Chromosomes

Genes are organized as linear structures called *chromosomes*, with many thousands of genes on each chromosome. Each chromosome has distinguishable sites that aid in cell division and in the maintenance of chromosome integrity. The *centromere* is visualized as the central constriction on a chromosome and it separates the p (short) and the q (long) arms from one another. The centromere enables correct segregation of the duplicated chromosomal material during meiosis and mitosis. *Telomeres* are present at both ends of the chromosome and are required for stability of the chromosomal unit.

Following cell culture, the arrest of cell division at metaphase (when the chromosomes have duplicated and condensed), and appropriate staining techniques, the chromosomes in a cell can be analyzed under the microscope. At this stage of the cell cycle, a chromosome has two double-stranded DNA molecules. Together, the strands are called sister chromatids. The sister chromatids are held together by the centromere. Photographs are magnified and the chromosomes are arranged into a *karyotype*. The normal human chromosome complement consists of 46 chromosomes, arranged in 23 pairs, with one member of each pair inherited from each parent (Fig. 2.7). The first 22 pairs, called *autosomes*, numbered 1 through 22, are arranged according to size and are the same in males and females. The pair of sex chromosomes generally predicts an individual's gender. Most females have two X chromosomes, while males have one X inherited from the mother and one Y chromosome inherited from the father. The gender of an individual is therefore determined by the father.

Because two copies of each chromosome are present in a normal somatic (body) cell, the human organism is *diploid*. In contrast, egg and sperm cells have *haploid* chromosomal complements, consisting of a single member of each chromosome pair. The correct number of chromosomes in the normal human cell was finally established in 1956, three years after the report of the double-helical structure of DNA was described, when Tjio and Levan demonstrated unequivocally that the chromosomal complement is 46 (Tjio and Levan, 1956).

Regions of chromosomes are defined by patterns of alternating light and dark regions called *bands*, which become apparent after a chemical treatment has been applied. One of the most common types of banding process, called Giemsa or G banding, involves digesting the chromosomes with trypsin and then staining with a Giemsa dye. G banding identifies late-replicating regions of DNA; these are the dark bands (Fig. 2.7). Other chemical processes will produce different banding patterns and identify unique types of DNA.

A specific genetic locus can be defined quite precisely along a chromosome, such as the gene *FRAXA* (fragile X syndrome), which is located on the X chromosome at band q27.3. Alternatively, its localization may be specified as an interval flanked by two genetic markers. Any two loci that occur on the same chromosome are considered to be *syntenic*. Two genes may be syntenic, yet far enough apart on the chromosome to segregate independently from one another. Thus, two syntenic genes may be unlinked. Two syntenic genes that fail to be transmitted to gametes in-

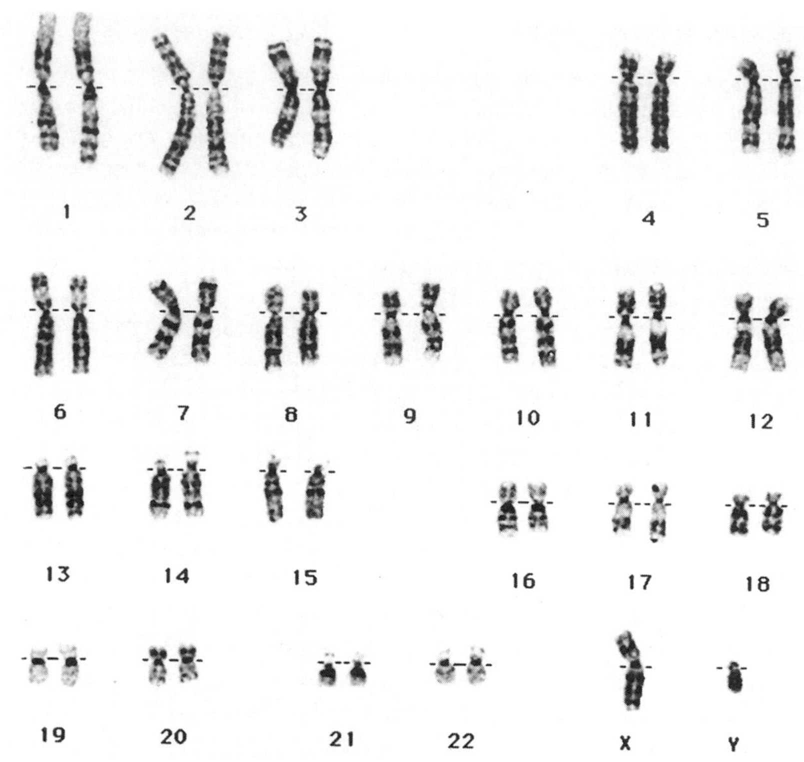

Figure 2.7. *A G-banded human male karyotype. (Courtesy of Mazin Qumsiyeh, Ph.D., Duke University Medical Center, Durham, NC.)*

dependently from one another are *linked* (Fig. 2.8). The location of two loci on the same arm of the chromosome is specified by their positions relative to each other and to the centromere. The gene closer to the centromere is termed *centromeric* or *proximal* to the other; similarly, the gene further from the centromere is *distal* or *telomeric* to the other (Fig. 2.9).

The X and Y chromosomes are very different in their genetic composition except for an area at the distal end of the p arm of each, termed the *pseudoautosomal region*. The pseudoautosomal region behaves similarly to the autosomes during meiosis by allowing for segregation of the sex chromosomes. Just proximal to the pseudoautosomal region on the Y chromosome are the *SRY* (sex-determining region on the Y) and *TDF* (testes-determining factor) genes, which are critical for the normal development of male reproductive organs. When crossing over extends past the boundary of the pseudoautosomal region and includes one or both of these genes, sexual development will most likely be adversely affected. For instance, the rare occurrences of chromosomally XX males and XY females are due to such aberrant crossing over.

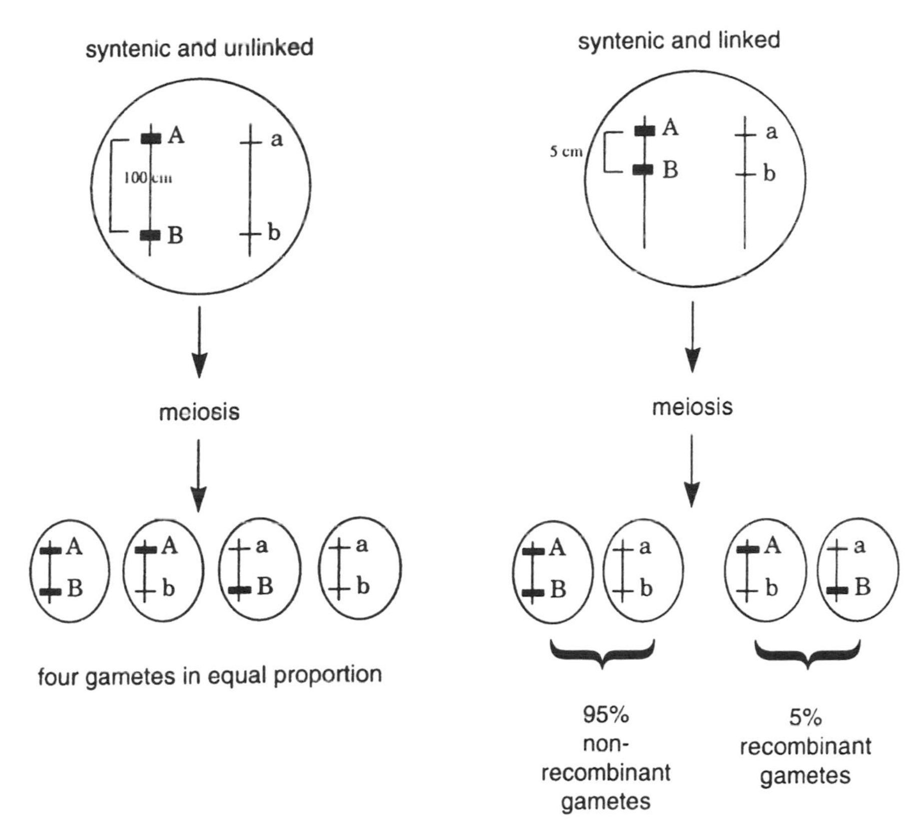

Figure 2.8. *Genes that are on the same chromosome (syntenic) may be linked or unlinked to one another.*

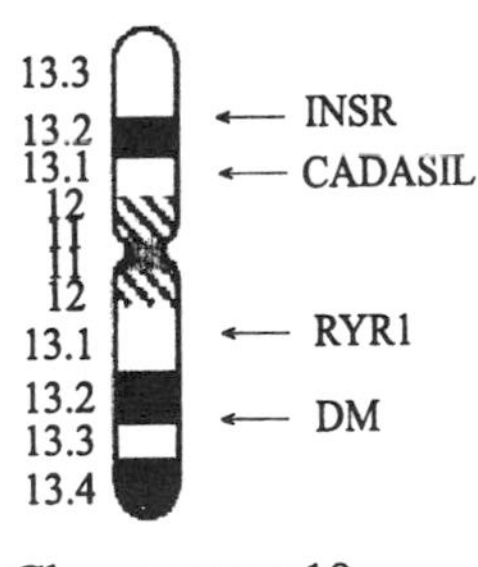

Figure 2.9. *The myotonic dystrophy (*DM*) and insulin receptor (*INSR*) genes are distal (telomeric) to the ryanodine receptor 1 and* CADASIL*, respectively.* RYR1 *and* CADASIL *are proximal (centromeric) to* DM *and* INSR*, respectively.*

Genes, Mitosis, and Meiosis

A cell's ability to reproduce itself is critical to the survival of an organism. This cell duplication process, utilized by somatic cells, is called mitosis. Similarly, an organism's ability to reproduce itself is critical to the survival of the species. In sexual organisms, the reproductive process involves the union of gametes (sperm and egg cells), which are haploid. Meiosis is the process by which these haploid gametes are formed.

Meiosis consists of two parts, meiosis I and meiosis II. In meiosis I, which is called a reduction division stage, each chromosome in a cell is replicated to yield two sets of duplicated homologous chromosomes. During meiosis I, physical contact between chromatids may occur, resulting in the formation of chiasmata. Chiasmata are thought to represent the process of *crossing over* or *recombination*, in which an exchange of DNA between two (of the four) chromatids occurs (Fig. 2.10). A chiasma occurs at least once per chromosome pair. Thus, a parental haplotype (the arrangement of many alleles along a chromosome) may not remain intact upon transmission to an offspring. Following crossing over, at least two of the four chromatids becomes unique, unlike those of the parent. The cellular division process that occurs ensures that one paternal homologue and one maternal homologue are transmitted to each of two diploid daughter cells. This cell division marks the end of meiosis I.

The process of genetic recombination helps to preserve genetic variability within a species. It allows for virtually limitless combinations of genes in the transmission

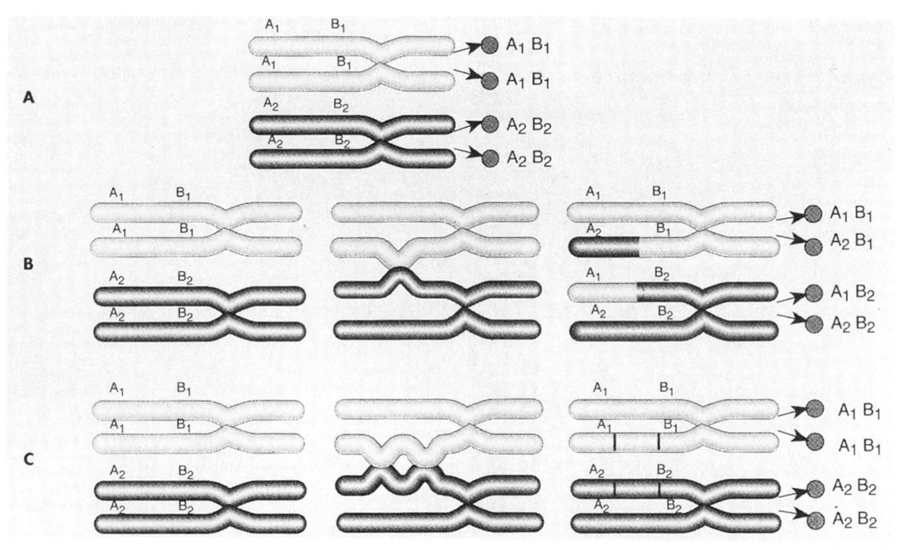

Figure 2.10. The genetic results of crossover: (A) no crossover: A, and B, remain together after meiosis; (B) crossover between A and B results in a recombination (A, and B, are inherited together on a chromosome, and A, and B, are inherited together on another chromosome): (C), double crossover between A and B results in no recombination of alleles: (Reprinted by permission from Jorde et al., eds., Medical Genetics; *C.W. Mosby, St. Louis, MO, 1995.)*

from parent to offspring (see Chapter 12 for a description of estimation of the recombination fraction). The frequency of recombination is not uniform through the genome. Some areas of some chromosomes have increased rates of recombination (*hot spots*), while others have reduced rates of recombination (*cold spots*). For instance, recombination frequencies may vary between sexes or may vary depending on whether the loci are at the telomere or centromere of the chromosome.

The second phase of meiosis is identical to a mitotic (somatic cell) division, in which genetic material is transmitted equally, identically, and without recombination to daughter cells. However, in contrast to a mitotic division, which yields two identical diploid daughter cells, the end result of the entire meiotic process in sperm cells is four haploid daughter cells with chromosomal haplotypes different from those originally present in the parent; in egg cells, the final outcome is a single haploid daughter cell, with the remainder of the genetic material lost because of the formation of nonviable polar bodies.

The fundamental differences between meiosis and mitosis are summarized in Table 2.1.

When Genes and Chromosomes Segregate Abnormally.

Failure of meiosis at either phase (meiosis I or II) is termed nondisjunction and leads to *aneuploidy*, or abnormal chromosomal complements. The most well-known aneuploidy is Down syndrome, caused by an extra copy of chromosome 21. Down syndrome is often called trisomy 21 because patients have a total of 47 chromosomes, with three copies of chromosome 21. A monosomy, or the absence of a second member of a chromosome pair, is rarely viable. A noted exception is Turner syndrome, in which a female has a total of 45 chromosomes, only one of which is an X chromosome.

Triplody or tetraploidy are the terms for the presence of one or two entire extra sets of chromosomes, leading to a total of 69 or 92 chromosomes. These anomalies,

TABLE 2.1 Differences Between Meiosis and Mitosis

	Meiosis	Mitosis
Purpose	Produce gametes; ensure or produce genetic variability through recombination	Replace somatic cells
Location	Gonads	Body cells
Number of cell divisions per cycle	Two: meiosis I and meiosis II (latter identical to mitosis)	One
Chromosome number in daughter cells	Halved from parental complement to produce gametes; resultant cells are haploid	Identical to parental complement; resultant cells are diploid
Recombination	Occurs in diplotene of prophase I of meiosis I	Occurs rarely, usually the result of an abnormality

which usually are inviable in humans, are due to errors in fertilization such as dispermy (two sperm fertilizing an ovum) or failure of the ovum's polar body to separate.

Segregation distortion, a phenomenon observed rarely in humans, is characterized by a departure from the 50:50 segregation ratio expected from normal meiosis. Specifically, one allele at a locus is transmitted to the gamete more than 50% of the time. Segregation distortion, also termed *meiotic drive*, has been described in many experimental systems. Myotonic dystrophy (Carey et al., 1994), an autosomal dominant muscular dystrophy, was among the first human disease alleles suggesting preferential transmission of an allele to offspring.

THE ORDERING AND SPACING OF LOCI BY MAPPING TECHNIQUES

The segregation of loci in meiosis provides the opportunity for assessment of Mendel's law of independent assortment. When two loci are unlinked to one another, this law holds true; however, the law is violated when two loci are linked to one another such that the transmission of one is not independent from transmission of the other. Estimating the distance between linked loci by assessing the frequency of recombination between them allows the development of an order of the markers relative to one another.

Once a gross localization for a disease or trait locus has been identified, either through linkage analysis or from some other approach (e.g., clues from chromosomal rearrangements), it is necessary to home in on the actual gene. This process is always complicated, but it can be simplified by the use of mapping resources, many of which were developed as a direct result of the Human Genome Initiative. No single mapping resource is best for all situations. Thus several different approaches to ordering pieces of DNA are in common use; the primary difference among them is in the way distance is measured. Regardless of the mapping approach utilized, the resultant locus order along a chromosome should be identical.

Genetic maps order polymorphic markers by specifying the amount of recombination between markers, whereas *physical maps* quantify the distances among markers in terms of the number of base pairs of DNA. For small recombination fractions (usually < 10–12%), the estimate of the recombination fraction provides a very rough estimate of the physical distance. In general, 1% recombination is equivalent to about a million base pairs of DNA and is defined as one centiMorgan (cM). Physical measurements of DNA are often described in terms of thousands of kilobases (10 kb of DNA is equivalent to 10,000 base pairs). A specific type of physical map, the radiation hybrid (RH) map (see below), also allows quantification of the length of a segment of DNA. RH map distance are measured in centirays (cR), and on average throughout the genome 1 cR is equivalent to about 30,000 bp of DNA, although this estimate varies according to radiation dose. Estimates of distance from physical and genetic maps of the identical region may vary dramatically (Table 2.2) throughout the genome.

TABLE 2.2 Estimated Physical and Genetic Lengths of Selected Chromosomes

Chromosome	Physical Length (Mb)[a]	Genetic Length (cM)[a]	Radiation Hybrid Length (cR)[b,c]
1	263	305	7894 (31.4)
2	255	271	6973 (34.4)
3	214	237	7785 (25.9)
4	203	244	2867 (24.4)
5	194	224	5611 (32.6)
6	183	207	6095 (28.3)
7	171	178	6606 (24.3)
8	155	172	3996 (36.6)
9	145	146	4513 (30.0)
10	144	181	5423 (25.1)
11	144	150	4858 (27.9)
12	143	160	5002 (26.9)
13	114	130	3306 (27.8)
14	109	122	3513 (25.0)
15	106	154	2822 (29.8)
16	98	157	2735 (34.4)
17	92	208	3039 (28.6)
18	85	143	2977 (26.8)
19	67	148	2122 (29.6)
20	72	122	2010 (33.8)
21	50	114	1562 (23.7)
22	56	81	1522 (26.9)
X	164	220	3644 (42.5)

[a]From Morton (1991).
[b]Numbers in parentheses give kilobases per centiray.
[c]Data extracted from the Stanford Human Genome Center (see Appendix for Web site).

Genetic Mapping

The study of human inherited disease has benefited from numerous experiments in other organisms. Although mapping in humans has a relatively recent history, the idea of a linear arrangement of genes on a chromosome was first proposed in 1911 by T. H. Morgan from his work in the fruit fly *Drosophila melanogaster*. The possibility of a genetic map was first formally investigated by A. H. Sturtevant, who ordered 5 markers on the X chromosome in *Drosophila* and then estimated the relative spacing among them.

In experimental organisms, genetic mapping of loci involves counting the number of recombinant and nonrecombinant offspring of selected matings (Table 2.3). Genetic mapping in humans is usually more complicated than in experimental organisms for many reasons, including our inability to design specific matings of individuals, which limits the unequivocal assignment of recombinants and nonrecombinants. Therefore, maps of markers in humans are developed by means of one of several statistical algorithms used in computer programs including CRIMAP, MAPMAKER (Lander and Green, 1987), CLINKAGE, and MULTIMAP (Matise et al.,

TABLE 2.3 Example Development of a Genetic Map using Four Linked Loci, A, B, C, and D, scored in 100 Offspring[a]

A. Score Recombination Events Between All Pairwise Combinations of the Four Loci

Loci Scored	Number of Recombinants	Frequency of Recombination
A–B	10	0.10
A–C	3	0.03
A–D	15	0.15
B–C	7	0.07
B–D	5	0.05
C–D	12	0.12

B. Process

1. Determine which two loci have the highest frequency of recombination between them; these two loci are the farthest apart on the map.
2. Fit the other loci into the map like pieces of a puzzle.

C. Resulting Genetic Map

Percent recombination		0.03		0.07		0.05	
loci order	A		C		B		D

[a]This example demonstrates no evidence for interference (see text). Positive interference would be manifested as a decrease in overall map length from what would be expected by adding the pairwise distances. In this example, if (A–C) + (C–B) + (B–D) > A–D, positive interference may be acting.

1994). Genetic maps can assume equal recombination between males and females or can allow for sex-specific differences in recombination. These maps are generally produced utilizing a single set of *reference pedigrees* such as the those developed by the Centre d' Étude du Polymorphisme Humain (CEPH; Dausset et al., 1990), which are mostly comprised of three-generation pedigrees with a large number of offspring (average = 8.5). Both sets of maternal and paternal grandparents are usually available, so that linkage phase (see Chapter 4) frequently can be established. The collection of CEPH pedigrees in its entirety consists of more than 60 pedigrees and includes more than 600 individuals; DNA from this valuable resource is available through the Coriell Institute for Medical Research. The complexity of the underlying statistical methods used to generate them renders genetic maps sensitive to marker genotyping errors, particularly in small intervals (Buetow, 1991), and these maps are less useful in small regions of less than about 2 cM. While marker order is usually correct, the genotyping errors can result in falsely inflated estimates of map distances. An example of a genetic map for chromosome 5 is shown in Figure 2.11.

Interference and Genetic Mapping Another factor complicating genetic mapping is *interference,* where the probability of a crossover in a given chromosomal region is influenced by the presence of an already existing crossover. In so-called positive interference, the presence of one crossover in a region decreases the probability that another crossover will change. Negative interference, the opposite of positive interference, implies that the formation of a second crossover in a region is made more likely by the presence of a first crossover. Most documented interfer-

Integrated Genetic Map of Chromosome 05

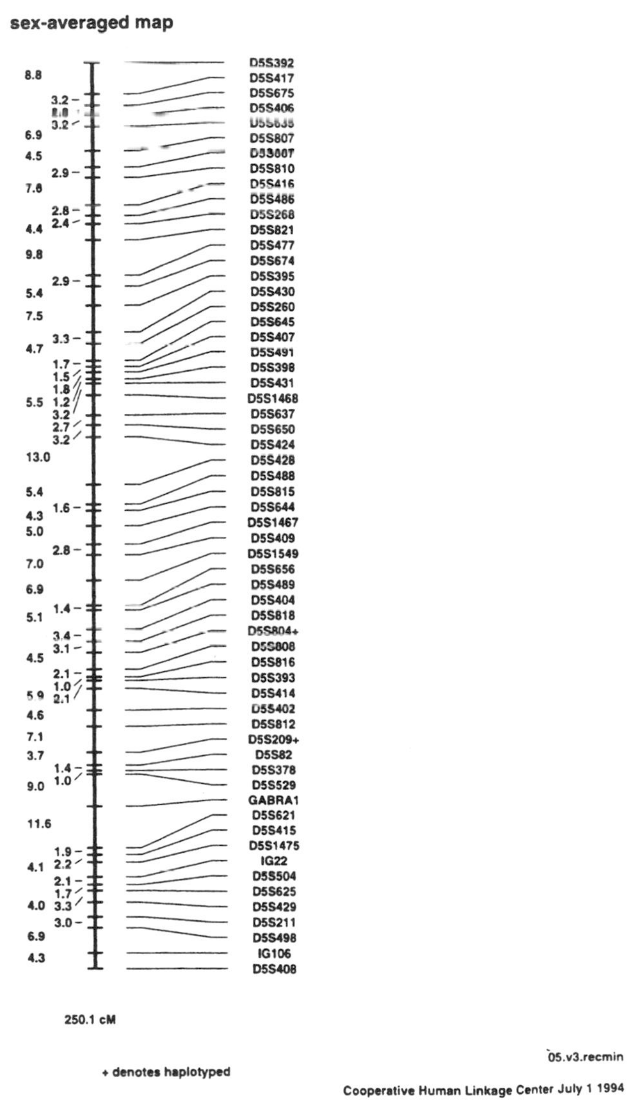

Figure 2.11. *Genetic map of chromosome 5 from the Cooperative Human Linkage Center (CHLC) database. Distance between markers is measured in sex-averaged recombination fractions.*

ence has been positive, but some reports of negative interference exist in experimental organisms. Interference is very difficult to measure in humans because exceedingly large sample sizes, usually on the order of 300–1000 fully informative meiotic events, are required to detect it (Weeks et al., 1994).

The investigation of interference is important because accurate modeling of interference will provide better estimates of true genetic map length, intermarker distances, and more accurate mapping of trait loci. Interference (*I*) is frequently measured in terms of the *coefficient of coincidence* (c.c.) in genetic crosses where three separate linked markers can be scored. The coefficient of coincidence is the ratio of the observed number of double crossovers to the expected number of double crossovers assuming no interference. When *I* is greater than 0, interference is present (positive) and when *I* is less than 0, interference is negative; when $I = 0$, there is no evidence for interference and recombination fractions across intervals are additive.

For example, assume three loci whose order is A-B-C. If the distance between A and B is 10 cM and between B and C is 5 cM, when $I = 0$, the distance between A and C is 15 cM. As noted earlier, the frequency of recombination in humans is generally decreased near the centromeric region of chromosomes, tends to be greater near the telomeric regions, and is increased in females when compared to males. It should be further noted that genetic map distance is not tied directly to physical map distance.

Several mathematical formulas have been developed to account for interference in predicting physical map distance from recombination frequencies in human linkage studies. These mapping functions include those developed by Kosambi (1944), Rao et al. (1977), and Haldane (1919). Haldane's map function assumes the absence of interference, while Kosambi's map function assumes that interference is large at small genetic distances but decreases as the genetic distance between two loci increases. A program in the LINKAGE utility package, MAPFUN, translates recombination frequencies into map distances and vice versa, under a variety of mapping functions. This package is available at Jurg Ott's Web site at Rockefeller University (see Appendix).

Meiotic Breakpoint Mapping

Meiotic breakpoint maps, an outgrowth of genetic maps, are graphical descriptions of critical, confirmed recombination events within reference pedigrees (Fig. 2.12). Once a meiotic breakpoint map for a region has been developed, a marker whose location is nearby can be genotyped in a subset of pedigrees in which critical recombination events have occurred. Limiting the genotyping efforts to a small number of specific individuals within pedigrees greatly increases genetic mapping efficiency.

Physical Mapping

The goal of developing a physical map of the genome is identical to that for the genetic maps: to order pieces of DNA and, subsequently, genes. However, the materials utilized and the average resolutions of various mapping methods differ. Some of the oldest available physical maps of the genome are *restriction maps*, which identi-

CEPH Meiotic Break Point Panel

Family	ID	D5S818	D5S804	D5S816	D5S812
1331	9	○	●	●	●
1331	10	●	○	○	○
1332	3	○	●	□	□
1332	12	○	□	●	●
1333	4	○	□	●	□
1333	9	○	□	●	□
1344	5	○	□	●	□
1347	6	○	○	○	●
1347	10	○	○	○	●
1362	9	□	○	○	●
1362	11	□	○	●	●
884	3	○	○	○	●
884	11	○	○	○	●

Figure 2.12. *Sample meiotic breakpoint map, from the data at the CHLC. Shaded circles indicate markers that recombine with those having open circles. Squares indicate marker is either uninformative or not genotyped. To test whether a marker was located between D5S818 and D5S804, an investigator would need to genotype individuals 9 and 10 in family 1331 and 3 in 1332, in addition to their parents. In typical practice, the laboratory is blinded to which individuals are recombinants to avoid potential bias. Consequently, two or three nonrecombinant siblings for each known recombinant individual are also genotyped.*

fy sites at which an enzyme cuts (*digests*) a specific sequence of DNA. For instance, DNA may be digested by a restriction enzyme such as *Not* I and the resultant pieces of DNA sized. *Contig maps* are developed by cloning pieces of DNA into vectors such as yeast artificial chromosomes (YACs) or cosmids and then ordered by their overlapping sequences (Chapter 17). *STS* (sequence tagged site) *maps* utilize unique stretches of DNA to identify particular clones. *EST* (expressed sequence tag) *maps* order the coding stretches of DNA. *DNA sequence maps* order specific stretches of DNA at the level of single base pairs (see Chapter 17).

A recent focus of the Human Genome Initiative has been the development of *radiation hybrid maps* (Lawrence et al., 1991). RH maps are developed by exposing DNA to high doses of radiation, thereby breaking the DNA into small pieces. The frequency with which particular markers are retained in a piece of DNA is scored, and this provides an estimate of the relative order and distance between markers in centirays. In the United States, a large portion of the RH mapping is performed at the Stanford Human Genome Center (SHGC). As of August 1997, SHGC had placed over 10,000 markers on RH maps for all chromosomes.

Through the SHGC RH server, a researcher may submit data obtained from hybridization of a marker to the publicly available (through Research Genetics, Inc.) G3 mapping panel, and mapping results will be returned. As with genetic maps, the basis of the RH maps is statistical: relative order of a marker is determined in a manner analogous to calculating lod scores (Falk, 1991; Boehnke, 1992, 1993; Boehnke et al., 1992) and is reported in terms of odds in favor of a placement relative to another placement.

Differences among these types of maps are summarized in Table 2.4.

TABLE 2.4 Summary Characteristics of Genome Maps of Selected Types

Map Type	Measurement of Distance	Material Needed	Caveats and Notes
Genetic maps	Recombination frequency (θ) or centimorgan (cM); 1% recombination is approximately equal to 1 cM.	Reference pedigrees; polymorphic markers.	Extremely sensitive to genotyping error; usually not useful in areas < 2 cM.
Meiotic breakpoint maps	Not applicable.	Reference pedigrees; polymorphic markers.	A good mechanism for minimizing genotyping to determine marker order using statistical techniques designed to minimize the number of recombination events.
Radiation hybrid maps	Centiray (cR): 1 cR represents 1% breakage between two markers.	Somatic cell hybrid panel; markers not necessarily polymorphic	Maps are developed using statistical techniques that assess the frequency of chromosome breakage
STS maps	Not applicable.	The clones from which the STSs are derived must be ordered.	Resulting maps have landmarks assayed by PCR, but markers are often not polymorphic.
Restriction maps	Tens of thousands of base pairs of DNA; 1000 bp of DNA is termed a kilobase.	Genomic DNA	
EST maps	Not applicable.	Genomic DNA	

GENES AND HUMAN DISEASE

Inheritance Patterns in Mendelian Disease

Two types of inheritance pattern conforming to Mendel's laws have already been described, dominant and recessive. Alleles whose loci are on an autosome can be transmitted in a dominant or recessive (or codominant) fashion; similarly, alleles having loci on the X chromosome are expressed and transmitted as either X-linked recessive or X-linked dominant disorders. These well-known patterns of inheritance are shown in Figure 2.13.

The hallmark of dominant inheritance, regardless of whether the underlying gene is located on an autosome or on an X chromosome, is that only a single allele is necessary for expression of the phenotype. In an autosomal recessive trait, two copies of a trait allele must be present for it to be expressed. In most cases, it is correctly assumed that each parent of an offspring with the trait carries a recessive allele at the locus. Rarely, an individual expressing an autosomal recessive disorder has inherited both abnormal recessive alleles from one parent (see Chapter 3); this phenomenon is called uniparental disomy.

To express an X-linked recessive disorder, a male needs only one abnormal allele; a female usually needs two abnormal alleles (one on each X chromosome) and this is generally rare. Thus, mostly males are affected. For X-linked dominant traits, both men and women can be affected since only one copy of the trait allele is necessary for phenotypic expression.

Alleles located on the Y chromosome are transmitted from affected males to all sons, and in each case the son's Y-linked phenotype will be identical to that of the

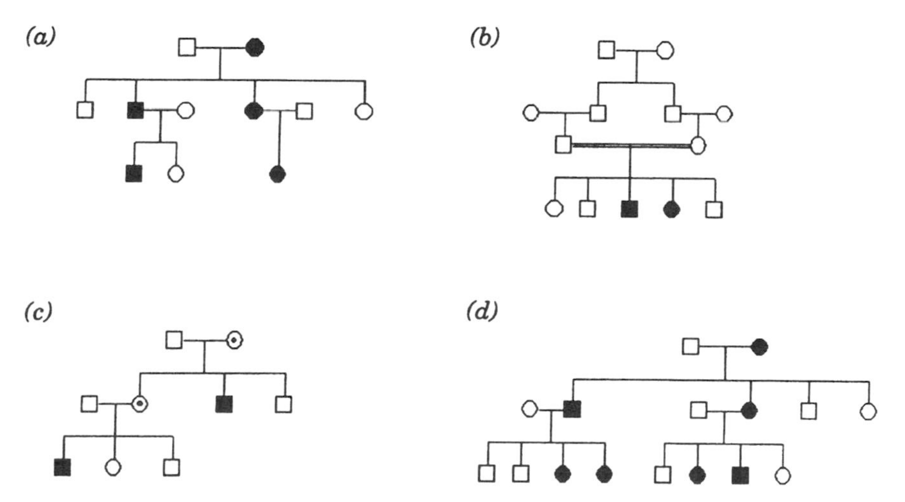

Figure 2.13. Pedigrees consistent with (a) autosomal dominant inheritance, (b). autosomal recessive inheritance, (c). X-linked recessive inheritance, and (d) X-linked dominant inheritance. Here and elsewhere: squares, males; circles, females; open symbols, unaffected individuals; solid symbols, affected individuals.

father; daughters of males with a Y-linked trait will not inherit the trait, since they receive their father's X chromosome. Very few expressed genes have been localized to the Y chromosome. Characteristics of each of the different Mendelian inheritance patterns are summarized in Table 2.5.

Genetic Changes Associated with Disease/Trait Phenotypes

Alterations or mutations in the genetic code can be neutral, beneficial, or deleterious. Changes in the genetic code can lead to trait and/or disease phenotypes; the pathology can be the result of either loss or gain of function of the gene product. Such changes can occur in a number of different ways.

Point Mutations A point mutation is defined as an alteration in a single base pair in a stretch of DNA, thereby changing the 3 bp codon. Since the genetic code is degenerate, many such changes do not necessarily alter the resulting amino acid; however, if the single base pair change leads to the substitution of one amino acid for another, the result can be devastating. Point mutations can be classified as transition mutations (a purine → purine or a pyrimidine → pyrimidine) or as the less common transversion mutations (purine → pyrimidine or a pyrimidine → purine). In general, transitions are less likely than transversion mutations to change the resulting amino acid. Five effects of point mutations have been defined: *synonymous* or *silent* mutations are single base pair changes in the DNA that do not affect the resultant amino acid; *nonsense mutations* result in a premature stop codon, leading to a polypeptide of reduced length; *missense mutations* lead to the substitution of one amino acid for another; *splice-site mutations* affect the correct processing of the mRNA strand by eliminating a signal for the cutting out of an intron; and *mutations in regulatory genes* alter the amount of material produced. Several examples of point mutations in human diseases are discussed in the subsections that follow.

Amyotrophic Lateral Sclerosis Approximately 10–15% of amyotrophic lateral sclerosis (ALS) patients have a positive family history consistent with autosomal dominant inheritance. This rapidly progressive neurodegenerative disorder has an average age of onset in the mid-40s and is usually fatal within a few years after onset. The gene responsible for about 15–20% of familial ALS cases has been identified as the cytosolic form of superoxide dismutase (Cu, Zn SOD) at 21q22.1. To date, 38 different point mutations have been identified. A recent study of clinical correlations associated with different mutations within the SOD1 gene (Juneja et al., 1997) found evidence for a significantly faster rate of progression (1.0 ± 0.4 year vs. 5.1 ± 5.1 years) in patients with the most common mutation, a transition mutation in which a thymine at codon 4 is substituted for a cytosine. This substitution changes the resultant amino acid from an alanine to a valine.

Sickle Cell Anemia Sickle cell anemia, an autosomal recessive disorder with a carrier frequency in African Americans of approximately 1/12, is a classic example of a point mutation leading to disease. Affected patients have the familiar phenotype

TABLE 2.5 Hallmarks of Mendelian Inheritance Patterns of Different Types

Inheritance Pattern	Examples	Gender Differences in Proportion of Affecteds?	Transmission Features	Recurrence Risks	Prevalence in Population	Other Critical Features
Autosomal dominant	Marfan syndrome; neurofibromatosis; myotonic dystrophy	No	Transmitted from affected parent to affected offspring (vertical transmission); male-to-male transmission.	For each offspring of affected parent, risk to child to inherit disease gene is 50%.	$p^2 + 2pq$	Reduced penetrance frequent; for a "true dominant," individuals heterozygous for the trait allele are no more severe than individuals homozygous for the trait allele.
Autosomal recessive	Sickle cell anemia; cystic fibrosis	No	Carrier parents generally unaffected.	For parents who have one affected child risk for each subsequent child is 25%.	q^2	Consanguinity frequent.
Sex-linked recessive	Duchenne muscular dystrophy; hemophilia	Males more frequently affected; carrier females generally unaffected; rare cases of nonrandom X-inactivation can lead to affected females.	Gene transmitted from unaffected carrier mother to affected son; no male-to-male transmission.	Carrier mother has 25% chance to have affected son and 25% chance to have carrier daughter; all daughters of affected males are carriers and no sons of affected males are affected.	Females: q^2 Males: q	Females affected in rare cases of nonrandom X inactivation.

(continued)

TABLE 2.5 *(continued)*

Inheritance Pattern	Examples	Gender differences in Proportion of Affecteds?	Transmission Features	Recurrence Risks	Prevalence in Population	Other Critical Features
Sex-linked dominant	Hypophosphatemic rickets; fragile X syndrome	No	Vertical transmission from mothers to both sons and daughters; fathers transmit to daughters only; no male-to-male transmission.	50% of offspring of affected mothers are affected (unless mother is homozygous for disease allele); all daughters of affected males are affected and no sons of affected males are affected.	Females: $p^2 + 2pq$ Males: p	Females are affected 3 times more frequently than males.

Y-linked	Genes *SRY* and *TDF*, important in sex determination, are on the Y chromosome; no known diseases are located on Y.	Yes; only males would express trait	Exclusively male-to-male transmission.	All sons of affected males are affected; no daughters of affected males are affected.	Females: 0 Males: q	Male-determining genes are located just proximal to the pseudoautosomal region on the Y chromosome, and faulty recombination in the pseudoautosomal region can lead to errors in sex determination.
Autosomal codominant	MN blood group; microsatellite repeat markers.	No	Each allele confers a measurable component to the phenotype.	Varies according to mating type.	Genotypes expected to occur in Hardy–Weinberg proportions of p^2, $2pq$, and c^2.	

of chronic anemia, sickle cell crises, and debilitating pain. Sickle cell anemia results from a single nucleotide substitution of an adenine to a thymine at position 6 in the β chain of hemoglobin. This changes the resultant amino acid from glutamine to valine. Interestingly, the carrier state for sickle cell may lead to a selective advantage in certain environments: carriers have a resistance to malaria that is useful in tropical climes.

Achondroplasia Achondroplasia, the most common type of short-limbed dwarfism, is an autosomal dominant disorder. About 85% of cases are the result of a new mutation. It has been observed that the rate of new dominant mutations increases with advancing paternal age (Penrose, 1955; Stoll et al., 1982). Achondrophasia is now known to result from mutations in the fibroblast growth receptor-3 gene (FGFR3), located on chromosome at 4p16.3. Interestingly, over 95% of the mutations are the identical G-to-A transition at nucleotide 1138 on the paternal allele (Rousseau et al., 1994; Shiang et al., 1994; Bellus et al., 1995). Other mutations in *FGFR3* are also responsible for hypochondroplasia and thanatophoric dysplasia, types of dwarfism that are clinically distinct from achondroplasia.

Deletion/Insertion Mutations Another class of mutations involves the deletion or insertion of DNA into an existing sequence. Deletions or insertions may be as small as one base pair, or they may involve one or many exons or even the entire gene. Even single base pair deletions or insertions can have devastating effects, frequently by altering the reading frame of the DNA strand.

Neurofibromatosis Neurofibromatosis type 1 (NF1) is an autosomal dominant disorder with variable expression; penetrance of the disorder is high, and some clinicians consider penetrance of this gene to be complete. The most common phenotypic manifestations are multiple café-au-lait spots and peripheral neurofibromatous skin tumors. Approximately 50% of all cases are due to new mutation, most frequently on the paternally inherited allele.

The gene for NF1, coding for a protein called neurofibromin, is located at chromosome 17q11.2 and has been cloned. Its function is thought to involve tumor suppression (DeClue et al., 1992; Li et al., 1992). To date, only a fraction of the mutations responsible for NF1 have been identified. Those identified include entire deletions of the gene (Wu et al., 1995), insertion mutations, and small and large deletions. Single base pair mutations leading to premature stop codons (Valero et al., 1994) and deletions, both leading to the production of a truncated protein, account for the majority of *NF1* mutations.

Duchenne and Becker Forms of Muscular Dystrophy Duchenne muscular dystrophy (DMD) is a severe, childhood-onset X-linked muscular dystrophy, and Becker muscular dystrophy (BMD) is its allelic, clinically milder variant. Boys with DMD develop normally for the first few years of life, after which rapidly progressive muscle deterioration becomes obvious. Affected males lose the ability to walk by age 10–12 years. The eventual loss of muscle strength in the cardiac and respiratory muscles leads to death in early adulthood. The gene coding for the protein dystrophin, which is abnormal in DMD/BMD, has been cloned (Koenig et al., 1988). Approximately two-thirds of mutations in this very large gene have been identified; the majority are deletions and duplications, although point mutations have also been identified. Correlations of the clinical phenotype with the molecular mutation have been complicated in DMD/BMD. The general deletions and point mutations leading to alterations in the reading frame of the DNA molecule (frameshift mutations) are more severe than those that do not alter the reading frame (in-frame mutations).

Cystic Fibrosis—Cystic fibrosis, an autosomal recessive disorder, is the most common hereditary disease among Caucasians, with a carrier frequency of between 1/20 and 1/30. The function of the pancreas, lungs, and sweat glands, among other organ systems, is affected. In American Caucasians, a single mutation called ΔF508 accounts for about 70% of the abnormal CF alleles. Three base pairs (codon 508) are deleted, and the resulting amino acid sequence is missing a phenylalanine. Over 300 other deleterious mutations have been identified throughout the world. The frequency of specific mutations differs among populations.

Novel Mechanisms of Mutation: Unstable DNA and Trinucleotide Repeats Dynamic mutations, or *unstable* DNA, have received considerable attention of late. Areas of the genome have variable numbers of dinucleotide or trinucleotide repeats. Most are not associated with expressed genes but can be exploited as markers, since they are highly polymorphic. A few areas of trinucleotide repeats are near or within genes and by expansion beyond a certain threshold, these disrupt gene expression and cause disease. To date, 11 disorders (7 autosomal and 4 X-linked) have been shown to be the result of expansion of these unstable triplet repeats (Table 2.6).

The phenotypes of myotonic dystrophy, Huntington disease, and Machado–Joseph disease, among others, are associated with *anticipation*, a clinical phenomenon in which disease severity worsens in each successive generation (see also Chapter 3). Because disease expression can be quite variable and difficult to measure, age of onset is frequently utilized as an analogue of severity, and anticipation is then observed as a decreasing mean age of onset with each passing generation. Since the discovery that expanding trinucleotide repeats may explain anticipation, investigators have reported clinical evidence for anticipation in many various disorders including bipolar affective disorder (McInnes et al., 1993), limb–girdle muscular dystrophy (Speer et al., 1998; familial spastic paraplegia (Raskind et al, 1997; Scott et al, 1997); and facioscapulohumeral muscular dystrophy (Tawil et al., 1996; Zatz et al., 1995). None of these disorders is proven to be caused by trinucleotide repeat expansions, and elucidation of their underlying defect will shed additional light on the phenomenon of anticipation.

Susceptibility Versus Causative Genes As the study of common and genetically complex human diseases identifies the significant contribution of heredity in their development, it is likely that more and more of these genes or genetic risk fac-

TABLE 2.6 Salient Features of Known Human Trinucleotide Repeat Diseases

Condition	Gene Symbol	Chromosome Location	Repeat Type	Repeat Localization	Repeat Number Abnormal Range	Clinical Features
Fragile X syndrome	*FRAXA*	Xq27.3	CGG	5′ untranslated region (premutations in range of 52–200)	200–1000	Moderate to severe mental retardation, macroorchidism, large ears, and prominent jaw. *FRAXA* accounts for about one-half of all X-linked mental retardation.
Fragile site mental retardation -2	*FRAXE*	Xq28	GCC	?	200–1000	Similar to fragile X syndrome phenotypically; cytogenetic evidence for Xq fragile site; negative for expansion in *FRAXA*.
Fragile site F	*FRAXF*	Xq28	$(GCCGTC)_n$ $(GCC)_n$	?	300–500	Cytogenetic evidence for Xq fragile site without molecular expansion at *FRAXA* or *FRAXE*; whether aberrant phenotype is associated with expansion at *FRAXF* is uncertain.
Fragile site 16q22	*FRA16A*	16q22	CCG	?	1000–2000	The expansion is the molecular explanation for the cytogenetic observation of a fragile site at 16q22; the expansion has been associated with infertility and spontaneous abortions.

Kennedy spinal and bulbar muscular atrophy (SMBA)	*SBMA*	Xq11-q12	CAG	Open reading frame	40–52	Caused by a defect in the androgen receptor gene, SBMA usually presents in midlife with bulbar signs and facial fasciculations.
Huntington Disease (HD)	*HD*	4p16.3	CAG	Open reading frame	37–100	Caused by a defect in the Huntington gene, HD is characterized by choreiform movements, rigidity, and dementia.
Spinocerebellar ataxia, type 1	*SCA1*	6p23	CAG	Open reading frame	< 100	Autosomal dominant ataxia with onset in 30s; upper motor neuron signs and extensor plantor responses.
Dentatorubropallido-luysian atrophy, Haw River syndrome	*DRPLA*	12p13.31	CAG	Open reading frame	< 100	Myoclonus epilepsy, dementia, ataxia, and choreoathetosis transmitted as an autosomal dominant.
Machado–Joseph (spinocerebellar ataxia type 3)	*MJD*; *SCA3*	14q24.3-q31	CAG	Open reading frame	61–84	Ataxia with onset usually in 40s; frequent dystonia and facial fasciculations
Myotonic dystrophy	*DM*	19q13.2-q13.3	CTG	3′ Untranslated region	200–4000	Myotonia, ptosis, characteristic cataracts, testicular atrophy, and frontal balding.
Spinocerebellar ataxia type 2	*SCA2*	12q24.1	CAG	5′ Coding region	36–59	Abnormalities of balance due to cerebellar dysfunction or pathology; clinically identical to *SCA1*.

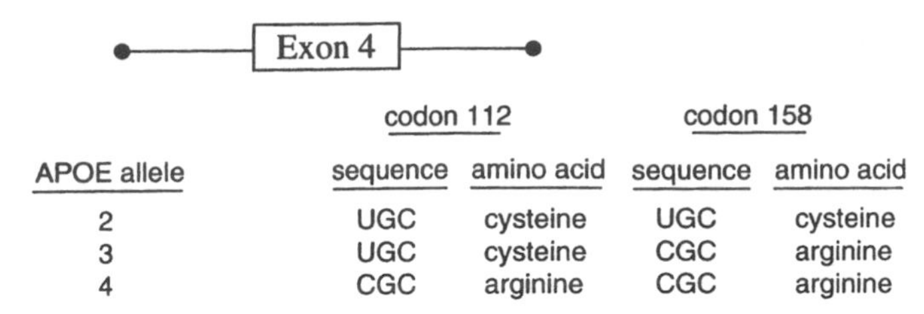

APOE allele	codon 112 sequence	codon 112 amino acid	codon 158 sequence	codon 158 amino acid
2	UGC	cysteine	UGC	cysteine
3	UGC	cysteine	CGC	arginine
4	CGC	arginine	CGC	arginine

Figure 2.14. *Single base pair changes in exon 4 of* APOE *define the 2, 3, and 4 alleles at this locus. [Modified from MA Pericak-Vance and JL Haines.* Trends Genet *11 (1995).]*

tors will be found to affect susceptibility to disease rather than the more traditionally considered causative genes (see also Chapter 18). Historically, one of the most widely investigated examples of susceptibility loci is the human leukocyte antigen (HLA) loci on the p arm of chromosome 6. Specific HLA antigens have been associated with various human diseases; for instance, the Bw47 antigen confers an 80–150-fold increased risk for congenital adrenal hyperplasia; the B27 antigen confers an 80–100-fold increased risk for anklylosing spondylitis; and the DR2 antigen confers a 30–100-fold increased risk for narcolepsy, a 3-fold increased risk for systemic lupus erythematosus, and a 4-fold increased risk for multiple sclerosis. Recently, a genome-wide scan for genes predisposing to multiple sclerosis confirmed the importance of the HLA region in the pathogenesis of the disease (Multiple Sclerosis Genetics Group, 1996).

A recent and well-characterized example of a susceptibility locus is that of the *APOE* gene and Alzheimer disease (AD). The *APOE* locus on chromosome 19 consists of three different alleles, scored as '2', '3', and '4', which occur with frequency 6, 78, and 16% in most European populations (e.g., Saunders et al., 1993). These alleles differ in their DNA sequence by only one base at codons 112 and/or 158 (Fig. 2.14). The *APOE* '4' allele increases risk and decreases age of onset in familial and sporadic late-onset AD and early-onset sporadic AD. The '2' allele has been shown to be protective to some extent for risk to develop AD (Corder et al., 1994, 1995a, 1995b Farrer et al., 1997). Interestingly, the '4' allele has been shown to be at lower frequency in the Indiana Amish, at least partially explaining the decreased frequency of Alzheimer disease in this inbred population (Pericak-Vance et al., 1996). The critical distinction for *APOE* and AD is that the '4' allele is not by itself sufficient or necessary for the development of AD; it is simply one of three alleles at the *APOE* locus, all of which exist with relatively high frequency in the general population, which increases susceptibility for AD.

SUMMARY

The study of genes, chromosomes, and patterns of transmission of human traits within families has led to remarkable discoveries that are useful in genetic counseling for recurrence risk, presymptomatic testing, and prenatal diagnosis (see Chapter

6), and in the understanding of the pathogenesis of diseases. The genetic basis of Mendelian disease is relatively straightforward, and in many cases is well understood. The situation in common complex disorders is markedly different from the study of Mendelian disease, since more than one gene with or without environmental factors likely will be associated with trait phenotype expression. Yet, many of the same principles hold true in complex disease: Mendel's laws regarding the transmission of genes and alleles at loci are as important to the study of resemblance between relatives in genetically complex disease as in Mendelian disease; the same holds true for the extent and result of the differing types of mutation. As the millennium approaches, the advances will accumulate more rapidly; human genomic study can only expand dramatically.

REFERENCES

Bellus GA, Hefferon TW, Ortiz de Luna RI, Hecht JT, Horton WA, Machado M, Kaitila I, McIntosh I, Francomano CA (1995): Achondroplasia is defined by recurrent G380R mutations of *FGFR3*. *Am J Hum Genet* 56:368–373.

Boehnke M (1992): Radiation hybrid mapping by minimization of the number of obligate chromosome breaks. *Cytogenet Cell Genet* 59:96–98.

Boehnke M, Lange K, Cox DR (1992): Statistical methods for multipoint radiation hybrid mapping. *Am J Hum Genet* 49:1174–1188.

Boehnke M (1993): Multipoint analysis for radiation hybrid mapping. *Ann Med* 24:383–386.

Buetow KH (1991): Influence of aberrant observations on high-resolution linkage analysis outcomes. *Am J Hum Genet* 49:985–994.

Burke JR, Wingfield MS, Lewis KE, Roses AD, Lee JE, Julette C, Pericak-Vance MA, Vance JM (1994): The Haw River syndrome: Dentatorubropallidoluysian atrophy (DRPLA) in an African-American family. *Nat Genet* 7:521–524.

Carey N, Johnson K, Nokelainen P, Peltonen L, Savontaus M-L, Juvonen V, Anvret M, Grandell U, Chotai K, Robertson E, Middleton-Price H, Malcolm S (1994): Meiotic drive at the myotonic dystrophy locus? *Nat Genet* 6:117–118.

Corder EH, Saunders AM, Risch NJ, Strittmatter WJ, Schmechel DE, Gaskell PC, Rimmler JB, Locke PA, Conneally PM, Schmader KE, Small GW, Roses AD, Haines JL, Pericak-Vance MA (1994): Apolipoprotein E type 2 allele decreases the risk of late onset Alzheimer disease. *Nat Genet* 17:180–184.

Corder EH, Saunders AM, Strittmatter WJ, Schmechel D, Gaskell PC, Rimmler JB, Locke PA, Conneally PM, Schmaer KE, Tanzi RE, Gusella J, Small GW, Roses AD, Pericak-Vance MA, Haines JL (1995a): Apolipoprotein E, survival in Alzheimer disease patients and the competing risks of death and Alzheimer disease. *Neurology* 45:1323–1328.

Corder EH, Saunders AM, Strittmatter WJ, Schmechel DE, Gaskell PC, Jr, Roses AD, Pericak-Vance MA, Small GW, Haines JL (1995b): The apolipoprotein E E4 allele and sex-specific risk of Alzheimer's disease. *JAMA* 273:373–374.

Dausset J, Cann H, Cohen D, Lathrop M, Lalouel J-M, White R (1990): Centre d' Étude du Polymorphisme Humain (CEPH): Collaborative genetic mapping of the human genome. *Genomics* 6:575–577.

DeClue JE, Papageorge AG, Fletcher JA, Diehl SR, Ratner N, Vass WC, Lowry DR. (1992):

Abnormal regulation of mammalian p21ras contributes to malignant tumor growth in von Recklinghausen (type 1) neurofibromatosis. *Cell* 69:265–273.

Falk CT (1991): A simple method for ordering loci using data from radiation hybrids. *Genomics* 9:120–123.

Farrer LA, Cupples LA, Haines JL, Hyman B, Kukull WA, Mayeux R, Pericak-Vance MA, Risch N, van Dujn CM. (1997): Effects of age, sex, and ethnicity on the associations between apolipoprotein E genotype and Alzheimer disease. A meta-analysis. APOE and Alzheimer Disease Meta Analysis Consortium. *JAMA* 278 (16):1349–1356.

Garrod AE (1902): The incidence of alkaptonuria: A study in chemical individuality. *Lancet* ii: 1616–1620.

Haldane JBS (1919): The combination of linkage values and the calculation of distances between the loci of linked factors. *J Genet* 8:299–309.

Harper PS, Harley HG, Reardon W, Shaw DF (1992): Anticipation in myotonic dystrophy: New light on an old problem. *Am J Hum Genet.* 51:10–16.

Juneja T, Pericak-Vance MA, Laing NG, Dave S, Siddique T. (1997): Prognosis in familial ALS: Progression and survival in patients with glu100gly and ala4val mutations in Cu, Zn superoxide dismutase. *Neurology* 48:55–57.

Kawaguchi Y, Okamoto T, Taniwaki M, Aizawa M, Inoue M, Katayama S, Kawakami H, Nakamura S, Nihimura M, Akiguchi I, Kumura J, Narumiya S, Kakizuka A (1994): CAG expansions in a novel gene for Machado–Joseph disease at chromosome 14q32.1. *Nat Genet* 8:221–228.

Koenig M, Monaco AP, Kinkel LM (1988): The complete sequence of dystrophin predicts a rod-shaped cytoskelital protein, *Cell* 53: 219–228.

Kosambi DD (1944): The estimation of map distances from recombination values. *Ann Eugen* 12:172–175.

Lander ES, Green P (1987): Construction of multilocus genetic linkage maps in humans. *Proc Natl Acad Sci USA* 84:2363–2367.

Lawrence S, Morton NE, Cox DR (1991): Radiation hybrid mapping. *Proc Natl Acad Sci USA* 88:7477–7480.

Mahadevan M, Tsilfidis C, Sabourin L, Shutler G, Amemiya C, Jansen G, Neville C, Narang M, Barcelo J, O'Hoy K, Leblond S, Earle-MacDonald J, deJohn PJ, Wieringa B, Korneluck RG (1992): Myotonic dystrophy mutation: An unstable CTG repeat in the 3′ untranslated region of the gene. *Science* 255:1253–1255.

Matise TC, Perlin M, Chakravarti A (1994): Automated construction of genetic linkage maps using an expert system (MultiMap): A human genome linkage map. *Nat Genet* 6:384–390.

McInnis MG, McMahon FJ, Chase GA, Simpson SG, Ross CA, DePaulo JR (1993): Anticipation in bipolar affective disorder. *Am J Hum Genet* 53:385–390.

Morton NE (1991): Parameters of the human genome. *Proc Natl Acad Sci USA* 88:7474–7476.

The Multiple Sclerosis Genetics Group (1996): A complete genomic screen for multiple sclerosis underscores a role for the major histocompatibility complex. *Nat Genet* 13:469–471.

Penrose LS (1955): Parental age and mutation. *Lancet* ii:312–313.

Pericak-Vance MA, Johnson CC, Rimmler JB, Saunders AM, Robinson LC, D'Hondt EG, Jackson CE, Haines JL (1996): Alzheimer's disease and apolipoprotein E-4 allele in an Amish population. *Ann Neurol* 39:700–704.

Rao DC, Morton NE, Lindsten J, Hulten M, Yee S (1977): A mapping function for man. *Hum Hered* 27:99–104.

Raskind WH, Pericak-Vance MA, Lennon F, Wolff J, Lipe HP, Bird TD (1997): Familial spastic paraparesis: Evaluation of locus heterogeneity, anticipation, and haplotype mapping of the *SPG4* locus on the short arm of chromosome 2. *Am J Med Genet* 74:26–36.

Rousseau F, Bonaventure J, Legeai-Mallet L, Pelet A, Rozet J-M, Maroteauz P, LeMerrer M, Munnich A (1994). Mutations in the gene encoding fibroblast growth factor receptor-3 in achondroplasia. *Nature* 371:252–254.

Saunders AM, Strittmatter WJ, Schmechel D, St George-Hyslop PH, Pericak-Vance MA, Joo SH, Rosi BL, Gusella JF, Crapper-MacLachlan DR, Growden J, Alberts MJ, Hulette C, Crain B, Goldgaber D, Roses AD (1993): Association of apolipoprotein E allele g4 with late-onset familial and sporadic Alzheimer's disease. *Neurology* 43: 1467–1472.

Scott WK, Gaskell PC, Lennon F, Wolpert C, Menold MM, Aylsworth AS, Warner C, Farrell CD, Boustany R-M, Albright SG, Boyd E, Kingston HM, Cumming WJK, Vance JM, Pericak-Vance MA (1997). Locus heterogeneity, anticipation, and reduction of the chromosome 2p minimal candidate region in autosomal dominant familial spastic paraplegia. J. *Neurogenetics,* In Press.

Sherman SL, Jacobs PA, Morton NE, Froster-Iskenius U, Howard-Peebles PN, Nielsen KB, Partington NW, Sutherland GR, Turner G, Watson M (1985): Further segregation analysis of the fragile X syndrome with special reference to transmitting males. *Hum. Genet* 69:3289–3299.

Shiang R, Thompson LM, Zhu Y-Z, Church DM, Fielder TJ, Bocian M, Winokur ST, Wasmuth JJ (1994): Mutations in the transmembrane domain of *FGFR3* cause the most common genetic form of dwarfism, achondroplasia. *Cell* 78:335–342.

Speer MC, Gilchrist JM, Stajich JM, Gaskell PC, Westbrook CA, Yamaoka LH, Pericak-Vance MA (1998). Evidence for anticipation in autosomal dominant limb–girdle muscular dystrophy. *J. Med. Genet.* 35:305–308.

Stoilov I, Kilpatrick MW, Tsipouras P (1995): A common *FGFR3* gene mutation is present in achondroplasia but not in hypochondroplasia. *Am J Med Genet* 55:127–133.

Stoll C, Roth M-P, Bigel P. (1982): A reexamination of parental age effect on the occurrence of new mutations for achondroplasia. In: Papadatos CJ, Bartscocas CS, eds. *Skeletal Dysplasias*. New York: Alan R. Liss, pp. 419–426.

Strachan T, Read AP (1996): *Human Molecular Genetics*. New York: Wiley-Liss.

Tavormina PL, Shiang R, Thompson LM, Zhu Y-Z, Wilkin DJ, Lachman RS, Wilcos WR, Rimoin DL, Cohn DH, Wasmuth JJ (1995): Thanatophoric dysplasia (types I and II) caused by distinct mutations in fibroblast growth factor receptor 3. *Nat Genet* 9:321–328.

Tawil R, Forrester J, Griggs RC, Mendell J, Kissel J, McDermott M, King W, Weiffenbach B, Figlewicz D, and the FSH-DY group (1996): *Ann Neurol* 39:744–748.

Tjio JH, Levan A (1956): The chromosome number of man. *Hereditas* 42:1–6.

Valero MC, Valasco E, Moreno F, Hernandez-Chico C. (1994): Characterization of four mutations in the neurofibromatosis type 1 gene by denaturing gradient gel electrophoresis (DGGE) *Hum Mol Genet* 3:639–641.

Verkerk AJMH, Pieretti M, Sutcliffe JS, Fu Y-H, Kuhl DPA, Pizzuti A, Reiner O, Richards R, Victoria MF, Zhang F, Eussen BE, van Ommen GJB, Blonden LAJ, Riggins GJ, Chastain JL, Kunst CB, Galjaard H, Caskey CT, Nelson DL, Oostra BA, Warren ST (1991): Identification of a gene (*FMR-1*) containing a CGG repeat coincident with a breakpoint cluster region exhibiting length variation in fragile X syndrome. *Cell* 65:905–914.

Watson JD (1968): *The Double Helix*. New York: Atheneum Publishers.

Weeks DE, Ott J, Lathrop GM (1994): Detection of genetic interference: Simulation studies and mouse data. *Genetics* 136:1217–1226.

Weinberg W (1908): Uber den nachweis der vererbung beim Menschen. Jahreshefte des vereins fur vaterlandische naturkunde in wurttemberg 64: 368–82.

Willems PJ (1994): Dynamic mutations hit double figures. *Nat Genet* 8:213–215.

Wu B-L, Austin MA, Schnieder GH, Boles RG, Korf BR. (1995): Deletion of the entire NF1 gene detected by FISH: Four deletion patients associated with severe manifestations. *Am J Med Genet* 59: 528–535.

Zatz M, Marie SK, Passos-Bueno MR, Vainzof M, Campiotto S, Cerquerira A, Wijmenga C, Padberg G, Frants R (1995): High proportion of new mutations and possible anticipation in Brazilian facioscapulohumeral muscular dystrophy families. *Am J Hum Genet* 56:99–105.

3

Defining Disease Phenotypes

Arthur S. Aylsworth
Division of Genetics and Metabolism
Department of Pediatrics and the Neuroscience Center
The School of Medicine
The University of North Carolina at Chapel Hill
Chapel Hill, North Carolina

INTRODUCTION

This chapter discusses some of the mechanisms underlying variation in clinical expression of disease traits, and considers some of the issues involved in choosing a consistent phenotype for study. The importance of phenotype definition and delineation cannot be overemphasized. For what we refer to now as "gene mapping" is actually "phenotype mapping," with identification of new gene loci merely a by-product in our current age of ignorance. After the year 2005, when most functional loci in the human genome will have been mapped and sequenced, there will still remain huge gaps in our knowledge about genotype–phenotype correlations. So as long as there is interest in identifying alleles that cause human phenotypes, there will be continuing, long-term interest in phenotype mapping.

The chapter provides a general overview of issues for consideration. Investigators involved in gene mapping studies typically come from a wide variety of medical, biological, and mathematical backgrounds. Therefore, those not currently working in the field of medical genetics may wish to use as background material one or more basic general medical genetics texts, such as Vogel and Motulsky

Approaches to Gene Mapping in Complex Human Diseases, Edited by Jonathan L. Haines and Margaret A. Pericak-Vance. ISBN 0-471-17195-6

(1997), Strachan and Read (1996), Seashore and Wappner (1996), Nora et al. (1994), Passarge (1995), Gelehrter et al. (1998), Jorde et al. (1995), and Thompson et al. (1991), and other medical genetics monographs, such as Rimoin et al. (1996), Scriver et al. (1995), King et al. (1992), Stevenson et al. (1993), Winter et al. (1988), Aase (1990), Jones (1997), Graham (1988), Cohen (1982), and Gorlin et al. (1990). A few journal references are given, primarily original articles and reviews for topics not covered in depth in the other, general texts. For descriptions and up-to-date references concerning specific diseases, traits, and genetic mechanisms, the reader is encouraged to utilize the online version of *Mendelian Inheritance in Man* (Online Mendelian Inheritance in Man, OMIM (TM), 1997.). Each OMIM entry has a unique six-digit number: the first digit indicates the mode of inheritance (1 = autosomal dominant, 2 = autosomal recessive, 3 = X-linked, 4 = Y-linked, and 5 = mitochondrial phenotypes; numbers beginning with 6 refer to autosomal loci or phenotypes created after May 1994). Many topics discussed here include references to pertinent "MIM numbers" (e.g., MIM xxxxxx for phenotypes and loci; MIM xxxxxx.xxxx refers to specific mutant alleles). Other topics with too many relevant entries to list can be searched in OMIM using the topical key words and Boolean operators. The OMIM database is an extremely valuable resource for those engaged in gene mapping studies. It provides for phenotype delineation the same kind of information offered by other public databases for identification of sequence homology, probes, STSs, ESTs, and so on. OMIM is updated frequently, and its use is strongly recommended for any investigator interested in human and medical genetics.

One of the goals of research in human genetics is to delineate the causal associations and pathogenetic processes that relate genotype to phenotype. Human pedigree patterns frequently vary drastically from the classic, simple Mendelian patterns described in Chapter 2. A number of mechanisms affect gene expression, causing clinical expression to be different from the deterministic, simple dominant and recessive patterns. Heterozygous mutant allele expression ranges from complete (the well-understood, traditional deterministic pattern usually called "Mendelian dominant" inheritance), to silent ("Mendelian recessive"), with clinical effects expressed only in the homozygote or mutant allelic compound. In between these two extremes lies a range of partial or intermediate expression where the phenotype produced by an allele may depend on one or more interactions with products of the other homologous allele, alleles at other loci, environmental influences, and/or chance. Investigators interested in mapping genes associated with specific phenotypes must understand these mechanisms of expression and consider them in each family being studied.

VARIATIONS ON AND EXCEPTIONS TO SIMPLE MENDELIAN INHERITANCE PATTERNS

Mendel spent years selecting the particular traits on which he finally collected extensive crossing data. This was necessary because most traits in plants and animals probably don't "Mendelize" (i.e., most do not follow the straightforward patterns of unitary inheritance described by Mendel). Rather, many pedigrees for common conditions in children and adults resemble that seen in Figure 3.1. Following are some examples of exceptions to the "typical" or simple Mendelian inheritance patterns described in Chapter 2.

Pseudodominant Transmission of a Recessive

"Vertical inheritance" (expression in subsequent generations) of a simple Mendelian recessive trait or condition usually occurs because of consanguinity; that is, a homozygous affected individual mates with someone who is heterozygous by virtue of descent from a common ancestor (Fig. 3.2). Note that if the disease allele frequency is relatively common, an affected individual's mate would not have to be related in order for the couple to have a child who expresses the same recessive phenotype as the parent.

Mosaicism and Pseudorecessive Transmission of a Dominant

A mutation that occurs after the formation of the zygote results in mosaicism; that is, the developing organism contains some cells with the mutation and some without. Mosaicism can be present in all tissues, or it may be predominant in some tissues and not others. We see specific phenotypic effects associated with *somatic mosaicism, germline mosaicism*, and *placental mosaicism*.

Somatic Mosaicism This condition can cause a very asymmetric pattern of phenotypic expression. For example, an individual with somatic mosaicism for a mutation in the gene that causes neurofibromatosis type 1 (MIM 162200) may have café au lait spots and neurofibromas that involve only one segment of the body (Fig. 3.3).

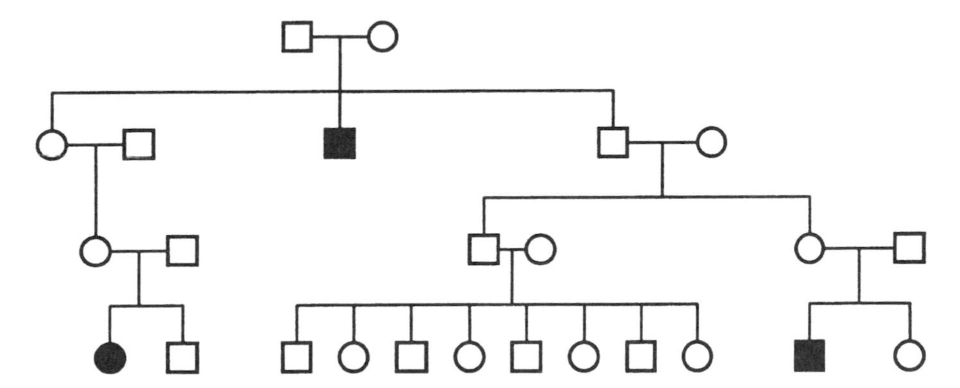

Figure 3.1. *Typical pedigree for many common disorders of children (birth defects, mental retardation) and adults (heart disease, cancer, diabetes, stroke, hypertension, arthritis, mental illness, etc.). Such disorders or traits appear to be familial, but have no obvious, straightforward Mendelian pattern.*

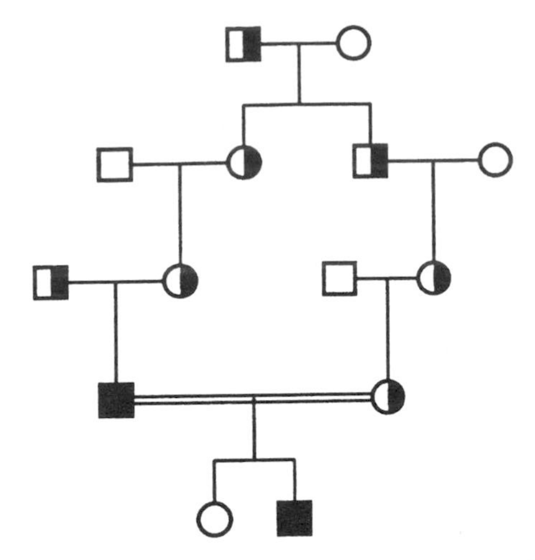

Figure 3.2. *Pseudodominant transmission of a recessive trait: solid symbols, homozygous and affected; half-open symbols, heterozygous and unaffected.*

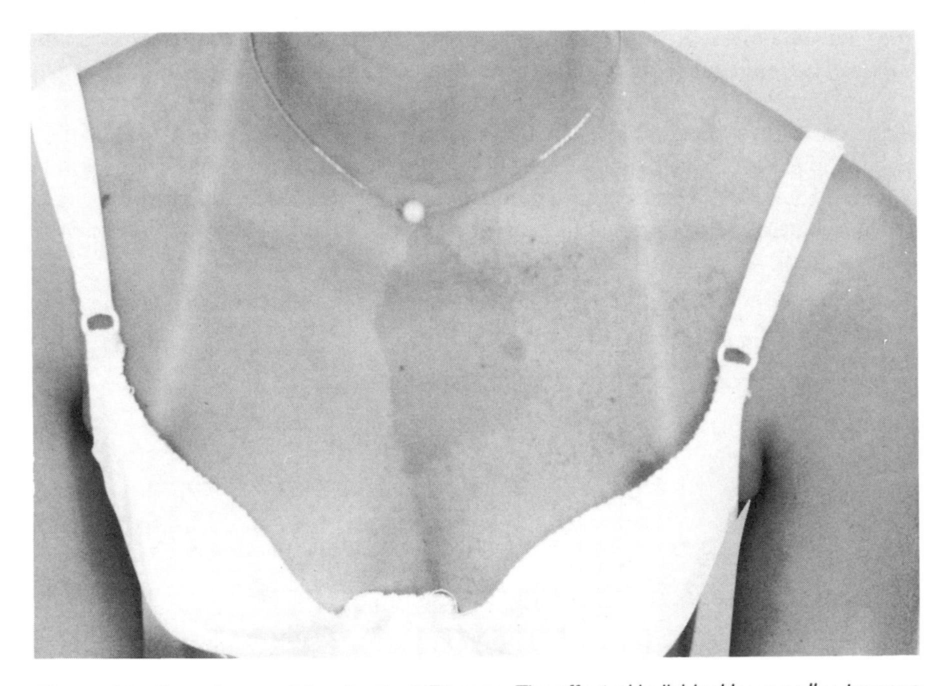

Figure 3.3. *Somatic mosaicism for the NF1 gene. The affected individual has small cutaneous neuorofibromas (not visible) and café au lait spots on her left arm and hand, shoulder, axilla, back, and chest. Note the large spot that stops at the midline. There were no Lisch nodules and no other features of NF1 elsewhere.*

Or, the result can be an unusually mild expression of the phenotype, as demonstrated by parents with phenotypes of Stickler syndrome (MIM 108300) or spondyloepiphyseal dysplasia congenita (MIM 183900), who have had children with the more severe condition known as Kniest dysplasia (MIM 156550, 120140.0012, 120140.0019). By this same mechanism, a parent with a mild form of osteogenesis imperfecta (OI) could have a child with a severe, lethal form of OI (MIM 166210). Normally, milder forms of OI tend to "breed true" within families, with affected relatives having approximately similar clinical manifestations. But it may be that a parent with mild OI, who is the first one in the family to be affected, is mildly affected because the mutation occurred after fertilization, hence is present in only a fraction of the body cells. Such a mutation produces a milder phenotype than one that is present in every cell.

Germline Mosaicism A postzygotic mutation may be present only in a very small proportion of body cells, resulting in minimal or no phenotypic expression. But if such a mutation is present in the germ line, the potential exists for transmission of the mutation with full phenotypic expression to multiple offspring. Therefore, in analogy with the "pseudodominant transmission of a recessive" mentioned above, this phenomenon would present as *pseudorecessive transmission of a dominant* (Fig. 3.4).

Osteogenesis imperfecta (see MIM 166210 and several dozen other related entries) is also a good example of the phenomenon of germ line mosaicism. The observed sib recurrence rate of severe OI (where neither parent is affected) was initially reported to be approximately 5–6%. This result was attributed to the distribution of cases: approximately 25% being autosomal recessive (with a 25% recurrence risk) and the remainder being new dominant mutations (with essentially a negligible recurrence risk). Now, however, it is recognized that most of these recurrences are probably due to germ line mosaicism in one of the parents, since most severe cases of OI have been shown to be heterozygous for mutations in the type I collagen genes, *COL1A1* (MIM 120150) and *COL1A2* (MIM 120160).

There are a number of explanations other than germ line mosaicism for the phenomenon of *pseudorecessive inheritance* when the phenotype of a dominant trait or syndrome appears in multiple offspring of unaffected parents as shown in Figure 3.4. These include mistaken paternity, misdiagnosis, and some of the phenomena considered below, including incomplete penetrance, genetic heterogeneity (a form of misdiagnosis), environmentally caused phenocopy, and genomic imprinting.

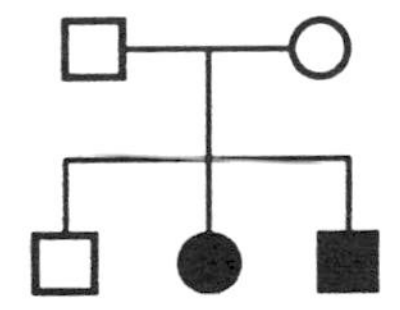

Figure 3.4. *Pseudorecessive transmission of a dominant.*

Mitochondrial Inheritance

General references: Wallace, 1995; Grossman, 1996.

A zygote receives its mitochondria from the egg, not the sperm. Therefore, expression of mitochondrial gene mutations follows a pattern of "maternal inheritance," where all the offspring of an affected woman receive copies of the mutant gene and may be affected (to varying degrees), but there is no transmission through affected males. Maternal mitochondria (and, therefore, traits encoded by mitochondrial genes) are transmitted to both sexes, and pedigrees may resemble the idealized one in Figure 3.5. At first glance, one might conclude that this pedigree is one of autosomal dominant inheritance, but on closer inspection, the key characteristic of exclusively maternal inheritance can be identified. In practice, there may be a great deal of variability of expression of mitochondrial mutations among affected individuals within families because of *heteroplasmy*, the existence of differing proportions of mutant and wild-type mitochondrial DNA in different relatives.

Interesting examples of human conditions with mitochondrial inheritance include several myopathic syndromes (MIM 530000, 540000, 545000), type 2 diabetes and deafness (MIM 520000), chloramphenicol resistance and toxicity (MIM 515000), aminoglycoside-induced hearing loss (MIM 580000), and others listed in Table 3.1. Investigators studying complex phenotypes should keep in mind that while most of the genes influencing their trait of interest are nuclear, one or more subgroups may have a significant genetic component that is mitochondrial. For example, there is good evidence for autosomal recessive inheritance of the Wolfram syndrome (MIM 222300), also referred to as DIDMOAD (**d**iabetes **i**nsipidus, **d**iabetes **m**ellitus, **o**ptic **a**trophy, and **d**eafness), with linkage to markers at 4p16.1. But evidence also exists for a rarer, mitochondrial form (MIM 598500). While many different regions of the nuclear genome have been implicated in the causation of idiopathic dilated cardiomyopathy (over a dozen OMIM entries), mitochondrial deletions in a mother-and-son pair suggest this may be responsible for a subset of cases (MIM 510000). Finally, mitochondrial genes have been implicated in the causation of both Alzheimer disease (MIM 502500) and Parkinson disease (MIM 556500),

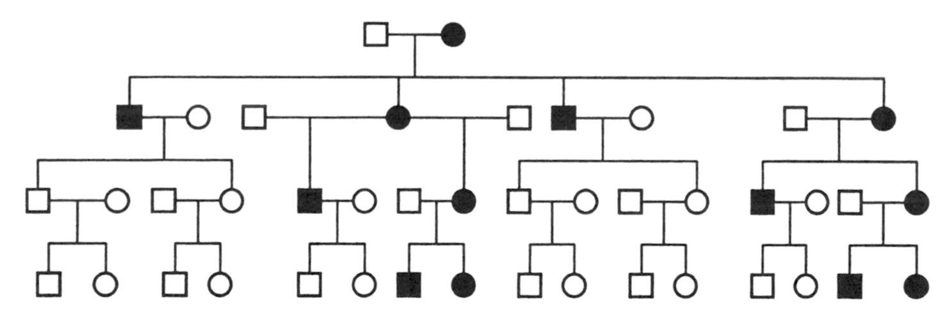

Figure 3.5. *Maternal inheritance of a trait determined by a mutation in mitochondrial DNA. The key features illustrated are that all offspring of affected women are affected, while none of the offspring of affected males are affected. Note that one woman in the F2 generation has had an affected son and daughter by two different mates.*

TABLE 3.1 Mitochondrial Phenotypes

MIM number	Phenotype
502000	Aging
502500	Alzheimer disease (AD)
510000	Cardiomyopathy, idiopathic dilated, mitochondrial
515000	Chloramphenicol toxicity
520000	Diabetes–deafness syndrome, maternally transmitted
520100	Diarrhea, chronic, with villous atrophy
530000	Kearns–Sayre syndrome (KSS)
535000	Leber optic atrophy
540000	MELAS syndrome
545000	MERRF syndrome
550000	Mitochondrial DNA breakage syndrome, secondary to nuclear mutation
550500	Myoglobinuria, recurrent
550900	Myoneurogastrointestinal encephalopathy syndrome
551000	Myopathy, mitochondrial, lethal infantile; limm
551200	Nephropathy, chronic tubulointerstitial
551500	Neuropathy, ataxia, and retinitis pigmentosa
553000	Oncocytoma
555000	Ophthalmoplegia, progressive external, and scoliosis
556500	Parkinson disease
557000	Pearson marrow–pancreas syndrome
560000	Renal tubulopathy, diabetes mellitus, and cerebellar ataxia due to duplication of mitochondrial DNA
580000	Streptomycin ototoxicity
598500	Wolfram syndrome, mitochondrial form

Source: Online Mendelian Inheritance in Man (see Appendix for URL), Sept 1997.

complex phenotypes of great interest to the medical community because of prevalence and morbidity.

Phenocopies and Other Environmentally Related Effects

The term "phenocopy" refers to an environmentally caused phenotype that mimics a trait or syndrome that is usually known or assumed to be of genetic causation. For example, while most cases of DiGeorge syndrome (MIM 188400, 600594) are associated with a chromosome 22q deletion, the phenotypic features can also be produced by prenatal exposure to 13-*cis*-retinoic acid. By the same token, some affected relatives in a family segregating an adult-onset disease such as cancer or diabetes may not share the predisposing gene of interest that is carried by most other affected individuals. Instead, they may have a condition due entirely to environmental factors (or other factors unrelated to the segregating susceptibility gene of interest), which would cause them to be coded (diagnosed) incorrectly in a linkage study.

Some suggest that the term "phenocopy" could also be generalized to include phenomena considered to be a part of the normal aging process, such as age-related macular degeneration, which can mimic autosomal dominant macular dystrophy. This would be appropriate if an environmental agent were to be identified as a major causative factor. But not everyone develops "age-related" degenerations, leaving open the very real possibility of genetic susceptibilities to many of these aging phenomena that are considered "normal." If other major genetic causes are found, these will be examples of "genetic heterogeneity." Use of the term "genetic phenocopy" to refer to "genetic heterogeneity" is discouraged: the term "phenocopy" should be reserved for instances of environmental factors that closely mimic the effects of mutant gene expression.

Note that environmentally caused phenotypes (whether or not they are phenocopies of known genetic syndromes) may also mimic genetic inheritance by being familial. For example, the detrimental effects of prenatal alcohol exposure (so-called fetal alcohol syndrome, or FAS) cause a syndrome of multiple congenital anomalies and mental retardation that can appear to be dominantly transmitted through a family if alcoholism occurs in several generations.

While the actual environmental exposure is presumed to be the major causative factor for conditions like FAS, one must also keep in mind the possibility that an underlying *genetic susceptibility* (embryo, fetus, or adult) may affect an individual's response to environmental factors. Environmental factors including chemicals, drugs, physical agents, and maternal disease may disrupt both prenatal morphogenesis and postnatal function.

While the list of human teratogens with well-documented effects is still relatively short, the environmental factors that influence both embryonic development and adult disease are currently of great research interest and controversy, especially with regard to the interaction of these environmental factors with an individual's underlying genetic susceptibility. For example, the risk of developing emphysema is greater in individuals with both an environmental exposure such as smoking *and* a genetic predisposition such as α_1-1-antitrypsin deficiency (MIM 107400) than it is for individuals with only one of these risk factors or none at all. Other examples include sun exposure and familial susceptibility to skin cancer, obesity and inherited susceptibility to type II diabetes, and subclinical maternal deficiency of folic acid (or other related metabolites) and mutations in genes expressed in the folate metabolism pathways that confer embryo susceptibility to neural tube defects. As pointed out by Khoury and Wagener (1995: see also the related editorial by Ottman, 1995), a future approach to the primary prevention of many diseases will probably be the identification of environmental "cofactors" that lead to clinical disease in individuals with "susceptibility genotypes." The translation of new genetic knowledge about susceptibility alleles into effective intervention programs, therefore, will require extensive population-based epidemiologic studies to characterize these cofactors and quantify their effects in genetically susceptible individuals.

Incomplete Penetrance and Variable Expressivity

Penetrance refers to the proportion of individuals with a particular mutant genotype that express the mutant phenotype. Penetrance is a proportion that ranges between 0 and 1.0 (or 0 and 100%). When 100% of mutant individuals express the phenotype, penetrance is *complete*. If some mutant individuals do not express the phenotype, penetrance is said to be *incomplete* or *reduced*. Dominant conditions with incomplete penetrance are characterized by "skipped" generations with unaffected, obligate gene carriers.

Penetrance, therefore, is the proportion of heterozygotes that express a dominant phenotype, or the proportion of homozygotes (or allelic compounds) that express the recessive phenotype. Conditions known for incomplete penetrance include split hand/foot malformation (MIM 183600) and hemochromatosis (MIM 235200). Note also that penetrance depends on definition of the phenotype. Therefore, carriers of the sickle cell gene (MIM 141900) express (i.e., are penetrant for) the trait of *in vitro* sickling, but do not express (i.e., are nonpenetrant for) the disease sickle cell anemia.

Expressivity refers to the variability in degree of phenotypic expression (i.e., severity) seen in different individuals with the mutant genotype. Expressivity may be extremely variable or fairly consistent, both within families and between families. *Intrafamilial* variability of expression may be due to factors such as epistasis, environment, chance, and mosaicism. *Interfamilial* variability of expression may be due to the preceding factors as well, but may also be due to allelic or locus genetic heterogeneity.

Incomplete penetrance and variable expressivity make it imperative for the investigator to *examine* carefully all relatives of probands rather than just take a verbal history. Failure to carefully examine each relative can result in the missing of minor expressions of anomalies or symptoms of late-onset disease.

Note that the term "variable penetrance" is commonly, but *incorrectly*, substituted for "variable expressivity." *Variable penetrance* is strictly appropriate only for conditions that have different penetrance figures for different populations, which are stratified by some other parameter such as age (see below), race, or pedigree position. But since misapplication of "variable penetrance" has robbed the term of its usefulness, alternative terms for the concepts of "variable penetrance" and "variable expression" are used below.

Incomplete or Semidominant Expression While some conditions we consider to be autosomal dominant may actually fit Mendel's original definition of "dominant" (i.e., where homozygous expression is the same as that of the heterozygote), many probably do not. Current evidence suggests that Huntington disease (MIM 143100) is a true dominant, but achondroplasia (MIM 100800), to name just one example, clearly is not. Homozygous achondroplasia is much more severe than heterozygous achondroplasia, with homozygotes usually dying in the neonatal period. Our knowledge of homozygous phenotypes is limited by the rarity of heterozygote matings for most "dominant" human conditions. Therefore, while the term "semidominant" is used in mouse and other experimental genetic systems, "dominant" is commonly used in human medicine for traits that show vertical transmission, regardless of whether the homozygous phenotype is the same as or more severe than the heterozygous phenotype. This is because, in most cases, only the heterozygous phenotype is known. Investigators studying large families should try to identify and define homozygous phenotypes and should be alert to the possibility of different phenotypic expression in these homozygous individuals. Note that X-linked gene expression is potentially "intermediate" and inherently variable in heterozygous females because of Lyonization. Also, the expression of a mitochondrial mutation in a pedigree with a maternal inheritance pattern is variable because of heteroplasmy.

Sex-Influenced Penetrance and Expression Expression may be affected by the sex of the individual carrying a particular gene. Examples include male pattern baldness (MIM 109200), susceptibility to breast cancer, and congenital anomalies of the internal and external genitalia such as hypospadias or vaginal septum.

Age-Related Penetrance and Expression (Variable Age of Onset) In all conditions of postnatal onset, the proportion of gene carriers who are affected varies with age. Therefore, one must specify age when describing penetrance. For some well-studied conditions such as Huntington disease, age-of-onset data are fairly good, but for many others such data are incomplete. Gene mapping studies should begin with a thorough search of the literature or one's own patient population to define, as well as possible, age-related penetrance values.

Transmission-Related Penetrance and Expression (Dynamic Mutations) (General references: Caskey et al., 1992; Hummerich and Lehrach, 1995; Monckton and Caskey, 1995; O'Donnell and Zoghbi, 1995; Warren and Ashley, 1995; Warren, 1996.) Expression becomes more severe and/or is characterized by earlier age of onset with transmission from generation to generation because of a phenomenon called *true anticipation.* In these disorders, a region of the gene is "unstable" because it contains a trinucleotide repeat. Normally, a few such repeats will be "stable" while a moderate number of repeats (the critical number varies with the disorder) will produce "instability" of the sequence, leading to a large number of repeats and disrupted gene function in subsequent generations. Examples include fragile X syndrome (MIM 309550), myotonic dystrophy (MIM 160900), Kennedy disease (MIM 313200), Huntington disease (MIM 143100), dentatorubro-pallidoluysian atrophy (DRPLA) (MIM 125370), Haw River syndrome (MIM 140340), spinocerebellar ataxias (SCA1, 2, 6, 7) (MIM 164400, 183090, 183086, 164500), and Machado–Joseph disease (MIM 109150).

Note that another explanation for *apparent anticipation* is biased ascertainment of patients. For conditions with a wide range of intrafamilial expressivity, one is much more likely to ascertain families where symptoms are more severe (or started earlier) in later generations than those in which the condition was much more severe in earlier generations. Biased ascertainment is still a possible explanation for apparent anticipation.

Genomic Imprinting (General references: Hall, 1990; Driscoll, 1994; Gold and Pedersen, 1994; Barlow, 1995; Cassidy, 1995; Feinberg et al., 1995; Latham et al.,

1995; Ledbetter and Engel, 1995; Sapienza, 1995.) Genetic contributions from both a mother and father are required for mammalian development. This requirement is related to mechanisms of genetic regulation that cause some normal genes to be expressed differently depending on the parent from which they are inherited. Imprinting implies suppression of expression. Such parent-of-origin differences in gene expression have been identified in humans in the past few years, largely through the study of genetic disorders that are influenced by this mechanism. Figure 3.6 shows how imprinting could result in distantly related individuals being affected despite an absence of affected relatives in between them. This pedigree could be produced by a dominantly expressed, autosomal mutation in a gene that undergoes maternal imprinting (i.e., the maternal allele is not expressed, thereby protecting the organism from expressing the deleterious mutation when it is maternally transmitted). On the other hand, the mutation is expressed when transmission occurs through the father, causing the phenotype to be expressed. The pattern shown represents *"maternal imprinting,"* where both gene expression and the phenotype are suppressed in maternal transmission. Note that layers of complexity can be added to the interpretation of these situations by considering the possibility of an imprinted repressor of an unimprinted mutant gene, or mutations that affect the imprinting process directly, resulting in functional uniparental disomic expression (see below).

Uniparental Disomy (UPD)

General references: Cassidy, 1995; Ledbetter and Engel, 1995.

Inheritance of both homologues of a chromosome pair from one parent with corresponding loss of the chromosomal contribution from the other parent is called uniparental disomy (UPD), an obvious exception to Mendel's "law" of segregation of alleles. (If this corresponding loss of a chromosome from the other parent did not occur, then the zygote would be trisomic.) Phenotypic expression of UPD may be due to homozygosity for a mutant allele that is carried by only one parent. Other mechanisms related to imprinting include gene product deficiency (two copies of an imprinted gene), or gene product overproduction (two copies of a normally ex-

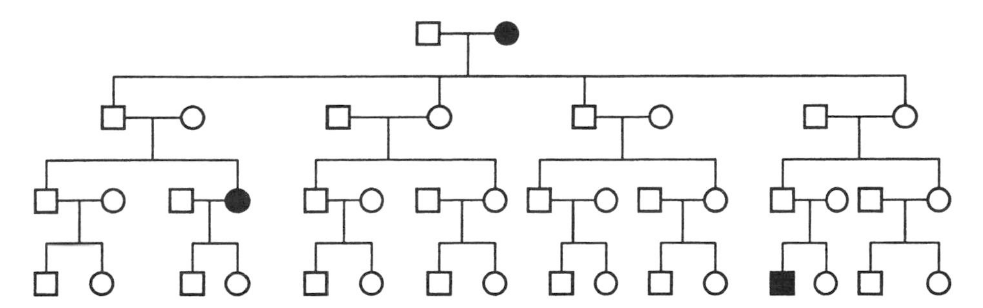

Figure 3.6. *Genomic imprinting: mutation in a gene that undergoes maternal imprinting, where the maternally inherited mutant allele is not expressed.*

printed locus). Note that uniparental disomy can cause male-to-male transmission of an X-linked trait if a son receives both his X and Y chromosomes from his father.

Multigenic Patterns

Contiguous Gene Mutations Chromosome deletions or partial duplications may be cytogenetically visible or submicroscopic [identifiable only by molecular techniques such as fluorescence *in situ* hybridization (FISH) or Southern blot] and may involve more than one gene. Strictly speaking, full trisomy could be listed here also. Small duplications and deletions can be vertically transmitted through families in a "dominant" inheritance pattern, sometimes with large variability in expressivity. Gene mappers are on the lookout for patients who have two major genetic phenotypes that are not usually related. Such findings suggest that the genes for these conditions are closely linked and are affected by a genomic rearrangement such as a submicroscopic deletion. If one of these conditions has already been mapped, one can look for the other in the same chromosomal region. For example, one patient had three different X-linked disorders, a form of chronic granulomatous disease (MIM 306400), Duchenne muscular dystrophy (MIM 310200), and retinitis pigmentosa (MIM 312610). A subtle interstitial deletion at Xp21 was evidence that in this individual, the three disorders were caused by closely linked, contiguous genes.

Translocations Unbalanced translocations may affect several distant individuals in the pedigree, who are related through balanced translocation carriers. Figure 3.1 showed the typical appearance of such a pedigree, with the solid symbols representing individuals with unbalanced karyotypes related through balanced translocation carriers. A number of genes have been mapped by finding balanced translocations in individuals that express the trait or condition of interest. Examples include neurofibromatosis type I (MIM 162200) and Duchenne dystrophy (MIM 310200). The translocation breakpoints are used to clone the gene of interest, which has been interrupted by the translocation. Therefore, one looks for families that both segregate the trait of interest and also have multiple individuals with features compatible with a major chromosome abnormality. Those features usually include growth retardation, mental retardation, and multiple birth defects. Individuals with unbalanced translocations often have a remarkable reproductive history including neonatal death, stillbirth, and miscarriage.

Polygenic/Multifactorial Determination For many, perhaps most traits, expression of an allele depends, at some level, on interactions with other genes and/or environmental factors, where these interactions may be genetically programmed or may be purely random (i.e., stochastic or chance events). Most of the common diseases of children (birth defects, mental retardation) and adults (heart disease, cancer, diabetes, stroke, hypertension, arthritis, mental illness, etc.) probably contain major subsets that fall into this category of causation. Even the phenotypic manifestations of conditions usually considered to be straightforward monogenic disorders may be the result of gene–environment interaction. For example, in the case of phenylketonuria (PKU), phenotypic expression (i.e., mental retardation) depends not only on genotype but also on exposure to the amino acid phenylalanine in dietary protein. If the amount of phenylalanine in our dietary protein was much less than it is, a deficiency of phenylalanine hydroxylase might be identified simply as a benign polymorphism in the human population rather than as a disease that causes mental retardation.

Single birth defects and common diseases of adult life are frequently familial, but available pedigrees, such as that in Figure 3.1, may not suggest a straightforward pattern of Mendelian inheritance. (Note that in such a family, evidence for straightforward Mendelian inheritance can be missed if all relatives are not carefully examined for manifestations of the trait or condition being studied.)

To explain observed "non-Mendelian" patterns of familial recurrence, mathematical models discussed by Falconer (1965) and Edwards (1969) were adapted. These assume a continuous distribution in the general population of "liability" to malformation or disease (Carter, 1969). "Polygenic" determination refers to the mathematical model in which a number of genes with small, additive effects provide an underlying genetic predisposition to malformation or disease. Some quantitative traits and clinical disorders in humans have been studied and found to be compatible with this mechanism of determination (Holt, 1961; Dragani et al., 1996). "Multifactorial" describes models in which environmental factors interact with genetic predisposition, which is frequently thought of as polygenic [i.e., the resulting (additive) effect of a large number of genes, each of which contributes a small effect]. In the case of quantitative traits such as blood pressure, weight, and height, a normal (or Gaussian) curve would represent the distribution of measurements in a population. This model was adapted to account for discontinuous traits by the addition of a threshold, the point within that liability distribution beyond which individuals are affected (Fig. 3.7). A number of observations in human populations and experimental animals have been reported as being consistent with this model of multifactorial inheritance (Carter, 1969, 1973, 1976; Fraser, 1976, 1980, 1981; Mitchell and Risch, 1993).

Monogenic Predisposition An alternative to the polygenic/multifactorial models is that underlying genetic liability is determined by a single gene with incomplete penetrance or variable expressivity (Biddle and Fraser, 1986; Temple et al., 1989; Moore et al., 1990; Hecht et al., 1991; Marazita et al., 1992). For many clinical conditions, the existence of less severely affected relatives seems consistent with the hypothesis of a single allele running through the family which *predisposes* carriers to develop the disorder. But penetrance and expressivity are phenomena that may be affected by or depend on the expression of other genes, environmental factors, time, and parent of origin. Since the factors that influence penetrance and expressivity may very well be a mix of genes and environmental factors, the monogenic model, in practice, may actually be a very complex "mixed" model of major single gene effects on a multifactorial background, where liability may be determined by more than one gene. Therefore, the term "multifactorial" would still ap-

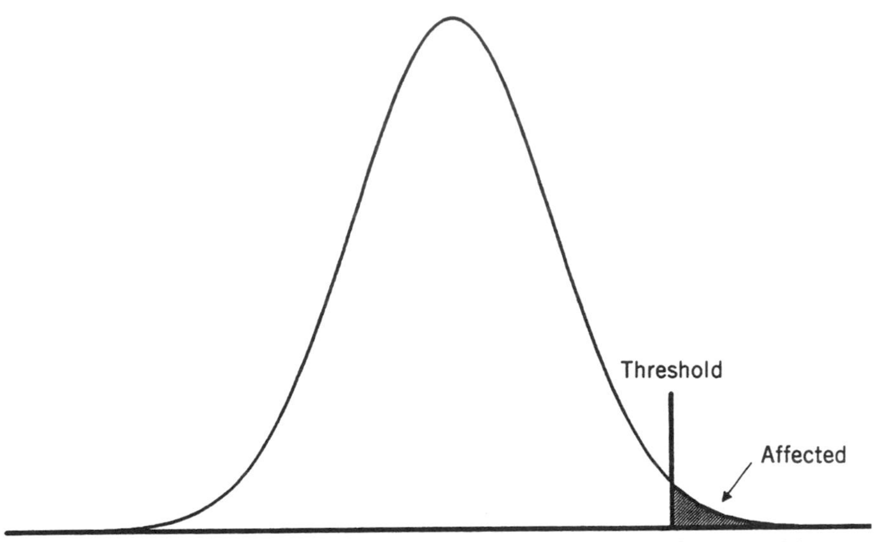

Figure 3.7. Multifactorial threshold model combining a "polygenic" predisposition with environmental effects.

pear to be an appropriate one, if it is not restricted to imply only a "polygenic" contribution without major single gene effects.

The Role of Chance (or, the Roll of the Dice) in Phenotype Expression

Another factor, traditionally ignored in discussions about the cause of genetic disease, is chance. Kurnit and colleagues (Kurnit et al., 1987) used computer modeling to show that some of the non-Mendelian familial clustering of anomalies usually attributed to concepts such as "reduced penetrance" and "multifactorial inheritance" may be accounted for by simple, random chance. Studies of this stochastic, probabilistic model allow us to conclude that we should think of incomplete penetrance in terms of single genes that predispose the organism to develop an anomaly or disease, but do not always result in an abnormal phenotype (Cohen, 1989). Central to this concept is the idea that biological processes (like many other phenomena we encounter daily) are error-prone because they are not unvarying or totally controlled by constants; that is, they are nonlinear systems. The complex processes of embryologic development may be very sensitive to initial, small perturbations (which are, by themselves, "within normal limits") that subsequently, *by chance,* are amplified by the randomness that is inherent in all such systems.

The relative merits of the popular, hypothetical models (polygenic, multifactorial-threshold, and major single gene with incomplete penetrance) have been debated vigorously (Fraser, 1976, 1980, 1981, 1989a, 1989b; Fraser, 1989a; Fraser, 1989b; Mitchell and Risch, 1992; Nemana et al., 1992). In general, it is probably reasonable to assume that most phenotypes can have many different causes, including stochastic combinations of genetic and environmental factors, where the genetic contribution may be determined by many genes, or by only one or a small number of genes.

HETEROGENEITY

Genetic Heterogeneity

A number of different genetic mutations may produce phenotypes that are identical or similar enough to have been traditionally lumped together under one diagnostic heading. In other words, a particular phenotype may be caused by more than one genotype. The conditions of "anemia," "mental retardation," or "dwarfism" are so obviously causally heterogeneous as to seem silly examples. On the other hand, conditions such as tuberous sclerosis (MIM 191092, 191100) and adult polycystic kidney disease (MIM 173900) were incorrectly thought of as genetically homogeneous disorders until relatively recently. Clearly, phenotype definition is a function of our depth of knowledge about a particular trait. But even with very refined, specific phenotype definition, we still find frequent genetic heterogeneity. There are two types of genetic heterogeneity, *allelic heterogeneity* and *locus heterogeneity*.

Locus Heterogeneity A phenotype may be caused by mutations at more than one gene locus; that is, mutations at different loci cause the same phenotype or a group of phenotypes that appear so similar that they had been wrongly classified as a single disease, clinical "entity," or diagnostic spectrum. An example is the Sanfilippo syndrome (MPS III A, B, C, and D; MIM 252900, 252920, 252930, 252940), where the phenotype is produced by four different enzyme deficiencies, each of which is important in the degradation of heparin sulfate, the substance stored in the tissues that causes the clinical disease manifestations. Other examples include polycystic kidney disease and the examples shown in Table 3.2.

Allelic Heterogeneity A phenotype may be caused by more than one allele at a specific gene locus; that is, different mutations at a single locus cause the same phenotype. It is now clear that one of the earliest studied genetic diseases, sickle cell disease (MIM 141900), a disorder caused by one specific base substitution, is the

TABLE 3.2 Examples of Locus Heterogeneity

Mutations in	Cause
Genes on 9q and 16p	Tuberous sclerosis
COL1A1 and *COL1A2*	Osteogenesis imperfecta
Genes on chromosomes 2p, 8, 14q, 15q, and Xq	Spastic paraplegia
FGFR1 and *FGFR2*	Pfeiffer syndrome
BRCA1 and *BRCA2*	Familial breast cancer
Genes on chromosomes 1, 14, 19, and 21	Alzheimer disease

exception rather than the rule. Instead, most disease phenotypes are more like the β-thalassemias, where a variety of different kinds of mutations throughout the β-globin locus cause the disease phenotype. Other examples are shown in Table 3.3.

Phenotypic Heterogeneity

More than one phenotype may be caused by allelic mutations at a single locus. Once again, the β-globin locus (MIM 141900) provides a good example. Different β-globin mutations cause sickle cell disease, hundreds of other hemoglobinopathies, numerous thalassemia variants, and methemoglobinemia. Other examples are shown in Table 3.4. Also, it should be noted that in different families with Crouzon syndrome (OMIM 123500) and Pfeiffer syndrome (OMIM 101600), identical mutations (i.e., involving the same codon or base pair) have been associated with these two very different phenotypes. Possible explanations include epistatic and stochastic factors that affect gene expression. OMIM entries include a subheading for "Allelic variants." These lists are increasingly illustrating the phenomenon of phenotypic heterogeneity by cross-referencing different phenotype entries that have mutations identified at a single locus.

PHENOTYPE DEFINITION

Classification of Disease

The classification systems traditionally used in medicine are somewhat arbitrary. Anatomic systems are commonly used by the surgical specialties, and physiologic or other functionally oriented classifications are frequently used by physicians who see the body as a collection of organ systems as they consider the treatment of symptom complexes. But a classification system based on *causal factors* or *pathogenetic mechanisms* is much more appropriate in genetic analysis. Each of the first two approaches is important in various aspects of medical practice, but the causal approach should be utilized for phenotype definition in gene mapping studies. In this approach, one attempts to classify phenotypes into categories that have etiologic and pathogenetic implications.

The study of congenital anomalies, for example, begins by deciding whether one is dealing with one or more deformations, disruptions, malformations, or an under-

TABLE 3.3 Examples of Allelic Heterogeneity

Different Mutations in	Cause
COL1A1	Osteogenesis imperfecta
CFTR	Cystic fibrosis
FGFR2	Crouzon syndrome
Dystrophin	Duchenne muscular dystrophy

TABLE 3.4 Examples of Phenotypic Heterogeneity

Different Mutations in	Cause
COL2A1	Hypochondrogenesis, Kniest dysplasia, spondyloepiphyseal dysplasia congenita; some cases of Stickler syndrome
CFTR	Cystic fibrosis and congenital bilateral absence of the vas deferens
FGFR2	Crouzon syndrome, Jackson–Weiss syndrome, Pfeiffer syndrome, Apert syndrome
Dystrophin	Duchenne and Becker muscular dystrophies
L1CAM	X-linked hydrocephalus, MASA syndrome, familial spastic paraplegia (SPG1)
FGFR3	Achondroplasia, thanatophoric dysplasia types I and II, hypochondroplasia, Crouzon syndrome with acanthosis nigricans, Muenke nonsydromic coronal craniosynostosis, skeleton-skin-brain syndrome

lying dysplasia (Spranger et al., 1982). This approach, while originally defined for congenital anomalies, can be expanded to apply to genetically caused or influenced functional disorders, including those of postnatal onset. For example, in defining a new or poorly delineated phenotype, it might be helpful to consider whether it represents a dysplasia, an inborn error of metabolism, susceptibility to environmental factors, or a degenerative or dystrophic process.

Major clinical features are clinical findings that are of major medical, surgical, functional, or cosmetic significance. On the other hand, minor clinical features are those that do not have significant medical, surgical, or cosmetic implications for the patient. It is important to recognize these minor features, however, because they may be important diagnostic handles in phenotype definition. A good example is the significance of Lisch nodules in the diagnosis of neurofibromatosis type I (MIM 162200). Also, it should be pointed out that many malformation syndromes are recognizable by their pattern of minor features, rather than by their major features. Down syndrome is a good example of this. Minor features and familial variants, which may provide clues about the cause or pathogenesis of major problems, are important to assess when one is identifying probands for study. The finding of minor features segregating with a major medical problem suggests pleiotropy, and these minor features may be of great help in phenotype definition. Therefore, one's hypothesis of underlying cause as well as the assignment of individuals to study groups should take into account minor as well as major clinical features.

Single, Major Phenotypic Manifestations

A condition with only a single, major clinical feature is "nonsyndromic." Examples include such common, isolated birth defects as spina bifida, cleft lip with or without cleft palate, and congenital heart disease, as well as conditions of postnatal onset (e.g., retinitis pigmentosa, deafness, spastic paraplegia) when they occur as isolated clinical findings.

Pathogenetic Sequences A "sequence" is a pattern of multiple clinical features that are all part of one pathogenetic *sequence* of events. A sequence may have many different potential primary causes and may comprise the only obvious clinical findings in an individual, or it may be part of a broader pattern of clinical expression as in a syndrome or pleiotropic phenotype where several pathogenetic sequences arise from a single, primary mutation. For example, a baby with spina bifida, hydrocephalus, and clubbed feet should not be thought of as having a "multiple congenital anomalies" syndrome. Rather, the hydrocephalus and clubbed feet are secondary manifestations, caused by the single primary malformation, the spinal neural tube defect. Classifying such an infant as having a "hydrocephalus and clubfoot syndrome" would be like classifying a familial cancer syndrome by the sites of metastases, rather than by the primary malignancy. Similarly, a stroke may be secondary to one or more underlying abnormalities including hypertension, atherosclerosis, and congenital vascular malformation. An attempt should be made to use the most primary cause as the basis for phenotypic classification.

Syndromic Phenotypic Expression

If more than one tissue, organ system, developmental field, or region of the body is dysmorphic, dysplastic, dysfunctional, or dystrophic, the collection of features can be considered *syndromic*. Used in this way, the term does not necessarily imply that the phenotype constitutes a recognizable or previously well-defined syndrome. The word *syndrome* means "a running together," and it is used to refer to a pattern of multiple abnormalities thought to be causally and/or pathogenetically related. Potential causes of syndromes include single alleles that manifest pleiotropy (e.g., a mutant allele having effects in multiple body sites, organs, or tissues), multigenic (genomic or contiguous gene) mutations, environmental factors, and conditions caused by a combination of genetic and nongenetic factors. For example, birth defects like spina bifida and cleft lip with or without cleft palate may be found in children who also have other primary malformations. Such individuals should be separated out from a study group targeting the nonsyndromic malformation. When retinitis pigmentosa (RP) is found in adults who also have severe, congenital sensory deafness, the condition should be classified as syndromic (see MIM 276900 and others) rather than as nonsyndromic RP. This is a good example of how the OMIM database can help one begin studying a trait of interest. The database contains 112 entries that mention "retinitis pigmentosa," including several dozen conditions in which RP is syndromic, associated with abnormalities of the nervous system, skeleton, kidneys, and other organs.

Associations and Syndromes of Unknown Cause

A recurrent and recognizable pattern of clinical features may be called a "syndrome" even though there is no suggestion of a genetic or environmental cause. Such *unknown genesis syndrome* phenotypes are initially reported as occurring sporadically (i.e., only once in each family). At this initial stage of phenotype delineation, one should keep in mind the probability of causal heterogeneity. Sporadic occurrence of a "genetically lethal" phenotype (where affected individuals do not reproduce) suggests the possibility of a new, dominantly expressed mutation in a single gene or a small chromosome deletion that is undetectable by routine standard cytogenetic banding techniques. Some phenotypes previously classified as "unknown genesis syndromes" are now seen to be caused by identifiable genetic or environmental factors. Examples include the Williams syndrome (MIM 194050) and the Prader–Willi syndrome (MIM 176270).

The term *association* refers to the observation that multiple primary clinical features not known to be related by common etiologic or pathogenetic mechanisms occur together significantly more often than expected by chance. An early example in the childhood malformation literature is the VATER association (MIM 192350), also sometimes called the VACTERL association, where the name is an acronym representing the areas involved in malformation (**v**ertebral, **a**nal, **t**racheo**e**sophageal fistula, and both **r**adial and **r**enal anomalies). Such an association is a *causally nonspecific category* used to keep track of a heterogeneous group of undelineated syndromes and sequences. Because they are statistical rather than biological entities, associations are not definitive diagnoses. Therefore, one cannot BEGIN with a strategy of mapping an "association" because of the assumed underlying causal heterogeneity. Rather, the goal should be to delineate causally specific syndromes and disease complexes from within association categories for eventual mapping studies. For example, there is evidence for Mendelian inheritance of VATER anomalies when these are associated with hydrocephalus (MIM 276950 and 314390).

Qualitative (Discontinuous) Traits

Most medical conditions and "abnormal" traits are recognized by their state of being significant deviations from the norm. Signs and symptoms indicate whether a disease state is present or absent. Traits with complete penetrance and easily defined patterns of phenotypic expression are optimal for linkage analysis. For example, mapping the achondroplasia locus (MIM 100800) was facilitated by existence of a readily recognizable phenotype in both children and adults. On the other hand, adult-onset disorders with variable expression and incomplete penetrance will be much more difficult to diagnose and assign to diagnostic groups.

Quantitative (Continuous) Traits

Many traits such as height, blood pressure, head circumference, or how well one scores on a standardized test can be quantified, with the resulting measurements being distributed in a continuous fashion across a population (see Polygenic/Multifactorial Determination, above). Conversion of continuous traits into discontinuous ones by defining a "threshold" for diagnosis as in Figure 3.7 (i.e., short stature, hypertension, microcephaly, or mental retardation, to use the examples mentioned above) may be used in linkage studies, but the criteria used for such conversion should be as biologically meaningful as possible, to minimize problems of underlying genetic heterogeneity. The true state of nature, however, is probably that quantitative or continuous traits are genetically complex, with potentially a large number of loci that determine or affect each trait. This high degree of genetic complexity is probably also true for most of the common medical conditions that afflict both children and adults.

A SUGGESTED APPROACH TO PHENOTYPE DEFINITION

Some general suggestions can be made about approaching phenotype definition in mapping studies.

1. Begin by searching for the trait or disease phenotype of interest in OMIM, using more than one search strategy to identify all possible entries. This approach may uncover a surprising amount of literature that otherwise would not have been considered. In addition, look for other model organisms, especially the mouse, that express a phenotype similar to the one of interest. One additional recommended resource for the human geneticist is the Jackson Laboratory and its online services, which include the Mouse Genome Database (accessible at *http://www.informatics.jax.org/mgd.html*). Mouse–human homologies are rapidly being delineated, and these should provide the human geneticist with a wealth of additional, useful mapping information.
2. Start with inclusive definitions. That is, if possible, assume that the phenotype of interest may be an association. Utilize the phenotypic spectrum of literature cases to assess the feasibility of defining the trait in the study population.
3. Remember that cytogenetic abnormalities can be of great help in identifying candidate regions in which to focus mapping studies. Try to ascertain patients or families that segregate the trait of interest (or some of the syndromic manifestations) *plus* other major or minor features, and look for a small deletion or translocation to identify a candidate region. To find these patients and families, be on the lookout for those with other major findings that are not usually associated with the trait or syndrome of interest, especially such apparently nonspecific findings as mental retardation, short stature, and dysmorphic features. Such minor findings reflect an altered pattern of embryonic or fetal morphogenesis that may be due to deletions or rearrangements of contiguous genes. Such patients should then be studied carefully for chromosomal rearrangements that might provide candidate regions for further mapping studies.
4. Consider carefully the biological significance of quantitative traits. In converting quantitative traits to qualitative traits, where possible, try to include other biologically significant criteria in deciding on threshold definitions. For example, if one wanted to map a gene that causes only mild or borderline short stature, one would try to combine a height threshold with one or more other significant (major or minor) clinical or radiographic features. The more overlap there is between the

general population and the affected population, the more important are these associated features in establishing criteria for a diagnostic classification.

5. Study apparently nonpenetrant individuals as carefully as possible and publish the clinical data.

6. Analyze the pedigrees in the study population for evidence of maternal inheritance and imprinting.

7. Use a checklist form for data collection. List definite and possibly pertinent clinical findings gleaned from literature review. Be inclusive. Collect data that can be used to better define the phenotypic features and natural history once the gene has been identified.

8. Just as it is important to have on the team one or more medical specialists with in-depth interest in and knowledge about any medical condition under study, it is also appropriate to include one or more clinical geneticists into the research protocol if they are not already part of the study. Every gene mapping project should also be a phenotype delineation study, since most phenotypes are not well defined in terms of the relationship between genotype and pathogenesis, variability of expression, penetrance, and natural history. Teams involved in family studies should collect data on intrafamilial and interfamilial variability of expression and longitudinal natural history. Funding agencies should recognize the importance of these clinical aspects of human genome research by emphasizing and supporting phenotype delineation in the same manner that they require and support expertise in genotyping.

REFERENCES

Aase JM (1990): *Diagnostic Dysmorphology*. New York: Plenum Medical Book Company.

Barlow DP (1995): Gametic imprinting in mammals. *Science* 270:1610–1613.

Biddle FG, Fraser FC (1986): Major gene determination of liability to spontaneous cleft lip in the mouse. *J Craniofac Genet Dev Biol* [Suppl] 2:67–88.

Carter CO (1969): Genetics of common disorders. *Br Med Bull* 25:52–57.

Carter CO (1973): Multifactorial genetic disease. In McKusick VA, Claiborne R, eds. *Medical Genetics*. New York: HP Publishing, Chapter 19, pp. 199–208.

Carter CO (1976): Genetics of common single malformations. *Br Med Bull* 32:21–26.

Caskey CT, Pizzuti A, Fu YH, Fenwick RG, Jr., Nelson DL (1992): Triplet repeat mutations in human disease. *Science* 256:784–789.

Cassidy SB (1995): Uniparental disomy and genomic imprinting as causes of human genetic disease. *Environ Mol Mutagen* 25:13–20.

Cohen MM, Jr (1982): *The Child with Multiple Birth Defects*. New York: Raven Press.

Cohen MM, Jr. (1989): Syndromology: An updated conceptual overview. VI. Molecular and biochemical aspects of dysmorphology. *Int J Oral Maxillofac Surg* 18:339–346.

Dragani TA, Canzian F, Pierotti MA (1996): A polygenic model of inherited predisposition to cancer. *FASEB* 10:865–870.

Driscoll DJ (1994): Genomic imprinting in humans. *Mol Gen Med* 4:37–77.

Edwards JH (1969): Familial predisposition in man. *Br Med Bull* 25:58–64.

Falconer DS (1965): The inheritance of liability to certain diseases, estimated from the incidence among relatives. *Ann Hum Genet* 29:51–76.

Feinberg AP, Rainier S, DeBaun MR (1995): Genomic imprinting, DNA methylation, and cancer. Monographs, National Cancer Institute, 21–26.

Fraser FC (1976): The multifactorial/threshold concept—Uses and misuses. *Teratology* 14:267–280.

Fraser FC (1980): Evolution of a palatable multifactorial threshold model. *Am J Hum Genet* 32:796–813.

Fraser FC (1981): The genetics of common familial disorders—Major genes or multifactorial? *Can J Genet Cytol* 23:1–8.

Fraser FC (1989a): Mapping the cleft-lip genes: The first fix? *Am J Hum Genet* 45:345–347.

Fraser FC (1989b): Research revisited. The genetics of cleft lip and cleft palate. *Cleft Palate J* 26:255–257.

Gelehrter TD, Collins FS, Ginsburg D (1998): *Principles of Medical Genetics*, 2nd ed. Baltimore: Williams & Wilkins.

Gold JD, Pedersen RA (1994): Mechanisms of genomic imprinting in mammals. *Curr Top Dev Biol* 29:227–280.

Gorlin RJ, Cohen MM, Jr, Levin LS (1990): *Syndromes of the Head and Neck*, 3rd ed. New York: Oxford University Press.

Graham JM, Jr (1988): *Smith's Recognizable Patterns of Human Deformation*, 2nd ed. Philadelphia: Saunders.

Grossman LI (1996): Mitochondrial mutations and human disease. *Environ Mol Mutagen* 25:30–37.

Hall JG (1990): Genomic imprinting: Review and relevance to human diseases. *Am J Hum Genet* 46:857–873.

Hecht JT, Yang P, Michels VV, Buetow KH (1991): Complex segregation analysis of nonsyndromic cleft lip and palate. *Am J Hum Genet* 49:674–681.

Holt SB (1961): Quantitative genetics of finger-print patterns. *Br Med Bull* 17:247–250.

Hummerich H, Lehrach H (1995): Trinucleotide repeat expansion and human disease. *Electrophoresis* 16:1698–1704.

Jones KL (1997): *Smith's Recognizable Patterns of Human Malformation*, 5th ed. Philadelphia: Saunders.

Jorde LB, Carey JC, White RL (1995): *Medical Genetics*. St. Louis: Mosby.

Khoury MJ, Wagener DK (1995): Epidemiological evaluation of the use of genetics to improve the predictive value of disease risk factors. *Am J Hum Genet* 56:835–844.

King RA, Rotter JI, Motulsky AG, eds. (1992): *The Genetic Basis of Common Diseases*. New York: Oxford University Press.

Kurnit DM, Layton WM, Matthysse S (1987): Genetics, chance, and morphogenesis. *Am J Hum Genet* 41:979–995.

Latham KE, McGrath J, Solter D (1995): Mechanistic and developmental aspects of genetic imprinting in mammals. *Int Rev Cytol* 160:53–98.

Ledbetter DH, Engel E (1995): Uniparental disomy in humans: Development of an imprinting map and its implications for prenatal diagnosis. *Hum Mol Genet* 4:1757–1764.

Marazita ML, Hu DN, Spence MA, Liu YE, Melnick M (1992): Cleft lip with or without cleft palate in Shanghai, China: Evidence for an autosomal major locus. *Am J Hum Genet* 51:648–653.

Mitchell LE, Risch N (1992): Mode of inheritance of nonsyndromic cleft lip with or without cleft palate: A reanalysis. *Am J Hum Genet* 51:323–332.

Mitchell LE, Risch N (1993): The genetics of infantile hypertrophic pyloric stenosis. A reanalysis. *Am J Dis Children* 147:1203–1211.

Monckton DG, Caskey CT (1995): Unstable triplet repeat diseases. *Circulation* 91:513–520.

Moore G E., Ivens A, Newton R, Balacs MA, Henderson DJ, Jensson O (1990): X chromosome genes involved in the regulation of facial clefting and spina bifida. *Cleft Palate J* 27:131–135.

Nemana LJ, Marazita ML, Melnick M (1992): Genetic analysis of cleft lip with or without cleft palate in Madras, India. *Am J Med Genet* 42:5–9.

Nora JJ, Fraser FC, Bear J, Greenberg CR, Patterson D, Warburton D (1994): *Medical Genetics: Principles and Practice*, 4th ed. Philadelphia: Lea & Febiger.

O'Donnell DM, Zoghbi HY (1995): Trinucleotide repeat disorders in pediatrics. *Curr Opini Pediat* 7:715–725.

Online Mendelian Inheritance in Man, OMIM (@TM). Center for Medical Genetics, Johns Hopkins University (Baltimore, MD) and National Center for Biotechnology Information, National Library of Medicine (Bethesda, MD), 1997.
http://www3.ncbi.nlm.nih.gov/omim/

Ottman R (1995): Gene–environment interaction and public health. *Am J Hum Genet* 56:821–823.

Passarge E (1995): *Color Atlas of Genetics*. Stuttgart and New York: G. Thieme Verlag; Thieme Medical Publishers.

Rimoin DL, Connor JM, Pyeritz RE, eds. (1996): *Emery and Rimoin's Principles and Practice of Medical Genetics*, 3rd ed. Edinburgh: Churchill Livingstone.

Sapienza C (1995): Genome imprinting: An overview. *Dev Genet* 17:185–187.

Scriver CR, Beaudet AL, Sly WS, Valle D, eds. (1995): *The Metabolic and Molecular Bases of Inherited Disease,* 7th ed. New York: McGraw-Hill.

Seashore MR, Wappner RS (1996): *Genetics in Primary Care and Clinical Medicine* Stamford, CT: Appleton & Lange.

Spranger J, Benirschke K, Hall JG, Lenz W, Lowry RB, Opitz JM, Pinsky L, Schwarzacher HG, Smith DW (1982): Errors of morphogenesis: Concepts and terms. Recommendations of an international working group. *J Pediatr* 100:160–165.

Stevenson RE, Hall JG, Goodman RM, eds. (1993): *Human Malformations and Related Anomalies*. New York: Oxford University Press.

Strachan T, Read AP (1996): *Human Molecular Genetics*. New York: BIOS Scientific Publishers Limited/Wiley-Liss.

Temple K, Calvert M, Plint D, Thompson E, Pembrey M (1989): Dominantly inherited cleft lip and palate in two families. *J Med Genet* 26:386–389.

Thompson MW, McInnes RR, Willard HF (1991): *Thompson & Thompson—Genetics in Medicine*, 5th ed. Philadelphia: Saunders.

Vogel F, Motulsky AG (1997): *Human Genetics: Problems and Approaches*, 3rd ed. Berlin: Springer-Verlag.

Wallace DW (1995): Mitochondrial DNA variation in human evolution, degenerative disease, and aging. *Am J Hum Genet* 57:201–223.

Warren ST (1996): The expanding world of trinucleotide repeats. *Science* 271:1374–1375.

Warren ST, Ashley CT, Jr (1995): Triplet repeat expansion mutations: The example of fragile X syndrome. *Annu Rev Neurosci* 18:77–99.

Winter RM, Knowles SAS, Bieber FR, Baraitser M (1988): *The Malformed Fetus and Still-birth. A Diagnostic Approach*. Chichester: Wiley.

4

Basic Concepts in Linkage Analysis

Margaret A. Pericak-Vance
Section of Medical Genetics
Center for Human Genetics
Duke University Medical Center
Durham, North Carolina

INTRODUCTION

Linkage analysis is just one of several methods that can be used to map genes. Laboratory mapping methods such as somatic cell hybrid studies and fluorescent in situ hybridization have been very successful in mapping genes that had already been identified. These conventional methods cannot always be applied, however, because for most diseases, the underlying biochemical defect remains unknown. Thus, statistical linkage techniques, which do not require this information, have become the key step in the positional cloning and gene identification process.

Application of molecular biology techniques in combination with statistical linkage analysis has led to rapid localization of genetic loci for many disorders including both those with a simple Mendelian mode of transmission (Gusella et al., 1983; Tsui et al., 1985; Vance et al., 1989; Siddique et al., 1991) and susceptibility loci for more complex diseases (Pericak-Vance et al., 1991; Jeunemaitre et al., 1992; Corder et al., 1993; Postma et al., 1995). Maps of highly polymorphic markers with spacing of 5–10 cM are now available for all chromosomes through numerous online databases (Chapter 16). These accurate and well-defined maps facilitate efficient

Approaches to Gene Mapping in Complex Human Diseases, Edited by Jonathan L. Haines and Margaret A. Pericak-Vance. ISBN 0-471-17195-6

screening for linkage, particularly in genetically complex disease (Davies et al., 1994; The Multiple Sclerosis Genetic Group; 1996, Pericak-Vance et al., 1997).

Beyond the initial localization of genes, genetic linkage analysis has also proven to be an extraordinary tool in the identification, cloning, and characterization of disease genes. As the field moves from the positional cloning to the positional candidate approach to identifying disease genes (Collins, 1995), linkage analysis will continue to play a key role in finding the genetic susceptibility loci for more common human afflictions. It will also help dissect the interactions with the other risk factors, both genetic and nongenetic, that represent the future of modern medicine

BACKGROUND

In linkage analysis cosegregation of two or more loci (genes or traits) is examined in a family to determine whether the two chromosomes demarked by differing alleles at those loci segregate independently according to Mendel's laws or tend to be inherited together more often that not, violating Mendel's laws. While there are several possible explanations for the latter result, the most likely is that the two (or more) genes reside in close physical proximity to each other. Alleles of genes residing on the same chromosome should segregate together at a rate that is related to the distance between them on the chromosome.

Recombination Fraction

The measure of genetic linkage is the recombination fraction, which is the probability that a parent will produce a recombinant offspring. Recombination occurs when homologous chromosomes cross over. A nonrecombinant offspring is one in which the parental type remains intact. Multiple crossovers can occur between two loci. If an even number of crossovers have occurred between the two loci, the resulting offspring looks like a nonrecombination between the two loci. However, multiple crossovers between closely linked genes are very rare. In humans the recombination fraction is referred to as theta (θ). Theta ranges from 0 for loci that are completely linked to 0.5 for unlinked loci on the same or different chromosomes.

The unit of measurement of genetic linkage is usually the centiMorgan (cM) (1/100 of a morgan). One map unit corresponds to 1 cM, which usually represents 1% recombination. Small values of θ are equivalent to the actual map distance (w) between loci, and thus recombination fractions are additive over small distances. The simplest case relating θ to w occurs when it can be assumed that multiple crossovers between two loci do not occur when the distance between two loci is very small, then $\theta = w$.

For larger distances, recombination fractions are not additive because multiple crossovers occur. When this is the case, mapping functions must be used to translate θ values into actual map distance. Some commonly used mapping functions are those of Kosambi (1944) and Haldane (1919). Haldane's mapping function assumes no interference, (i.e., crossing over is evenly distributed over the entire chromosome). The Kosambi mapping function takes into account interference and is the mapping function most often used in humans. Other mapping functions that also model interference include those of Rao (1977) and Carter and Falconer (1951).

Examples of Phase and Informativeness

To determine whether a pedigree will be useful for linkage analysis, two questions must be considered. First, are critical individuals in the pedigree doubly heterozygous ("informative") at the loci under investigation? An individual is considered informative for linkage analysis if his or her offspring can be scored as to whether or not they represent recombination events between the loci. Second, with what degree of certainty can these offspring be scored as recombinants or nonrecombinants? This certainty is determined by whether linkage phase can or cannot be established. The linkage phase of two loci is the linear arrangement of the alleles at the loci on the chromosomes. In experimental organisms, linkage phase is often described in terms of whether two alleles are in coupling (on the same chromosome) or repulsion (on different chromosomes). "Linkage phase known" and "linkage phase unknown" pedigrees generate different amounts of information for a linkage analysis.

In Figure 4.1, assume that a rare, autosomal dominant disorder is segregating. This assumption of a rare disorder is frequent in Mendelian disease and is important because it allows the investigator to assume that unaffected spouses (individuals 2 and 4) are not carriers for the disease gene. It further allows the assumption that the affected individuals (individuals 1, 3, 5, 6) are heterozygous for the disease allele, since the frequency of individuals homozygous for the disease allele is low enough

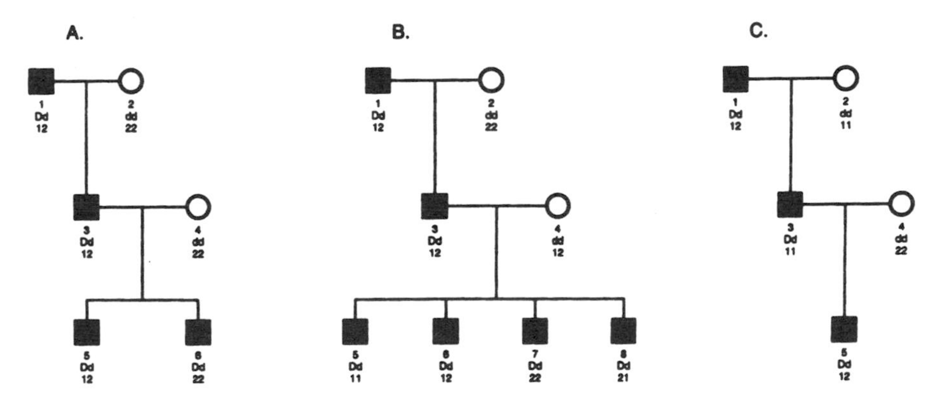

Figure 4.1. *Examples of fully penetrant, autosomal dominant pedigrees.* (A) *Linkage phase known, fully informative mating between individuals 3 and 4 with one recombinant offspring and one nonrecombinant offspring.* (B) *Linkage phase known mating between individuals 3 and 4. Note that at the marker locus, individuals 3 and 4 are identically heterozygous. Thus, only offspring who are homozygous at the marker locus contribute full information to the lod score. The heterozygous offspring contribute some small amount of information to the lod score, depending on the assumed recombination fraction.* (C) *Although linkage phase is known, individual 3 is homozygous at the marker locus, and this mating is uninformative.*

to be negligible. Consequently, affected individuals can be assigned the genotype *Dd* at the trait locus, where *D* represents the disease allele and *d* represents the normal allele. In Figure 4.1*A*, the affected grandfather, individual 1, has the genotype "12" at the marker locus, so he is heterozygous at both the disease and marker loci under question. Therefore, this individual meets the minimal criteria for being informative for linkage analysis, recognizing that several assumptions about genotypes at the disease locus have been made. Linkage phase is established in this pedigree by noting that the affected grandfather has transmitted the disease allele with the '1' allele at the marker locus to his affected son. Frequently, the knowledge of linkage phase in such a pedigree is designated as *D1/d2*, with the slash distinguishing the two chromosomes. Any offspring of individual 3 who have inherited his *D1* genotype are scored as nonrecombinants (individual 5) and those who have inherited a *D2* genotype are scored as recombinants (individual 6). In Figure 4.1*B*, since the affected father (individual 3) and his unaffected spouse (individual 4) are both identically heterozygous for the marker locus (an "intercross mating"), the pedigree is considered only partially informative. Offspring who are homozygous at the marker loci can be scored as recombinants or nonrecombinants after determination of linkage phase, while the heterozygous offspring cannot be scored unambiguously. Note that 50% of offspring of an intercross mating are heterozygous, in most cases leading to a significant loss of information. In Figure 4.1*C*, although individual 3 is heterozygous at the disease locus, he is uninformative for linkage analysis because he is homozygous at the marker locus, regardless of his spouse's genotype. Subsequent to the availability of highly polymorphic microsatellite repeat markers, the frequency of intercross matings and critical affected individuals who are homozygous at the marker locus has decreased considerably.

In a recessive disease, parents of an affected offspring are both carriers for the trait allele and are therefore heterozygous at the trait locus. Meioses from both parents will contribute segregation information, differing from studies in dominant disorders where only meioses of the affected parent contribute. Several examples are shown in Figure 4.2, in which the pedigrees are segregating for a rare autosomal recessive disorder. For example, the pedigree in Figure 4.2*A* is fully informative. Both parents are heterozygous at each of the marker loci under consideration, and the investigator can determine with certainty that the carrier father transmits the '1' allele with the disease allele and the carrier mother transmits the '3' allele with the disease allele. Once individual 3 has been used to establish the linkage phase as described above, it can be determined that individual 4 inherited a nonrecombinant gamete from his carrier father and similarly a nonrecombinant gamete from his carrier mother. The pedigree in Figure 4.2*B*, however, is only partially informative. Here, the mother is homozygous at the marker locus, so gametes transmitted from her cannot be scored as recombinant or nonrecombinant. Similarly, the pedigree in Figure 4.2*C* is partially informative. Although each parent is heterozygous at both the disease and marker loci, the mother and father are identically heterozygous at the marker locus. Thus, only offspring who are homozygous at the marker locus can be scored unequivocally as recombinants or nonrecombinants. (It should be noted that in both the dominant and recessive cases, the heterozygous offspring of such intercross matings do contribute a very

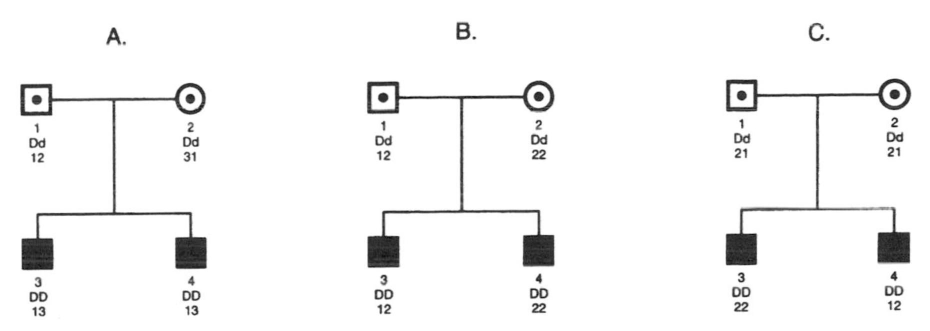

Figure 4.2. *Autosomal recessive pedigrees.* (A) *Fully informative mating: both parents are heterozygous, without identical genotypes at the marker locus, at the disease locus and at the marker locus.* (B) *Partially informative mating: individual 2 is homozygous at the marker locus and thus her gametes cannot be scored as recombinant or nonrecombinant.* (C) *Partially informative mating: the parents are identically heterozygous at the marker locus; therefore, only offspring who are homozygous at the marker locus contribute fully to the lod score.*

small amount to the lod score after establishment of linkage phase. This result is not intuitive, but can be shown utilizing likelihood theory).

Tests of Genetic Linkage

Two genes are said to be linked if the recombination fraction between them is significantly less than 0.5. Simple tests of linkage can be performed using a chi-square test comparing the number of expected number of recombinant or nonrecombinant offspring to the observed number. This is the approach used in most animal systems such as mice and *Drosophila*. Analysis of linkage in humans is more complex owing to the variability of family structure, the limitations in family size, and the inability to always determine phase in the doubly heterozygous parent. Thus these simple mathematical solutions were replaced by more complex analyses that involve the calculation of likelihoods. Over 50 years ago, Haldane and Smith (1947) developed a forerunner of the methods used commonly today. They proposed the probability ratio test for linkage. This ratio is the probability of the data at some given value of θ divided by the probability of nonlinkage ($\theta = 0.5$). To add the resulting scores across families, the log of the probability ratio or a lod score was used (Barnard, 1949).

The likelihood ratio approach to linkage analysis remains the most powerful linkage method available. In developing this technique, Morton (1955) integrated the probability ratio test of Haldane and Smith (1947) with the sequential sampling analysis developed by Wald (1947). The application of sequential analysis to the linkage test added a third option to the usual dualistic acceptance or rejection of the null hypothesis of no linkage—namely, that of testing additional data before a final decision is made regarding linkage or nonlinkage. Use of a sequential test is appropriate for pedigree studies because in such studies, data are usually ascertained on a family-by-family (i.e., sequential) basis.

LOD SCORES

The lod score as first defined by Morton (1955), represents $\log_{10}$ of the odds for linkage. The likelihood (L) of observing a particular configuration of a disease and a marker locus in a family is calculated assuming no linkage, hence free recombination between the two loci (i.e., θ, the recombination fraction, is 0.50). This likelihood is then compared with the likelihood of observing the same configuration of the two loci within the same family, assuming varying degrees of linkage over a selected range of recombination fractions (i.e., θ ranges from 0.0 to 0.5). Log_{10} of the ratio of these likelihoods is then determined for each value of θ within the range, and each of the resulting numbers is referred to as a lod score, $z(x)$, where x represents a particular value of θ within the range of recombination fractions:

$$z(x) = \log_{10}\left[\frac{L(\text{pedigree given } \theta = x)}{L(\text{pedigree given } \theta = 0.5)}\right] \tag{4.1}$$

The value $z(x)$ is referred to as a two-point lod score, since it involves linkage between only two loci (i.e., a disease locus and a marker locus). Equation (4.1) assumes equal recombination for both sexes in the pedigree under consideration, although differences in recombination fractions between sexes are well documented.

Two-Point Lod Scores

As previously outlined, the likelihood (L) of the pedigree is determined at x, a given value of θ, and compared to $\theta = 0.5$, the null hypothesis. The null hypothesis (H_0) assumes no linkage, the alternate hypothesis (H_A) assumes that the disease and the marker locus are linked. To demonstrate linkage, there must be evidence of cosegregation that can support rejection of the null hypothesis

$$L = \theta^R (1 - \theta)^{NR} \tag{4.2}$$

of no linkage. For the example in Figure 4.3, the likelihood (4.2) of observing the linkage data is where θ is the recombination fraction, R the number of recombinants, and NR the number of nonrecombinants. Thus, $N = NR + R$, where N is the total number of offspring. Therefore, the lod score for these data at $\theta = 0.05$ based on the general equation (4.1) is calculated as

$$z(x) = \log_{10}\left[\frac{\theta^R(1-\theta)^{NR}}{(0.5)^R(0.5)^{NR}}\right]$$

$$z(0.05) = \log_{10}\left[\frac{(0.05)^1(0.95)^7}{(0.5)^8}\right] = 0.9513$$

$$z(0.05) = 0.95 \tag{4.3}$$

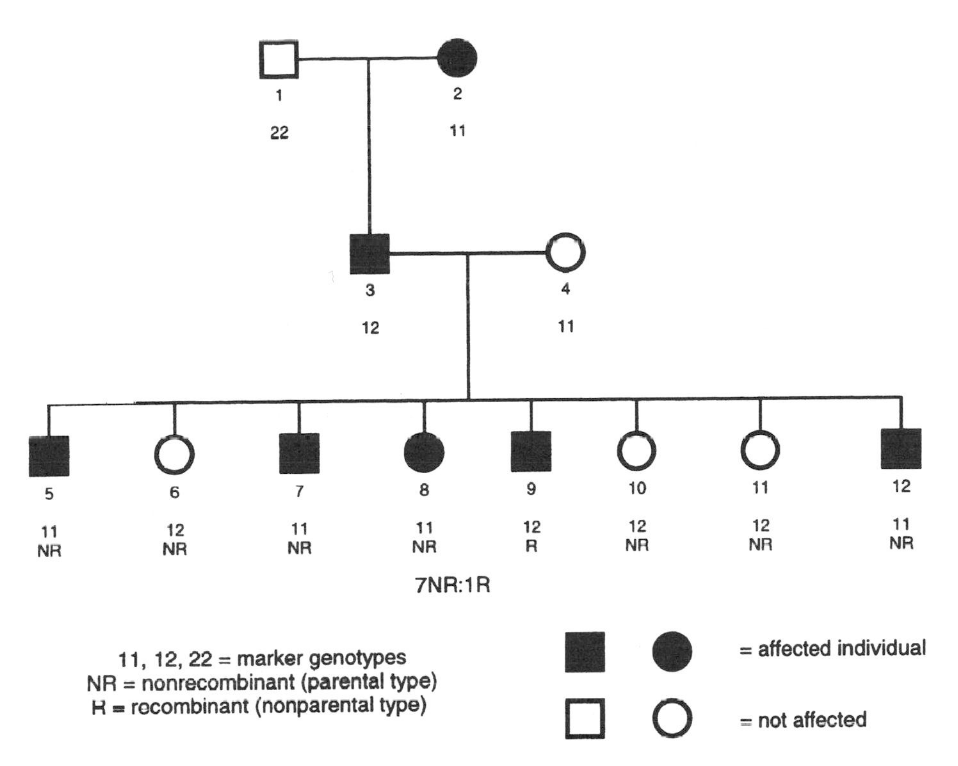

Figure 4.3. *Example of a phase-known autosomal dominant pedigree showing evidence of linkage.*

Thus, the lod score for these data is 0.95 at $\theta = 0.05$. This represents odds of 8.9:1 (e.g., $10^{0.95}$:1) in favor of linkage. Similar calculations are performed over a range of θ from 0.0 to 0.49. This single, small pedigree is not sufficient by itself to give significant evidence for linkage; additional families would need to be examined before a final conclusion could be reached. Although it may seem obvious, it should be noted that as new families are added to the analysis, lod scores across families can be summed only at the same recombination fraction.

The example above was straightforward: the pedigree was phase-known because the grandparental genotypes were available. When only one or two generations of data are available, the pedigree will be phase-unknown, indicating that the disease and marker alleles could be on the same or different chromosomes. In this case, the appropriate formula is:

$$z(x) = \log_{10}\left\{\frac{1}{2}\left[\frac{\theta^{R}(1-\theta)^{NR}}{(0.5)^{R}(0.5)^{NR}}\right] + \frac{1}{2}\left[\frac{\theta^{NR}(1-\theta)^{R}}{(0.5)^{NR}(0.5)^{R}}\right]\right\} \tag{4.4}$$

In equation (4.4), there are now two terms, one for each of the two possible phases of marker and disease allele. Since there is no a priori reason to assign either phase, each is given a probability of ½.

Once linkage has been established, it is often of interest to determine the maximum likelihood estimate of the recombination fraction θ [i.e., $\hat{\theta}$ the value of θ at which z is largest: $z(\hat{\theta})$]. The quantities $\hat{\theta}$ and $z(\hat{\theta})$, as they are often referred to, should be reported in lod score tables, together with a range of lod score values calculated on the basis of θ values from 0.00 to 0.49. Confidence limits on the maximum likelihood estimate of the recombination fraction are usually calculated by subtracting 1.0 from the maximum lod score and identifying the corresponding values of θ on the lod-score curve (1-lod-unit down method). Guidelines are established for consistent reporting of lod scores and their confidence curves (Conneally et al., 1985; Ott, 1992).

Calculation of lod scores using a sex-specific recombination fraction is a simple extension of the lod score calculation above. Differences in recombination fraction by sex have been noted for many chromosomes, and if they are large, they should be included in the analysis. In general, female recombination is greater than male recombination, except in the telomeric regions of some chromosomes. All current linkage analysis programs provide for calculation of lod scores using sex-specific recombination fraction as an option.

As stated previously, because of the sequential test approach, lod scores can be summed across families. Sums of lod scores of 3.0 or more (1000:1 odds in favor of linkage) are indicative of linkage and –2.0 or less are indicative of nonlinkage (Morton, 1955). Values between –2.0 and 3.0 are inconclusive and require collection of additional family data before a conclusion can be made. These criteria were originally suggested by Morton (1955) and are based on the probabilities of type I (α) and type II (β) errors. In this case, (α) refers to the probability of concluding that there is linkage between the loci tested when in fact linkage does not exist (i.e., rejection of a true null hypothesis—a false positive result); and (β) refers to concluding that there is no linkage when in fact linkage does exist (i.e., acceptance of a false null hypothesis—a false negative result). A conservative value of $z \geq 3.00$ for concluding linkage was chosen by Morton (1955) to guard against false positive evidence for linkage. Hence, if the total lod score over all families is 3.0 or more, we can conclude that there is significant evidence for linkage in these data. If the total is –2.0 or less, we can conclude that linkage is excluded for those values of θ where this is the case.

Calculation of lod scores is facilitated by use of computers. A commonly used program is the LINKAGE package (Lathrop et al., 1984). LINKAGE, based on the algorithm of Elston and Stewart (1971), allows calculation of two-point lod scores but extends analysis to multiple linked marker loci on a genetic map (i.e., allows calculation of multipoint location score: ($\log_{10}$; multipoint lod score). LINKAGE is just one of several linkage analysis programs that can implement this approach. Other programs in general use that can be used in lod score calculations include MENDEL (Lange et al., 1988) for multipoint and two-point analysis, LODLINK for two-point analysis, and VITESSE (O'Connell and Weeks, 1995) software for multipoint analysis.

Genetic Heterogeneity

An important phenomenon to consider when performing linkage analysis is genetic heterogeneity. As described in Chapter 3, genetic locus heterogeneity exists when two or more genes act independently to cause an identical trait. Thus clinically identical forms of the same disease phenotype can be caused by different genetic etiologies. This is an important consideration in genetic linkage studies because heterogeneity can confound the analysis. The earliest disease gene linkages (e.g., Huntington disease, cystic fibrosis) were found to be homogeneous, and nonallelic heterogeneity was thought initially to be a rare phenomenon. Recently, however, the more general case seems to be substantial locus heterogeneity (e.g., Charcot–Marie–Tooth disease, familial spastic paraparesis, the hereditary ataxias, Alzheimer disease). When one is using lod scores to perform a linkage study heterogeneity can be tested with relative ease.

In many instances, an investigator may not suspect heterogeneity because of the absence of distinguishing clinical or pathological features between different families with the disease. This is the case, for example, in Charcot–Marie–Tooth disease type 2, tuberous sclerosis, and amyotrophic lateral sclerosis. Other disorders present prior to initiation of the linkage study, with evidence that more than one disease locus may exist. Two examples of such diseases include Alzheimer disease (Goate et al., 1991; Levy-Lahad et al., 1995; Sherrington et al., 1995; Rogaev et al., 1995) and familial breast cancer (Hall et al., 1990). Stratifying the families based on some *a priori* characteristic before the analysis in linkage data greatly facilitated establishing linkage in these disorders. However, this is not always the case, and this approach does not guarantee elimination of the heterogeneity problem. Another approach to minimize the effect of heterogeneity is to analyze a single large family where a single major gene accounts for the disease. However, successful documentation of linkage under these conditions does not allow for immediate generalization of that linkage information to other pedigrees or the general population (Speer et al., 1995).

In many cases when heterogeneity is present, the overall lod score obtained is not 3.0 or more. Thus care must be taken to inspect not just the overall summary for the whole data set, but also individual family results. In particular, formal statistical tests can be performed. Since we are testing a null hypothesis that all families are linked (homogeneity), we consider this a test of homogeneity. Rejection of the null hypothesis (e.g., a significant result) indicates that heterogeneity is likely to exist. Formal tests of homogeneity include the M, β, and admixture tests.

The M Test Morton (1956) was the first to examine this phenomenon directly. The test statistic can be readily evaluated using standard chi-square statistics. It assumes a separate estimate of θ for each family, hence allowing for the potential that each family is segregating the disease at a different locus. The M test can be used to divide lod scores from N families into any number of classes from 2 to N. The key to the appropriate use of the M test is that the classes must be defined before the analysis begins. Classes can be based on criteria various types, including different methods of ascertainment, different clinical criteria, data from different research groups, and families each with a different value. If the number of classes is small, it is often called the predivided sample test.

The β Test A more powerful approach using these same principles is the β test developed by Risch (1988). The primary difference is that the β test compares its results to the β distribution, a somewhat more accurate distribution for the lod scores. It follows a chi-square distribution with one degree of freedom. Both the M and the β test are conservative and readily applicable to family data.

The Admixture Test Today the most commonly used methodology to test for heterogeneity is admixture analysis. Unlike the M and β tests, admixture analysis, first described by Smith (1963), assumes a mixture of families: one group linked and the other group unlinked to a marker of interest. These tests are available in the HOMOG program package (Ott, 1992). HOMOG examines heterogeneity with respect to single or multiple disease loci. Either two-point or multipoint lod scores between the disease and marker loci can be used as input into HOMOG to evaluate heterogeneity in the data set in question. The null hypothesis (H_0) assumes homogeneity and no linkage. This null hypothesis is compared to a series of alternative hypotheses: H_1 assumes linkage and homogeneity, and H_2 assumes heterogeneity with two family types, one type linked and one type unlinked to the marker locus in question. Additional alternatives include H_3, which considers differences in recombination between the sexes but assumes homogeneity overall, and H_4, similar to H_3, but allowing for two family types.

The HOMOG program package, in addition to the standard test just described, includes several other options: HOMOG1, HOMOG2, HOMOG3, and HOMOG3R. These additional programs permit examination of the possibility of multiple family types (i.e., multiple linkages), including linkages on two different chromosomes.

Multipoint Lod Scores

When an interesting result is found through two-point analysis, multipoint linkage analysis is often employed to maximize linkage information and/or to localize the disease gene more precisely on an established map of markers. For sublocalization of the disease gene, the most likely position is estimated by comparisons of logarithms of the likelihoods for the gene being at different positions on the genetic map.

For results in the promising range, multipoint analysis will, in essence, form an extended haplotype, thereby making family data more informative and allowing more accurate odds with which to exclude or confirm linkage. Researchers should keep in mind that if the pedigree data are fully informative in the two-point analysis (i.e., all meiotic information is realized), multipoint analysis will not add additional information to the linkage results. In other words, the maximum two-point lod score possible for a set of data is equal to the maximum multipoint lod score ($\log_{10}$) possible.

Power Studies

Power studies are discussed in detail in Chapter 7, hence only a brief overview is presented here. Prior to generation of genotype data, and prior to collection of DNA from pedigree members, it is necessary to evaluate pedigrees for their power to detect linkage. Power is the probability of correctly concluding that a linkage exists. Power studies are useful in linkage studies because they predict the sample size needed to detect linkage. Power analysis allows the researcher to determine whether the data available will be sufficient to produce a successful outcome that is, whether or not it is possible to establish linkage to the disease in question. Power studies can also be used to decide when enough data has been collected and genotyping should begin. In addition to the power to detect linkage, power studies can give information regarding the probability of generating a false positive or false negative result, the average exclusion region possible in a given data set, and the importance of specific individuals to the lod score outcome.

If locus heterogeneity exists for the disease in question, it will affect the power of the data set to detect linkage. More data will be needed to detect linkage with heterogeneity than without heterogeneity. Because estimates of the level of heterogeneity are difficult to determine prior to the establishment of a successful linkage, it is best to examine the data assuming homogeneity, followed by a series of analyses at different levels of heterogeneity.

ANALYSIS OF COMPLEX TRAITS

The term "complex genetic disease" refers to a trait phenotype with an unknown mode of inheritance. The same confounding factors that influence monogenic linkage analysis also complicate the complex trait mapping situation. Genetic heterogeneity, misclassification, penetrance, and so on all have an effect on mapping studies in a complex phenotype.

Complex genetic diseases are traits that show no clear pattern of Mendelian inheritance such as dominant or recessive but generally have moderate to high evidence of genetic involvement and exhibit familial aggregation of cases. They are common disorders that are either polygenic (multiple genes) or multifactorial (multiple genes interacting with environment) in nature. Examples of complex diseases are Alzheimer disease, diabetes, various cancer syndromes, psychiatric disorders, stroke, Parkinson disease, and cardiovascular disease. Unlike monogenic traits, where genetic variation in phenotype is due primarily to the variation of alleles at a single locus, phenotype expression in these polygenic traits is due to the actions of two or more loci. The interaction of these multiple disease loci and their resulting effect on risk and phenotypic expression is usually explained by either additive or multiplicative genetic models. In other words, the effects of two or more loci can equal the sum of their independent effects (additive) or the loci can interact in such a way that results in an even greater risk than that generated independently by each locus (multiplicative).

The genes involved in a polygenic or multifactorial disorder are often referred to as susceptibility genes in an effort to distinguish them from causative genes. A causative gene is a gene in which a mutation directly leads to the disease phenotype, as in many Mendelian disorders. However, it should be pointed out that even simple Mendelian diseases can exhibit complex monogenic inheritance. For example, the age-dependent penetrance found in Huntington disease and amyotrophic lateral sclerosis (ALS), the extensive genetic heterogeneity found in Charcot–Marie–Tooth (CMT) disease or in the hereditary ataxias, and the variation in expression in myotonic muscular dystrophy or tuberous sclerosis gene carriers are all examples of single gene disorders that have confounding variables in expression.

Even in a complex trait, a subset of families could be due to a single Mendelian factor. Often the subset that is attributable to a monogenic effect is small. Examples of this situation are familial breast cancer (*BRCA1* and *BRCA2*), Alzheimer disease (AD) [amyloid precursor protein (APP) and the presenilin 1 and 2 (*PSEN1*, *PSEN2*) genes), and ALS (the *SOD1* gene). In breast cancer, these Mendelian forms represent about 20% of all cases, and in AD these Mendelian forms represent less than 5% of all AD. In AD these families were distinguished early because they exhibited very early age of onset (< 60 years) and were usually found in large multigenerational pedigrees, in contrast to small family aggregates or as isolated cases.

Genetic Model–Dependent Versus Genetic Model-Independent Analysis

When research was concentrated solely on mapping Mendelian diseases, linkage analysis was virtually defined as lod score analysis. However, other approaches to linkage analysis also exist. With the movement toward studies of genetically complex diseases, these alternative approaches have become more widely used. All the approaches to linkage analysis can be divided into two types: genetic model dependent and genetic model independent.

Genetic Model–Dependent Analysis The most powerful method for detecting linkage is the parametric (genetic model–dependent) approach. This approach requires an assumption of a genetic model as well as specification of certain parameters (e.g., mode of inheritance of a disease and the markers, disease and marker allele numbers and frequencies, mutation rate, and penetrance). The parametric approach provides an estimate of the recombination fraction and makes it possible to test data statistically for evidence of genetic heterogeneity. The parametric approach is most commonly referred to as lod score analysis (Morton, 1955). As already mentioned, the lod score or pedigree likelihood method is the most powerful method of analysis, provided the genetic model parameters are specified correctly. Power is decreased, often severely, when the wrong assumptions about the genetic model are made.

Two of the most common problems in genetic model dependent analysis are the misspecification of the mode of inheritance of the trait locus and the misspecification of the marker allele frequencies (Chapter 12).

Genetic Model–Independent Analysis Genetic model–independent methods of linkage analysis do not require prior knowledge of the parameters that define the disease gene model, although they do require accurate definition of the disease phenotype (Chapter 3). Linkage is measured by allele sharing between a known marker and a disease where the mode of inheritance is unknown. Perhaps the most familiar model-independent method is the sib pair method (Penrose, 1935; Haseman and Elston, 1972; Risch, 1990a). In sib pair methods, identity by descent (IBD; Chapter 13) relationships are used in examining allele sharing among sibs. Other examples of genetic model independent methods are the SimIBD method (Davies et al., 1995) and the NPL method (Kruglyak et al., 1996). These method use other affected relative pairs (avuncular and cousin pairs), as well as sib pairs to test for linkage (Chapter 14). Although genetic model–independent methods are less powerful in their ability to detect linkage than genetic model–dependent approaches (under optimal conditions), when a probable disease gene model cannot be assumed, as in genetically complex disorders, genetic model–independent methods of analysis are a reasonable alternative.

Allelic Association "Allelic association" refers to a significantly increased or decreased frequency of occurrence of a marker allele with a disease trait and represents deviations from the random occurrence of the alleles with respect to disease phenotype. Allelic association can be due to either linkage disequilibrium or a true biological association. Linkage disequilibrium is allelic association due to tight linkage. Allelic association due to linkage disequilibrium represents another nonparametric approach to mapping disease gene loci. It is a useful and often necessary tool in identifying disease gene loci, particularly susceptibility genes in complex diseases. Allelic association can be explained either by direct biological action of the polymorphism (e.g., the *APOE-4* allele in Alzheimer disease) or by linkage disequilibrium with a nearby susceptibility gene (Chapter 15).

REFERENCES

Barnard GA (1949): Statistical inference. *J R. Stat Soc* B11:115–135.

Carter TC, Falconer DS (1951): Stocks for detecting linkage in the mouse and the theory of their design. *J Genet* 50:307–323.

Collins FS (1995): Positional cloning moves from perditional to traditional. *Nat Genet* 9:347–350.

Conneally PM, Edwards JH, Kidd KK, Lalouel J-M, Morton NE, Ott J, White R (1985): Report of the Committee on Methods of Linkage Analysis and Reporting. *Cytogenet Cell Genet* 40:356–359.

Corder EH, Saunders AM, Strittmatter WJ, Schmechel D, Gaskell P, Small GW, Roses AD, Haines JL, Pericak-Vance MA (1993): Apolipoprotein E4 gene dose and the risk of Alzheimer disease in late onset families. *Science* 261:921–923.

Davies JE, Kawaguchi Y, Bennett ST, Copeman JB, Cordell HT, Pritchard LE, Reed PW, Gough SC, Jenkins SC, Palmer SM (1994): A genome-wide search for human type 1 diabetes susceptibility genes. *Nature* 371:130–136.

Davis S, Schroeder M, Foldin LR, Weeks DE (1996): Nonparametric simulation-based statistics for detecting linkage in general pedigrees. *Am J Hum Genet* 58:867–880.

Elston RC, Stewart J (1971): A general model for the genetic analysis of pedigree data. *Hum Hered* 21:523–542.

Goate AM, Chartier-Harlin MC, Mullan MC, Brown J, Crawford F, Fidani L, Guiffra L, Haynes A, Irving N, James L, Mant R, Newton P, Rooke K, Roques P, Talbot C, Pericak-Vance M, Roses A, Williamson R, Rosser M, Owen M, and Hardy J (1991): Segregation of a missense mutation in the amyloid precursor protein gene with familial Alzheimer's disease. *Nature* 349:704–706.

Gusella JF, Wexler NS, Conneally PM, Naylor SL, Anderson MA, Tanzi RE, Watkins PC, Ottina K, Wallace MR, Sakaguchi AY, Young AB, Shoulson I, Bonilla E, Martin JB (1983): A polymorphic DNA marker genetically linked to Huntington's disease. *Nature* 306:234–238.

Haldane JBS (1919): The combination of linkage values and calculation of distance between the loci of linked factors. *J Genet* 8:299–309.

Haldane JBS, Smith CAB (1947): A new estimate of the linkage between the genes for colour-blindness and hemophilia in man. *Ann Eugen* 14:10–31.

Hall JM, Lee MK, Newman B, Morrow JE, Anderson LA, Huey B, King MC (1990): Linkage of early-onset familial breast cancer to chromosome 17q21. *Science* 250:1684–1689.

Haseman JK, Elston RC (1972): The investigation of linkage between a quantitative trait and a marker locus. *Behav Genet* 2:3–19.

Jeunemaitre X, Soubrier F, Kotelevtsev YV, Lifton RP, Williams CS, Charru A, Hunt SC, Hopkins PN, Williams RR, Lalouel JM (1992): Molecular basis of human hypertension: Role of angiotensinogen. *Cell* 71:169–180.

Kosambi DD (1944): The estimation of map distances from recombination values. *Ann Eugen* 12:172–175.

Kruglyak L, Daly MJ, Reeve-Daly MP, Lander ES (1996): Parametric and nonparametric linkage analysis: A unified multipoint approach. *Am J Hum Genet* 58:1347–11363.

Lange K, Weeks D, Boehnke M, (1988): Programs for pedigree analysis: MENDEL, FISHER, and dGENE. *Genet Epidemiol* 51:235–249.

Lathrop GM, Lalouel J-M, Julier C, Ott J (1984): Strategies for multilocus linkage analysis in humans. *Proc Natl Acad Sci USA* 81:3443–3446.

Levy-Lehad E, Wasco W, Pourkaj P, et al (1995): Candidate gene for the chromosome 1 familial Alzheimer's disease locus. *Science* 269:970–973.

Morton NE (1955): Sequential tests for the detection of linkage. *Am J Hum Genet* 7:277–318.

Morton NE (1956): The detection and estimation of linkage between the genes for elliptocytosis and the Rh blood type. *Am J Hum Genet* 8:80–96.

Multiple Sclerosis Genetics Group (1996): A complete genomic screen for multiple sclerosis underscores a role for the major histocompatibility complex. *Nat Genet* 13:469–471.

O'Connell JR, Weeks DE (1995): The VITESSE algorithm for rapid exact multilocus linkage analysis data via genotype set-recording and fuzzy inheritance. *Nat Genet* 11:402–408.

Ott J (1992): Strategies for characterizing highly polymorphic markers in human gene mapping. *Am J Hum Genet* 41:283–290.

Penrose LS (1935): The detection of autosomal linkage in data which consists of pairs of brothers and sisters of unspecified parentage. *Ann Eugen* 6:133–138.

Pericak-Vance MA, Bebout JL, Gaskell PC, Yamaoka LA, Hung W-Y, Alberts MJ, Walker AP, Bartlett RJ, Haynes CS, Welsh KA, Earl NL, Heyman A, Clark CM, Roses AD (1991): Linkage studies in familial Alzheimer's disease: Evidence for chromosome 19 linkage. *Am J Hum Genet* 48:1034–1050.

Pericak-Vance MA, Pritchard ML, Yamaoka LH, Gaskell PC, Scott WK, Terwedow HA, Menold MM, Conneally PM, Small GW, Vance JM, Saunders AM, Roses AD, Haines JL (1997). Complete genomic screen in late-onset familial Alzheimer disease: Evidence for a new locus on chromosome 12. *JAMA* 278:1237–1241.

Postma DS, Bleeker ER, Amelung PJ, Holroyd KJ, Xu J, Panhuysen CI, Meyers DA, Levitt RC (1995): Genetic susceptibility to asthma—Bronchial hyperresponsiveness coinherited with a major gene for atopy. *New Engl J Med* 333:894–900.

Risch N (1988): A new statistical test for linkage heterogeneity. *Am J Hum Genet* 42(2):353–64.

Risch N (1990): Linkage strategies for genetically complex traits. I. Multilocus models. *Am J Hum Genet* 46:222–228.

Rao DC, Morton NE, Lindsten J, Hulten M, Yee S (1977): A mapping function for man. *Hum Hered* 27:99–104.

Rogaev E, Sherrington R, Rogaeva EA, Levesque G, Ikeda M, Liang G, Chi H, Lin C, Holman K, Tsuda T (1995): Familial Alzheimer's disease in kindreds with missense mutations in a gene on chromosome 1 related to the Alzheimer's disease type 3 gene. *Nature* 376:775–778.

Sherrington R, Rogaev EI, Liang Y, Rogaeva EA, Levesque G, Ikeda M, Chi H, Lin C, Li G, Holman K, Tsuda T, Mari L, Foncin J-F, Bruni AC, Montesi MP, Sorbi S, Rainero I, Pinessi L, Nee L, Chumakov I, Pollen D, Brookes A, Sanseau P, Polinsky RJ, Wasco W, Da Silva HAR, Haines JL, Pericak-Vance MA, Tanzi RE, Roses AD, Fraser PE, Rommens JM, St George-Hyslop PH (1995): Cloning of a gene bearing missense mutations in early-onset familial Alzheimer's disease. *Nature* 375:754–760.

Siddique TS, Figlewicz D, Pericak-Vance MA, Haines JL, Roos RP, Rouleu G, Williams D, Watkins PC, Jeffers A, Sapp P, Hung W-Y, Bebout J, Noore FR, Nicholson G, Reed R, McKenna-Yasek D, Deng G, Brooks BR, Festoff B, Antel JP, Tandan R, Munsat TL, Mulder DW, Laing NG, Halperin J, Norris FH, van den Bergh R, Swerts L, Tanzi R, Jubelt B, Matthews KD, Bosch EP, Horvitz HR, Gusella JF, Brown RH Jr, Roses, AD (1991): Linkage of a gene causing familial amyotrophic lateral sclerosis to chromosome 21 and evidence for genetic locus heterogeneity. *New Engl J Med* 324:1381–1384.

Smith CAB (1963): Testing for heterogeneity of recombination fraction values in human genetics. *Ann Hum Genet* 27:175–182.

Speer MC, Tandan R, Yamaoka LH, Fries TJ, Stajich J, Lewis K, Stacy R, Pericak-Vance MA (1995): Bethlem myopathy is not allelic to limb–girdle muscular dystrophy, type 1A. *Am J Med Genet* 58:197–198.

Tsui-L-C, Buchwald M, Barker D, Braman JC, Knowlton R, Schumm JW, Eiberg H, Mohr J, Kennedy D, Plavsic N, Zsiga M, Markiewicz D, Akots G, Brown V, Helms C, Gravius T, Parker C, Rediker K, Donis-Keller H. (1985): Cystic Fibrosis locus defined by a genetically linked polymorphic DNA marker. *Science* 230: 1054–1057.

Vance JM, Nicholson GA, Yamaoka LH, Stajich J, Stewart CS, Speer MC, Hung W-Y, Roses AD, Barker D, Pericak-Vance MA (1989): Linkage of Charcot–Marie–Tooth neuropathy type 1a to chromosome 17. *Exp Neurol* 104:186–189.

Wald A (1947): *Sequential Analysis*. New York: Wiley.

Weeks DE Lange K (1988): The affected-pedigree member method of linkage analysis. *Am J Hum Genet* 42:315–326.

5

Determining the Genetic Component of a Disease

Lindsay A. Farrer

Department of Medicine, Neurology, Epidemiology, and Biostatistics
Boston University School of Medicine
Boston, Massachusetts, and
Department of Neurology
Harvard University School of Medicine,
Boston, Massachusetts

L. Adrienne Cupples

Department of Epidemiology and Biostatistics,
Boston University School of Medicine,
Boston, Massachusetts

INTRODUCTION

According to McKusick (1994), more than 7000 human diseases are inherited in a Mendelian fashion. Individually, these conditions are usually rare, most occurring in less than 0.01% of the population. Despite the limited availability of patients for any single investigation, more than 1000 genes associated with these disorders have been localized to a chromosomal region by linkage analysis. By comparison, relatively few genes for common diseases have been similarly mapped. Moreover, several putative linkages reported for this latter group of disorders were not confirmed by later research (Egeland et al., 1987; Kennedy et al., 1988; Sherrington et al., 1988; Detera-Wadleigh et al., 1989; Kelsoe et al., 1989). To understand the factors

Approaches to Gene Mapping in Complex Human Diseases, Edited by Jonathan L. Haines and Margaret A. Pericak-Vance. ISBN 0-471-17195-6

that may have contributed to these false starts, it is important to distinguish Mendelian diseases from other familial disorders.

The occurrence of a rare disease in multiple biological relatives is likely to have arisen from a single cause, whereas familial aggregation of a common disease might be due to multiple causes co-occurring by chance and does not imply that the etiology is the same in all members. For example, genetic linkage studies revealed that some instances of congenital sensorineural deafness in a large inbred family were caused by a defect other than the mutation linked to the *DFNB4* locus on chromosome 7q (Baldwin et al., 1995). Hearing impairment in these unlinked cases might be due to a defect in another gene. This type of genetic heterogeneity is often difficult to detect by inspection of the pedigree. However, there are recognized nongenetic causes of congenital deafness such as perinatal exposure to the cytomegalovirus. Therefore, repeated maternal infection might result in multiple affected offspring, mimicking a genetic pattern of inheritance.

Many common diseases are believed to have an underlying genetic predisposition, but most do not have a clearly recognizable mode of inheritance and are therefore considered to be genetically complex. Complex traits do not manifest at the same rate or at the same time in all people, perhaps because of chance, but also because of various modifying effects such as age, gender, or other genetic or environmental factors. Thus, to understand the genetic component to such traits, it is necessary to clarify the extent of phenotypic variability that may be under genetic control as well as to identify modifying factors that may simulate or obscure patterns of genetic transmission.

This chapter reviews several approaches for determining whether genes play a role in the risk or expression of a disease. Particular emphasis is placed on study design issues, especially ascertainment and phenotype definition, and on incorporating the information from these analyses into gene mapping studies. To maintain the gene mapping focus of this text, some approaches for assessing familial aggregation are omitted and mathematical formulations for the statistical models are not reviewed in detail. In-depth treatments of these subjects and related issues can be found elsewhere (Khoury et al., 1993). Succinct reviews of the genetic basis of many common diseases have been published in monograph form (RA King et al., 1992).

ASCERTAINMENT

Many studies aimed at understanding the genetic basis of a disease are biased because subjects are selected according to the presence or absence of the trait in at least one family member. Persons through whom the rest of the pedigree is *ascertained* (i.e., brought to the attention of the investigators) are called *probands*. Sampling through affected persons considerably alters the expected ratio of affected to unaffected persons (see Example subsection below for illustration of this bias); moreover, affected persons ascertained in this manner may not be representative of affected cases in the general population or even of the segment of the population from which they were culled.

In what ways might probands for a genetic study be nonrepresentative? If sampling were done through specialty clinics or a disease registry, there might be an enrichment of cases with more severe forms of the disorder or unusual presentation of symptoms or natural history. Aggregation in families of these patients is often more severe perhaps, but not necessarily, because disease expression may be correlated with genetic liability. Alternatively, patients obtained from these sources may tend to have a more pronounced family history of the disorder because of self-selection or selective referral by physicians.

Example Illustrating the Effect of Incomplete Ascertainment on Genetic Modeling Studies

Suppose one wished to determine the mode of inheritance for a disease. One relatively straightforward method is to compare the observed number of affected individuals with the number expected, assuming a particular mode of inheritance. A trait under the control of a single gene will segregate in families, and the frequencies of phenotypes should be consistent with Mendelian ratios. To illustrate the effect of ascertainment bias, consider the case of cystic fibrosis (CF), which is inherited in an autosomal recessive fashion. Persons homozygous for the CF allele, *f*, are affected and persons homozygous for the normal allele, *c*, or CF carriers (genotype *cf*) are normal. To evaluate segregation of CF, say we sampled all families with three children in which both parents were CF carriers (i.e., parental mating type of *cf* × *cf*). The distribution of affected children in such families would be as follows:

Number Affected	*Probability Observed*
0	$(3/4)^3 = 27/64$
1	$3(1/4)(3/4)^2 = 27/64$
2	$3(1/4)^2(3/4) = 9/64$
3	$(1/4)^3 = 1/64$

However, the families with zero affected offspring will not be ascertained (assuming that carrier parents are not identified by population-based DNA genotyping). Thus, the relative **observed** proportions of families with 1, 2, and 3 affected children will be 27/37, 9/37, and 1/37, respectively. The proportion of affected children in this group of families will be:

$$\frac{1(27) + 2(9) + 3(1)}{3(37)} = \frac{48}{111} \sim 50\%$$

The observed proportion is more in line with an autosomal dominant mode of inheritance, not the expected autosomal recessive proportion of 25%. The 25% proportion was not obtained because the families with zero affected children were excluded. When these families are included the proportion becomes:

$$\frac{0(27)+48}{3(37)} = \frac{48}{192} = 25\%$$

This is an example of *incomplete ascertainment*. Because families included in such studies are usually ascertained on the basis of having at least one affected member, it is important to apply an ascertainment correction (see Emory, 1986, for details). In most situations, this entails excluding the probands from the analysis. Thus we see that failure to recognize how families were ascertained or failure to apply an appropriate correction to the selection method may jeopardize the entire study, regardless of the statistical methods employed to analyze the data. In most linkage studies, ascertainment is *not* well defined.

CHOOSING A SAMPLE TO STUDY THE GENETIC BASIS OF A DISEASE

Perhaps the most crucial step in a study aimed at determining the genetic component of a disease is defining the parameters of the subject population. If the disease is rare, it is not feasible to enlist a sufficient number of patients from population- or community-based samples, which are carefully selected subsets of the population generally used for epidemiological studies. Solicitation by advertising or public announcement may net the requisite sample size relatively quickly, but the familial aggregation patterns in these samples will not be representative of most cases in the population. Clinical diagnostic and treatment centers are a better source of subjects with rare disorders. However, this is true only if attendance at the center is for reasons other than those related to family history or other risk factors. To further ensure random ascertainment, one should strive to capture consecutive patients. These considerations become magnified when it is necessary to choose subjects for studies of common diseases, owing to the added importance on having an unbiased sample with respect to environmental risk factors. Thus, for common diseases, such clinic-based population samples are usually less desirable than community-based or population-based samples for studies of common diseases. A health maintenance organization (HMO) roster is a viable alternative to a traditional clinic sample. HMO participants usually represent an unbiased cross section of the population with respect to familial aggregation of disease and factors that may be associated with disease (e.g., gender, socioeconomic status, lifestyle).

Some genetic designs require the use of unrelated controls. Criteria for selection of controls may depend on the type of study. Some studies consider the phenotype of the control subjects in the analysis. In such situations it is important to select controls who are demographically matched for relevant characteristics (e.g., age, gender). Convenient samples such as spouses or patients with other diagnoses ascertained through the same sources may not be appropriate because they are unmatched (e.g., for gender in the case of spouses) or overmatched (e.g., for environmental risk factors perhaps in the case of clinic controls). Although these concerns might be lessened somewhat when information about the relatives of controls is the focus of the study, one should still be cautious about employing what is often called a *heterogeneous sampling design*, in which the controls are not representative of the population from which the cases were drawn.

APPROACHES TO DETERMINING THE GENETIC COMPONENT OF A DISEASE

There are three major and sequential steps involved in the identification of genetic mechanisms (MC King et al., 1984):

1. Determination of familial aggregation.
2. If there exists evidence for familial aggregation, the discrimination among environmental, cultural, and/or genetic factors that may contribute to the clustering.
3. If there exists evidence for genetic factors, the identification of specific genetic mechanisms.

Documentation of familial aggregation is essential before laboratory studies are undertaken to identify specific genetic mechanisms. Without familial aggregation, the expense and effort involved in further studies would be wasted.

There are four major and complementary approaches for determining the genetic component of a complex disease. The use of any particular method will depend on the nature of the trait. For example, congenital diseases and diseases manifesting late in life require studies of different types. Some methods that are appropriate for evaluating quantitative traits such as blood pressure, weight, or skin color are not as efficacious when applied to discrete traits (i.e., traits that are either present or absent). The main approaches are outlined briefly in this section and are explained in detail by use of specific examples in the succeeding sections.

Family Studies

Studies of family members constitute the basic sampling design in determining the genetic etiology of a trait. The *family study* approach entails capturing information about the disease in specific relative sets (e.g., sibs, parent–offspring, first cousins, half-sibs), with twins being a special case (see below), or in entire pedigrees (nuclear families or extended kindreds).

Discrete Traits. There are several types of study design in which familial aggregation and ultimately the genetic component of a trait can be assessed for a discrete (usually dichotomous) trait such as a disease. Khoury et al. (1993) describe three types of sampling design, each of which begins with a case–control study design. These are the abbreviated, detailed, and full-fledged approaches.

Abbreviated Family History Approach—Here the disease is the trait of interest, and controls are selected and interviewed to ascertain whether biological relatives have the disease. Familial aggregation would be supported if the proportion of cases having relatives with the disease were found to be greater than that of the controls. In this type of study, family history is often reduced to a yes/no response for the cases and controls and is then typically treated as simply another potential exposure, such as age or cigarette smoking. There are a number of problems with this study design. For example, the chance that a case or control has a family history is dependent on how many relatives each has. This sampling design clearly is not appropriate for linkage analyses, since no detailed information about the relatives is collected.

Detailed Family History Approach—Here the relatives are no longer lumped into a single group, and each relative is a unit of observation and information is recorded specifically on them, albeit usually through report from the cases and controls. In this type of study design, prevalence and/or incidence of disease in the relatives can be estimated.

Full-Fledged Family Study Approach—Here the relatives are actually contacted, and detailed information on disease and environmental exposures, documented by medical records and/or direct interview, is collected. While this study has the most accurate information on relatives, it may be subject to selection bias (since relatives of cases and of controls may differentially participate in the study). It is, however, the design best suited for linkage analysis.

Quantitative Traits. For quantitative traits such as body weight or blood pressure, a case–control design is not applicable, since the trait is not dichotomous. Hence, the design of such studies takes on a different form. One approach that mimics the case–control design is to select subjects who are at the extremes of the distribution of the quantitative trait, such as those with high systolic blood pressures and a comparison group with low systolic blood pressures. Sometimes, subjects will be selected on a disease trait, such as the presence or absence of coronary heart disease. Then the research can proceed in a fashion similar to the three case–control designs described by Khoury et al. (1993). For example, analogous to the abbreviated family study design, information on blood pressure levels would be reduced a question of how many relatives have high blood pressure. However, in a full-fledged family study design, blood pressures would be measured on each relative and the results would be analyzed to evaluate these quantitative traits of the family members. In this situation, the blood pressure levels of those selected for high values would be compared with the levels of those selected for low values. If family members of those selected for high levels also had higher levels on average than those selected for low levels, the results would suggest some degree of heritability or at least sharing of environmental behavior.

Another study design that is also employed for quantitative traits is to randomly select families and to measure the quantitative trait in these family members as well as spouses (if possible). In this study design, the correlations of the quantitative traits between relatives of different types (e.g., parents and offspring; siblings; second- and third-degree relatives) will be calculated and compared, especially with data from spouses who are not biologically related. If the correlations among first-degree biological relatives are stronger than those between spouses, the results suggest that the trait is heritable. Furthermore, the mode of inheritance can be evaluated by examining the magnitude of the correlations for first-, second-, and third-degree relatives. The random sample approach is preferred because it avoids some of the biases of other designs; however, examining samples of subjects with high and low values is often a more powerful approach, especially for common and complex traits where the necessary sample sizes are easier to obtain.

Recurrence Risks—Familial aggregation of dichotomous traits is typically evaluated in family studies by examining the proportion of relatives of the probands (themselves affected with the trait) who also have the trait and comparing it with the proportion of relatives of control subjects who have the trait. If the proportion is greater among the relatives of the probands, then the trait is said to aggregate within families. For example, information collected from 1952–1987 shows the point prevalence in Denmark for nonsyndromic cleft palate to be 5.8/10,000 live births, whereas among first-degree relatives of an affected proband the point prevalence was 254/10,000 (Christensen and Mitchell, 1996). This contrast strongly suggests that there is familial aggregation of nonsyndromic cleft palate.

However, it is possible that this aggregation is simply due to similar environments shared by those living together, not to a genetic etiology. To determine whether a genetic etiology may be involved, additional comparisons are necessary. In particular, the pattern of familial recurrence risks is important (Risch, 1990). We define λ_R as the ratio of the risk to relatives of type R (e.g., sibs, parents, offspring, etc.) compared to the population risk ($\lambda_R = k_R/k$, where k_R is the risk to relatives of type R and k is the population risk). The most frequent studies examine λ_s, the recurrence risk to siblings of probands versus the general population risk, since information about sibs is often readily available from prior epidemiological studies. The higher the value of λ, the stronger the genetic effect. For example, for a Mendelian recessive trait such as cystic fibrosis the λ_s is about 500 (0.25/0.0004); for a dominant such as Huntington disease, it is about 5000 (0.50/0.0001). For complex traits the estimates are much lower. Some examples of estimates in complex traits include a λ_s of about 20–30 for multiple sclerosis, 4–6 for osteoarthritis, 4–5 for Alzheimer disease, 15 for insulin-dependent diabetes, and 75–150 for autism. In general, any $\lambda > 2.0$ is thought to indicate a significant genetic component.

What Does the Risk Ratio (λ) Really Tell You?

The risk ratio to relatives (e.g., sibs) is a powerful approach to judging the strength of an underlying genetic effect and its ease of detection. However, this property should not be confused with the importance or pervasiveness of the effect. Just because a disease trait has a λ of 3 or 4 does not mean that genes are not as important in that trait as in a trait with a λ of 30 or 40. Because λ is a ratio, both the prevalence in the relatives **and** the prevalence in the general population affect its size. Thus a strong effect in a very common disease will have a smaller λ than the same strength of an effect in a rare disease. A good example is Alzheimer disease. Even though the sibling risk is 30–40% by age 80, at least 10% of the population is also affected by age 80, leading to a λ_S estimate of 3–4. Despite this relatively low λ_S , a single allele, the *APOE-4* allele, is responsible for approximately 30–50% of all AD.

The dropoff in λ_{R-1} as a function of the degree of relationship R is indicative of the possible genetic modes of transmission (Chapter 13). For example, if a single major gene (with any number of alleles) determines the trait, then λ_{R-1} will decrease by a factor of 2 for each decreasing degree of unilineal relationship, using the parent–offspring relationship for the first-degree relative. If two major unlinked loci determine the trait, the dropoff in λ_{R-1} will be more rapid if the effects of the loci are multiplicative, but will behave as for a single locus if the effects are additive. Hence, the information on relatives of varying degree of relationship can provide not only evidence that a genetic component exists, but also evidence on how those genes might be acting.

For example, in the cleft palate study from the Danish Facial Cleft Registry, Christensen and Mitchell (1996) found that the risk to first-, second-, and third-degree relatives was 2.54, 0.28, and 0%, respectively. Since these recurrence risks do not follow a predictable pattern for a single major locus, with the risk to first-degree relatives nearly 10 times that of second-degree relatives rather than twice, the data suggest that a more complex genetic model is implicated in the expression of this trait. The authors examined several additional models, including a multifactorial model and a number in which two or more loci are involved in the expression

TABLE 5.1 Recurrence Risks for Cleft Palate in a Danish Registry

	λ Values[a]			
	λ_1	λ_{MZ}	λ_2	λ_3
Observed	47.2	1149	4.8	0
Predicted values from Single major locus		93	24.1	12.6
Multifactorial threshold		564	9.2	3.3
$\lambda_{11} = 1.5$		1980	7.0	2.7
$\lambda_{11} = 3.0$		1238	7.9	3.0
$\lambda_{11} = \lambda_{21} = 1.5$		1760	7.1	2.7
$\lambda_{11} = \lambda_{21} = 2.0$		1253	7.7	2.9
$\lambda_{11} = \lambda_{21} = \lambda_{31} = \lambda_{41} = 1.5$		1391	7.4	2.8
Infinite loci		2228	6.9	2.6

[a]λ_{iR} is the relative increase in risk to relatives of type R that is attributable to the ith locus; λ_R is the relative increase in risk to relatives of type R compared to the population prevalence. In models with more than one locus, λ_R is calculated as the product of the λ_{iR}, as described by Risch (1990) for a multiplicative model.
Source: Christensen and Mitchell (1996, Table 3).

of the trait (Table 5.1). Using the methods described by Risch (1990), the authors calculated λ_{iR} which is the relative increase in risk to relatives of type R that is attributable to the ith locus. Fitting various models with 1, 2, or more loci contributing to the expression of the trait, the authors determine which model comes closest to that observed in the Danish Registry. The models are based on a heritability of 85% and a fixed relative risk of 47.2 for siblings and offspring relative to the population prevalence (2.74/.058). The single major locus model underestimates the risk to monozygotic (MZ) twins and overestimates the risk to second-degree relatives. While a multifactorial model provides an improvement, it still seriously underestimates the risk to MZ twins. Better fits are provided by models that assume several loci of modest effect, such as the model in which four loci of relative increase of 1.5 are postulated. The authors conclude that the data "are compatible with several multilocus, multiplicative models of inheritance." Analyses such as these can provide a preliminary investigation into a suitable genetic model for a trait.

Quantitative Traits. For quantitative traits, correlations are typically calculated among relatives of different types (e.g., parents and offspring or sets of siblings) and compared with correlations between biologically unrelated individuals such as spouses. Correlations among biologically related individuals that are greater than those between unrelated individuals provide evidence that the trait aggregates within families. For example, in the Framingham Heart Study, correlations for high density lipoprotein cholesterol of 0.26 between parents and offspring, 0.25 among siblings and 0.09 between spouses suggest that this trait aggregates in families (Cupples et al., 1995). As with dichotomous traits, comparison of correlations among relatives of different types lends information to the genetic etiology of the trait.

Coaggregation. Evaluation of familial aggregation has been extended to two traits that are thought to possibly have a common genetic etiology. For a description of these approaches as applied to quantitative traits, see Lange and Boehnke (1983a) and Boehnke et al. (1986), who employed a bivariate approach to evaluate the aggregation and coaggregation in families of total serum cholesterol and serum triglyceride measured in 95 pedigrees with 771 individuals. For dichotomous traits, see Merikangas et al. (1988).

Heritability Estimation

For quantitative traits, another generalized approach to identifying a genetic component is to calculate heritability. The total phenotypic variance of a quantitative trait (σ^2_T) can be partitioned into genetic (σ^2_G) and nongenetic (σ^2_E) proportions. The genetic variance can be further divided into contributions from major genes and the genetic background (polygenes). Heritability, abbreviated as the symbol h^2, is the proportion of the phenotypic variance due to the additive effects of many genes, that is, the polygenic component (σ^2_A) (We will ignore the notion of heritability in the *broad sense*, which encompasses the proportion of variance among individuals due to differences at major gene loci). The greater the value for h^2, the greater the genetic contribution to the etiology. Because the genetic component measured by heritability is polygenic, heritability estimates are meaningful only if no major genes contribute to the disorder (Emory, 1986).

However, abnormal results arising from heritability estimations may help substantiate the possibility of major gene involvement, a requisite before one embarks on a search for such genes. For example, if a dominant gene contributes significantly to risk of the disorder, then the estimated heritability may exceed 100%. Evidence for higher heritability estimates obtained from sibs than from parents and offspring is consistent with the contribution from a major recessive gene. Such estimates are, of course, wrong and must be recalculated using a model that allows for both major gene and polygene components.

There are several other cautions to be observed in the interpretation of heritability estimates. First, such estimates assume that there is no heterogeneity with regard to mode of inheritance. Second, heritability estimates, particularly when derived from sibs, may be inflated because of shared environmental factors. To avoid this bias, heritability estimates also should be derived from other relative sets, including parents and offspring, and more distantly related individuals.

To illustrate the merits and limitations of using this approach to demonstrate the existence of a major gene, consider age at onset of Huntington disease (HD), a dominantly inherited neurodegenerative disorder caused by the expansion of the trinucleotide repeat CAG, located at one end of the HD gene on chromosome 4p16 (Huntington Disease Collaborative Research Group, 1993). Onset of the disease usually occurs between 30 and 50 years of age, although symptoms may appear at any time, from infancy until the ninth decade of life (Conneally, 1984). Studies of age at onset done prior to the identification of the mutation causing HD support a role for genetic modifiers (Myers et al., 1982; Pericak-Vance et al., 1983; Farrer et al., 1984); however, the existence of a major gene modifier was not clearly indicated. Heritability estimates may be useful in this situation and can be derived for a specified relative set by computing the intraclass correlation r and dividing it by the degree of relationship. Table 5.2 shows h^2 values for various relative sets obtained

TABLE 5.2 Estimates of Heritability for Age at Onset of Huntington Disease

Relative Set	Number of Pairs	Kinship Coefficient	r	h^2
Parent	2341	0.5	0.64	128
Sib	2532	0.5	0.64	128
Half-sib	40	0.25	0.80	320
Grandparent	648	0.25	0.48	192
Avuncular	2164	0.25	0.43	172
First cousin	1127	0.125	0.47	376
Great-grandparent	81	0.125	0.02	16
Great-avuncular	313	0.125	0.29	232
Half-avuncular	33	0.125	0.49	392

from data in the National HD Roster at Indiana University. These estimates, which are consistently higher than 100% and show similar heritability in parents and sibs, favor the hypothesis for a major dominant gene. Subsequent studies now show that about half of the total variation in age at onset is explained by the autosomal dominant HD mutation itself. The length of the CAG repeat tract is inversely related to age at onset, the correlation ranging from –.69 to –.75 (Duyao et al., 1993; Snell et al., 1993).

Heritability is generally associated with quantitative traits whose distribution in families, and the population can be studied by analysis of variance. However, heritability can also be estimated for discrete traits (a more complete description of the statistical approach can be found in Khoury et al., 1993). Many discrete traits including clubfoot, schizophrenia, and cleft lip and/or palate are thought to have complex etiology involving perhaps major gene, polygene, and environmental components. Therefore, these traits are *multifactorial*. Although the phenotype is dichotomous (i.e., correlates with presence or absence of disease), liability or predisposition to the trait (or disease) is measured on a quantitative scale. Assuming a *threshold model*, the trait manifests in those who have a liability beyond a set value (Falconer, 1965; see Fig. 5.1). This model predicts that the proportion of affected relatives will be highest among severely affected persons whose liability is presumably further beyond the threshold than that for mildly affected persons. Similarly, if a multifactorial trait is more frequent in one sex than in the other, the risk is higher for relatives of patients of the *less* susceptible sex. Furthermore, the risk of a multifactorial trait is sharply lower for second-degree than for first-degree relatives.

For gene mapping purposes, it is important to establish that the trait being studied has a substantial major gene component. As a rough approximation, one can dis-

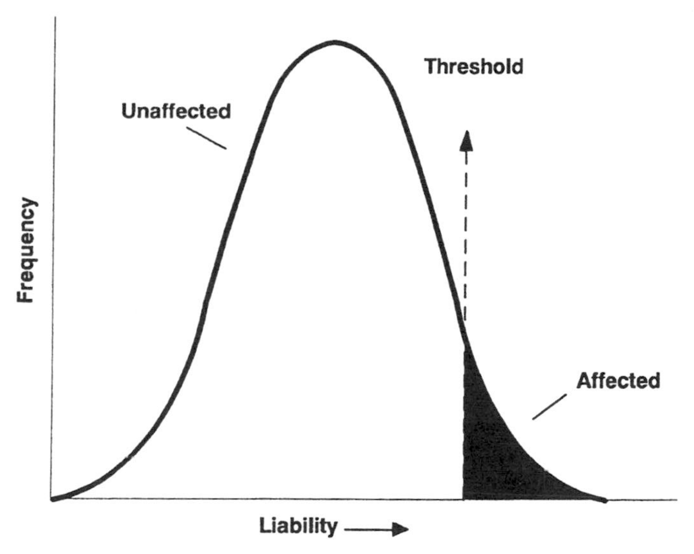

Figure 5.1. *The multifactorial threshold model. Liability to a trait is distributed normally, with a threshold dividing the population into unaffected and affected groups.*

criminate modes of transmission for a complex trait by comparing observed relative frequencies with expected values under specified inheritance models. For example, if x is the prevalence of a disease in the general population, among sibs the expected relative frequencies for the disorder are $1/2x$, $1/4x$, and $1/\sqrt{x}$ assuming autosomal dominant, autosomal recessive, and multifactorial inheritance, respectively (Emory, 1986).

Segregation Analysis

Once familial aggregation with a probable genetic etiology has been established for a trait, one may consider using segregation analysis to evaluate whether a major gene contributes to the expression of the phenotype. Segregation analysis is one of the most established methods for this purpose. In experimental animals, studies of this type are easily facilitated by controlling the mating types of the parents to examine segregation patterns for the trait of interest in their offspring. Obviously, in humans such studies must take advantage of natural mating types. Segregation analysis aims to determine the transmission pattern of the trait within families and to test this pattern against predictions from specific genetic models. For example, if a single major gene is the underlying cause for the trait, then segregation analysis attempts to identify the particular Mendelian model that determines the transmission of the trait. This can be an important step in the search for specific genes, since linkage studies will likely fail if factors other than a major gene account for familial clustering of the disease. Moreover, some linkage procedures (especially the parametric lod score approach, Chapter 12) are highly dependent on a valid specification of the transmission model. Thus one advantage of segregation analysis in genetically complex traits is that in addition to evidence for a major gene in the trait of interest, one obtains estimates for various parameters associated with the model (i.e., gene frequency, penetrance) that can be used in the parametric lod score approach.

Segregation analysis entails fitting a variety of models (both genetic and nongenetic) to the data obtained from families and evaluating the results to determine which model best fits the data. Since most family data are obtained from families with at least one "affected" person or through individuals with unusually high or low values of the quantitative trait, ascertainment corrections are necessary in these analyses. A variety of ascertainment corrections are possible depending on the way in which the sample was selected. See Khoury et al. (1993, pp. 274–276) for an in-depth discussion of this topic.

An important hallmark of segregation analysis is that it involves the estimation of many parameters, typically accomplished through maximum likelihood estimation procedures (Elston and Stewart, 1971; Cannings et al., 1976; Lange and Boehnke, 1983b). Most algorithms limit the major gene component to a biallelic system (e.g., alleles A and B) and assume Hardy–Weinberg equilibrium, although a number of the computer programs permit more general models. Estimable quantities (parameters) associated with the major locus include the allele frequencies, transmission parameters [t_1 (or t_{AA}) = probability that a subject with the AA genotype will transmit an A allele to offspring, t_2 (or t_{AB}) = probability that a subject with the AB genotype will transmit an A allele to offspring, t_3 (or t_{BB}) = probability that a subject with the BB genotype will transmit an A allele to offspring], and susceptibility parameters, which link each genotype to the phenotype (i.e., the chance that an individual with a particular genotype will express the trait).

If the trait is the result of polygenes, it becomes necessary to define a risk function. In this case, most segregation analyses apply Falconer's model of genetic liability, in which the trait is assumed to be the result of many genes that have additive effects, the sum of which follows a normal distribution (Falconer, 1965; Curnow and Smith, 1975; see preceding section on estimating heritability). For quantitative traits, the penetrance parameters define a distribution for expression of the trait for each genotype of the single major locus. Often, the normal distribution is assumed, for which the parameters are the mean and variance. An investigator can assume common variance with different means for different genotypes, or can permit different variances as well as different means.

As a preliminary step to segregation analyses of quantitative traits, investigators sometimes undertake a *commingling analysis*, the purpose of which is to determine whether the distribution of the phenotype comes from multiple populations of individuals, hence represents a mixture of those populations. If the data suggest that the phenotype is the result of a mixture of two or three populations, a possible conclusion is that these populations are determined by different genotypes. Hence, an evaluation of segregation of the phenotype by examining different transmission models is warranted. If the data suggest that the phenotype is drawn from only one population (possibly with a skewed distribution), then segregation analysis may not be warranted. In the latter case, an investigator should be concerned about the power of the data to detect multiple populations. More complete descriptions of how to conduct a commingling analysis can be found in Khoury et al. (1993) and Morton et al. (1983).

Major locus models for quantitative traits assume that each genotype corresponds to an array of phenotypes, or that, conversely, a phenotype may result from more than one genotype. If the trait is either autosomal dominant or recessive, there are two distributions because the distribution for the heterozygote is the same as the distribution for one of the homozygotes. For an additive model, the mean of the heterozygote population lies exactly at the midpoint between the means of the two homozygote populations. In many instances, known physical or environmental factors influence the expression of quantitative traits. For example, body weight and age are known to affect blood pressure and lipid levels. Investigators often adjust for these known effects by multiple linear regression techniques prior to the segregation analysis and then evaluate the standardized residuals in the segregation analysis. In some situations it is possible to adjust for these covariates in the segregation analysis. Since, however, these adjustments add considerable computing time to the analyses, this approach is not advisable unless there are only a few covariates.

Segregation analysis permits the evaluation of a variety of hypotheses regarding the transmission of the trait, including but not limited to no transmission, no major gene, all transmission due solely to shared environmental factors, and various modes of Mendelian inheritance. These hypotheses specify constraints on the parameters. For example, a Mendelian model would require that $t_1 = 1$, $t_2 = 0.5$, and $t_3 = 0$. Typically, these models are tested against a general model that permits a more complicated mode of transmission, such as a mixed model with a single major gene and a multifactorial/polygenic component.

Frequently, maximum likelihood estimation is employed to obtain estimates of the parameters as well as their standard errors. The likelihood represents a probabilistic model for observing the data given a certain set of parameters. As the parameters vary, the likelihood changes also, and the goal is to maximize the likelihood. In the most general model, all parameters are estimated, and this model will have the largest likelihood among all models tested. Other models restrict some parameters to specific values, thereby potentially lowering the likelihood. The likelihood for a single model considered by itself has little meaning, but likelihoods from different models can be compared to see which model yields a better fit of the data. If the two models are hierarchical (meaning that one model is nested in another, hence has a reduced set of parameters), a comparison can be made using a likelihood ratio test, which computes the difference between the –2 ln (likelihood) of the more unrestricted model and the –2 ln (likelihood) of the more restricted model. For large samples, this statistic is approximately distributed as a chi-square statistic with degrees of freedom equal to the difference in the number of estimated parameters for each model. Comparison of other models can be accomplished by examining Akaike's information criterion (–2 ln likelihood + 2 number of parameters estimated), though no formal test of hypothesis can be performed (1974).

While early segregation analyses considered single major locus models or polygenic models (in which the trait is the result of the cumulative effect of many genes, each having a small effect), more recent approaches have considered a mixed model, in which the phenotype may be the result of a combination of these two (Morton and MacLean, 1974; Lalouel et al., 1983). Initially, limited to analysis of nuclear families, the mixed model was later expanded to more general pedigrees by Hasstedt (1982). More recently, Bonney proposed the "regressive" model, which permits simultaneous adjustment for covariates while estimating the parameters of the various models in a segregation analysis (Bonney, 1984, 1986).

It is important to point out that designing and performing a segregation analysis is often a long and expensive proposition. It is critical that care be taken in clearly defining the ascertainment of each pedigree, and that appropriate corrections for the ascertainment scheme be used. In addition, the power of any segregation analysis to discriminate between models is critically dependent on the number of parameters that are being estimated and the size of the dataset. For example, many more families will have to be examined to discriminate between two models when 10 parameters are being estimated than when only 5 parameters are being estimated.

A more complete description of segregation analysis methods can be obtained in Khoury et al. (1993). Many of these models have been implemented in several computerized genetic epidemiology packages, among which are POINTER (Lalouel and Morton, 1981), SAGE (Statistical Analysis for Genetic Epidemiology: Bailey-Wilson and Elston, 1992), PAP (Pedigree Analysis Package: Hasstedt, 1993), and

PPA (Package for Pedigree Analysis, which includes MENDEL and FISHER: Goradia et al., 1992). These models are capable of demonstrating a single major gene (or effect) if such exists, but are limited for evaluation of oligogenic models in which two or more major loci are involved in the etiology of the phenotype.

Twin and Adoption Studies

Twin and adoption study methods are two approaches uniquely suited for neuropsychiatric diseases and other traits such as obesity, where the disentanglement of nature (genetics) and nurture (environment) is extremely difficult. The twin approach relies on biology—MZ twins are genetically identical, whereas DZ (dizygotic) twins share, on average, one-half of their genes. If both twins of a pair are affected, they are said to be concordant. By estimating and comparing concordance rates among MZ and DZ twins (Allen et al., 1967), one can get an idea of the role of genetic factors in the etiology of the disease. If the concordance rate is 100% in MZ twins and between 25 and 50% in DZ twins, it may be concluded that the disease is strictly genetic and probably due to a single recessive (if the concordance rate is ~25%) or dominant (concordance rate ~50%) gene. For a disorder in which genetic factors are important in etiology but not the sole factor, the concordance rate for MZ twins will still be greater than for DZ twins, and the concordance rate for MZ twins reared apart (if such difficult studies have been done) will be about the same as for MZ twins reared together. For example, the MZ twin concordance rate in multiple sclerosis is approximately 30%, which may seem low, and does indicate a substantial environmental influence. However, the DZ concordance rate is only 2–4%, and the difference between the two indicates a strong genetic component. Low concordance in MZ twins or equal concordance between MZ and DZ twins suggests a strong environmental influence on development of the disease.

Several complexities are inherent in the twin study design. Concordant twin pairs are more likely to come to attention and be published in the literature precisely because they are rare, both for being twins and for having the disease. In addition, most diseases have variability in age at onset. Unless the age at onset is tightly controlled by the genes involved, twin pairs may become concordant only over time. Thus follow-up of twins is critical. Finally, most twin studies are small, and any estimates derived from them have necessarily large confidence limits or standard errors.

The basic premise of adoption studies is that if a trait has a genetic influence, the risk of illness should be higher in biological relatives (parents, siblings, or offspring) than in adopted relatives living in the same household. This is a particularly useful design for diseases with a large known or suspected environmental component, as in many neuropsychiatric diseases such as alcoholism and bipolar disease. This design controls for the environmental component, since that is the only factor acting in the adopted relatives, while the biological relatives generally do not share the same environment. As with the twin design, there are several complexities that must be considered. First, obtaining an appropriate sample size can be difficult, particularly in countries where adoption records are generally unavailable for study.

Second, the environments of the adopted and biological relatives may not vary substantially, since adoption agencies often strive to match socioeconomic backgrounds during the adoption process.

EXAMPLE OF A DICHOTOMOUS TRAIT: ALZHEIMER DISEASE

In this example we first provide a brief description of Alzheimer disease (AD) and then describe the results of studies using several of the aforementioned methods to test the existence of a genetic component. AD was chosen because there are significant difficulties in applying any of the methods above, thus demonstrating many of the problems likely to be encountered when studying other disease traits.

Alzheimer disease is a neurodegenerative dementing illness characterized by pervasive deficits in cognition and by personality changes. It is a progressive disease and the fourth leading cause of death in adults. The basic etiology of AD is unknown, although advanced age and a positive family history of dementia are two prominent risk factors. The prevalence increases with age from about 3% of 65-year-old individuals to as much as 50% among individuals surviving to age 85 years (Evans et al., 1989; Mullan et al., 1994).

Numerous factors including gender, head trauma, aluminum, zinc, parental age, and nonsmoking have been implicated in the etiology of AD, but none shows a consistent or large effect on disease risk. The reasons for this are unclear, but may be related to the method of ascertaining cases and controls, how the risk factors were assessed, or cross-cultural differences in susceptibility to dementia. It is also possible that some of these factors are confounded, or interact with each other or genetic factors. Distinction between genetic and nongenetic models for AD is further complicated by its usual onset in very late life.

Family Studies

Published Pedigrees. A genetic component to AD was suggested as early as 1929 (Flugel, 1929). Over the next half-century a considerable number of families with multiple affected members have been reported (reviewed by St. George-Hyslop et al., 1989). In the vast majority of these families the mode of inheritance appears to be autosomal dominant. However, it is important to understand that biases introduced either by the manner of ascertainment or by failure to consider the ages of unaffected family members relative to the age at onset in affected members make this conclusion tenuous. Moreover, because nearly all these cases had onset of symptoms at an unusually early age (before 65 years), when familial clustering may be recognized more easily, genetic patterns in such families may be atypical. In fact, the vast majority of AD cases have no apparent family history. Not surprisingly, gene mapping strategies applied to these early-onset families have identified three early-onset AD loci: *APP* on chromosome 21 (Goate et al., 1991), *PS1* on chromosome 14 (Sherrington et al., 1995), and *PS2* on chromosome 1 (Levy-Lahad et al., 1995; Rogaev et al., 1995).

Identifying genetic mechanisms for the more common late-onset form of AD, which typically occurs after age 65, has been more elusive. Pedigree studies of AD have been hampered by several properties specific to AD, namely diagnostic uncertainty, using life-to-date information (censoring) of at-risk individuals, small effective family size, and incomplete family histories. Despite these problems, one genetic risk factor for late-onset AD has been identified. Linkage studies in late-onset families with apparently dominantly transmitted late-onset AD (Pericak Vance et al., 1991) led to the identification of apolipoprotein E genotype as a significant risk factor for the disorder (Strittmatter et al., 1993).

Risks to Relatives. One of the difficulties in any late-onset disease is accurately determining the risk to relatives. Even with the recent advances in the epidemiology and molecular genetics, this question is difficult to answer for AD. An associated question is whether the risk of AD increases unabated with age. That is, would all people develop AD if they lived long enough? During the last decade, several investigators addressed these questions by examining incidence of disease in relatives of AD probands using survival analysis methods (Kaplan and Meier, 1958; Chase et al., 1983; Cupples et al., 1989, 1991). Although all studies agree that first-degree relatives of a patient have a substantially higher risk of disease than unrelated individuals, they differ by as much as 25% in the estimated lifetime risk to relatives of AD cases. Predictably, authors finding a lifetime risk near 50% concluded that AD is most likely transmitted as an autosomal dominant trait with age-dependent penetrance (Breitner and Folstein, 1984; Breitner et al., 1986, 1988; Mohs et al, 1987; Huff et al., 1988; Martin et al., 1988), whereas others observing a lifetime risk below 40% conclude that the genetic mechanism is complex or heterogeneous (Farrer et al., 1989; Sadovnick et al., 1989; van Duijn et al., 1993; Silverman et al., 1994). Variability in lifetime risk estimates in earlier studies is due in part to the paucity of data among persons older than 85 years. The reported confidence intervals for risks in this portion of the age spectrum are very large.

Lautenschlager et al. (1996) investigated the relationship between risk of disease, age, and sex in 12,971 first-degree relatives of 1694 rigorously diagnosed probands. Risks of dementia and the age at onset distribution for first-degree relatives of the AD probands were estimated using a maximum likelihood procedure (Cupples et al., 1991). The lifetime risk for dementia to first-degree relatives of all AD probands was 39.0 ± 2.1% at age 96 years (Fig. 5.2). Further inspection of the age-specific risks for AD among relatives of probands revealed that cumulative risk increased by an average of 1.90 ± 0.79% per year between the ages of 80 and 90. During the tenth decade, the cumulative risk increased by 1.00 ± 0.60% per year. The significantly lower rate among the nonagenarians ($p = .034$) and the observation that 61 persons survived to the oldest onset age (96 years) apparently without developing symptoms of AD are consistent with the hypothesis that cumulative risk may be approaching an asymptote. Extrapolation of the cumulative risk curve to the maximum age in the sample of 104 suggests that, at most, the lifetime risk would increase to approximately 42%.

The risk of developing AD was significantly higher in female relatives than male

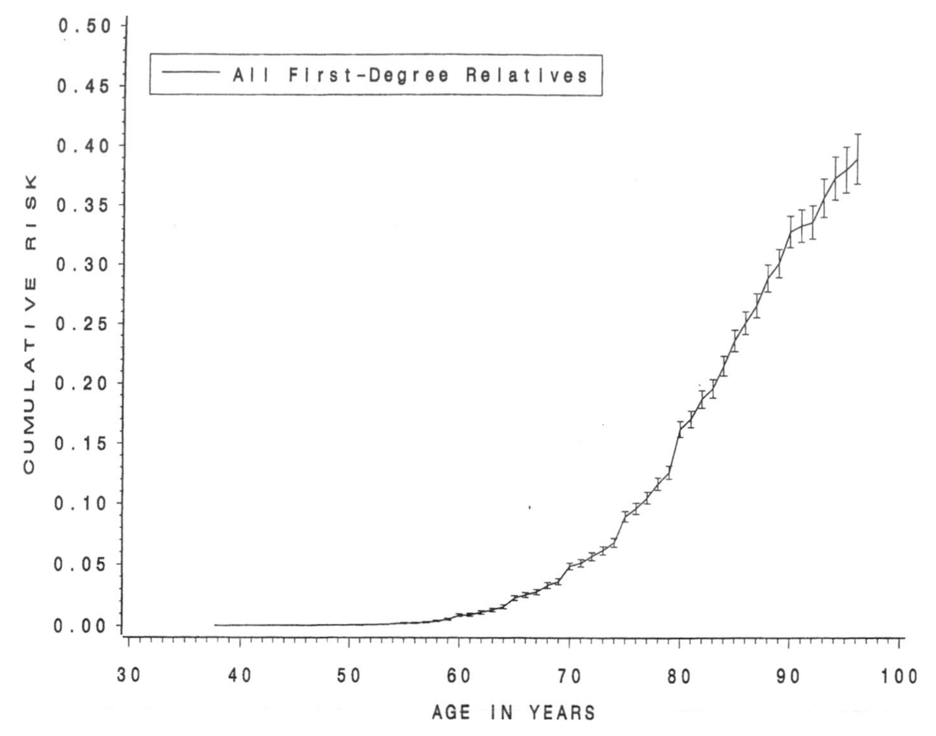

Figure 5.2. *Estimated lifetime incidence of Alzheimer disease in first-degree relatives of AD probands. Vertical lines show standard errors at each onset age in affected relatives. (From Lautenschlager et al. 1996; reprinted with permission.)*

relatives at all ages (Fig. 5.3). By age 93, women have a 13% higher risk than men of developing AD. At age 80, sibs of 21 probands having two affected parents had a risk of AD that was at least 32.3% greater than the risk among sibs having fewer than two affected parents (Fig. 5.4). At age 85, sibs of 124 probands having an affected father had twice the risk of developing AD than sibs of 1129 probands whose parents were normal (p = .0022) and a 1.4 times greater risk than sibs of 271 probands having an affected mother (p = .13). Notably, the sex ratio among affected sibs was the same whether the father (0.37) or the mother (0.30) was affected (P = .40). The risk of AD among sibs having an affected mother was not significantly different from the risk among sibs having two normal parents.

Several conclusions regarding genetic mechanisms can be gleaned from the survival distributions. The 39% risk of AD by age 96 years among first-degree relatives, which is approximately twice the estimated cumulative incidence of AD in the general population, supports the well-established hypothesis that AD has a strong genetic component, but it is unlikely that autosomal dominant or codominant inheritance can fully explain aggregation of disease in these families because the risk, which was adjusted for age-dependent penetrance, is significantly less than 50%. Data in figure 5.4 showing that by age 80 sibs of probands born to conjugal AD

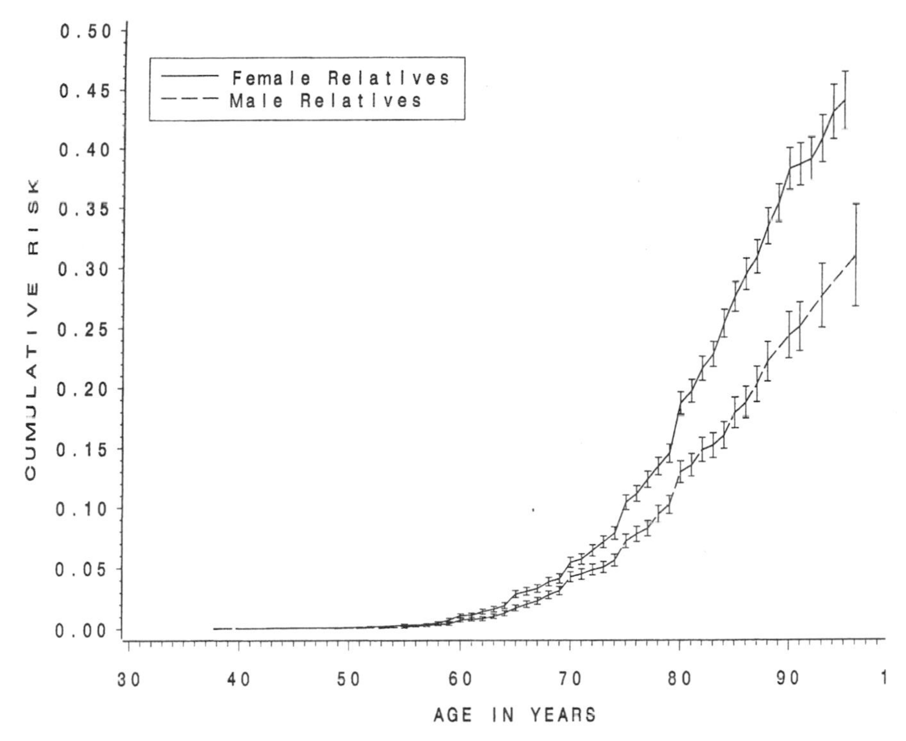

Figure 5.3. *Estimated lifetime incidence of Alzheimer disease in male and female first-degree relatives of AD probands. (From Lautenschlager et al., 1996; reprinted with permission.)*

couples have a risk (54%) that is greater than the sum of the risks to children having affected mothers or fathers (37%) and nearly 5 times greater than the risk to children having normal parents are consistent with an additive model such as the one describing a dose effect of the *APOE-4* allele on risk of AD (Corder et al., 1993); however, multifactorial or polygenic inheritance patterns cannot be ruled out based on these data alone. Bird et al. (1993) showed that children of conjugal AD couples have an increased risk of AD, but meaningful empirical risk estimates could not be gleaned from their sample because the majority of children were younger than 55.

There are several caveats to the interpretation of survival analysis data associated with diagnostic uncertainty, study design, and limitations of the statistical methods. First, the majority of relatives in this study (and in most family studies) were not examined in a manner as rigorous as the probands, and thus there is the possibility of misclassification. Informants may tend to underreport dementia in very old relatives because of preconceptions of the normal aging process. To guard against such errors, one should engage multiple informants and carefully review medical records, for these steps have been shown to increase substantially the classification of relatives in AD studies (Silverman et al., 1986). Second, results based largely on patients attending specialized AD clinics may not be representative of AD cases in

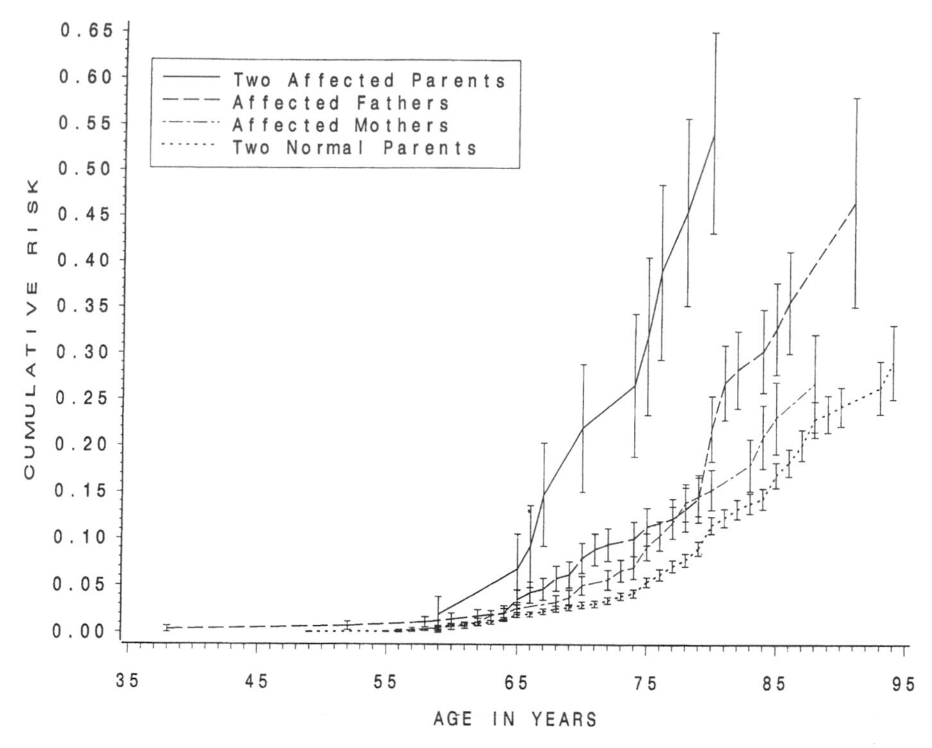

Figure 5.4. *Estimated lifetime incidence of Alzheimer disease in siblings of AD probands having two affected parents, cognitively normal parents, an affected father, or an affected mother. (From Lautenschlager et al., 1996; reprinted with permission.)*

the general population. However, a population-based study of early-onset AD in the Netherlands reported an identical lifetime risk of 39% among relatives of 198 patients (van Duijn et al., 1993). Third, nonparametric life table methods do not estimate risk of disease for ages in which there are no incident cases beyond the maximum age at onset in the sample. Finally, it is not feasible to discern specific disease transmission mechanisms from survival analyses. For example, although the risk curve for a simple age-dependent, fully penetrant autosomal dominant trait should be a sigmoid curve asymptotic to a final risk of 50%, failure to obtain this pattern does not dismiss dominant inheritance and presence of this pattern does not prove its existence.

Applying the formulations for risk of AD among sibs (Lautenschlager et al., 1996; see Fig. 5.2) and assuming an empirical estimate of 0.5% (Schoenberg et al., 1987) for prevalence, the observed relative frequency in sibs is 62 (Table 5.3). This risk does not match exactly the expected relative frequencies of any of the three modes of inheritance considered; however, it is clearly too large to fit either the recessive or multifactorial model. A dominant model would be an adequate explanation only under the assumption of reduced penetrance (by factors other than age).

TABLE 5.3 Comparison of the Observed Relative Frequency of AD Among Sibs with the Relative Frequencies Expected Under Various Modes of Inheritance.

Frequency		Relative Frequency			
			Expected		
General Population, x	Sibs, s	Observed s/x	Dominant $1/2x$	Recessive $1/4x$	Multifactorial $1/\sqrt{x}$
005	.31	62	100	50	14

Alternatively, a more complex model such as epistasis (i.e., one gene masking the effect of another) or one involving the interaction of a gene with an environmental factor could account for the sibling recurrence data. In any event, these data are consistent with the existence of a major gene for AD.

Segregation Analysis

Unlike lifetime risk analysis or recurrence risk studies, in which risk of illness is estimated for a fixed set of relatives, segregation analysis entails assessment of patterns of disease transmission from parents to offspring. The efficacy of this approach in AD (and most other late-onset disorders) is limited by changes in the definition of the disorder over time and by difficulty in accessing pedigree and medical history information for a disproportionately large number of deceased persons, including nearly all parents and many sibs of probands. Understandably, reliable data on three or more informative generations are seldom available (children of probands are considered to be uninformative because they are collectively too young). This limitation and the extremely late onset of the disorder impose considerable computational constraints. Thus, an extremely large data set, at least 200 families or more, is needed to obtain meaningful results.

Early attempts to elucidate mechanisms of AD transmission by complex segregation analysis using the mixed model approach (Morton and MacLean, 1974) implemented in the POINTER computer program yielded concurrence that there is a major dominantly transmitted susceptibility gene for AD (Farrer et al., 1991; van Duijn et al., 1993), but not all the parameter estimates from the best-fitting models were easily interpretable. Refinements to the genetic model and evidence for differential transmission of AD among early-onset and late-onset families were provided by Rao et al. (1994), who carried out segregation analyses in more than 400 families using logistic regressive models (Bonney, 1984, 1986) implemented in the REGTL program of SAGE. Table 5.4 shows their results in 198 early-onset families.

In this analysis, AD was treated as a dichotomous trait with age-dependent penetrance, and the major gene component was modeled as a diallelic locus. Age at onset was assumed to follow a logistic distribution constrained to cumulative incidence values of 0.2 for women and 0.11 for men by the age of 102 years (the oldest age in the sample) which were extrapolated from population incidence data

TABLE 5.4 Segregation Analysis of Alzheimer Disease in Early-Onset Families[a]

Models[b]	q_A	τ_{AA}	τ_{AB}	τ_{BB}	μ	γ_{AA}	γ_{AB}	γ_{BB}	γ_{AA}	γ_{AB}	γ_{BB}	$-2 \ln L$	n_e	p
Non-Mendelian														
1. General	.063	1.0[c]	.113	.030	73.7	1.0[c]	1.0[c]	.090	1.0[c]	.898	.001	753.5	10	
2. No major gene					72.2	.200			.110			772.6	2	.02
3. Sporadic	.015	.030[d]	.030[d]	.030[d]	72.4	.800	.620	.187	.890	.598	.095	772.3	8	< .005
4. Environmental	.040[e]	.040[e]	.040[e]	.040[e]	72.2	.810	.660	.162	.900	.600	.069	772.6	7	< .005
Mendelian														
5. Arbitrary major gene	.015	[1.0]	[0.5]	[0.0]	74.2	1.0[c]	1.0[c]	.176	1.0[c]	1.0[c]	.083	756.2	7	.47
6. Dominant	.015	[1.0]	[0.5]	[0.0]	74.2	1.0[c]	1.0	.176	1.0[c]	1.0	.083	756.2	5	> .995
7. Recessive	.288	[1.0]	[0.5]	[0.0]	74.2	.790	.147	.147	.776	.050	.050	761.5	5	.07
8. Additive	.052	[1.0]	[0.5]	[0.0]	74.0	1.0[c]	.578	.156	1.0[c]	.531	.061	759.6	5	.25

[a]Conventions: q_A, frequency of AD allele, A; τ_{AA}, τ_{AB}, τ_{BB}, probability of transmitting allele *A* with genotypes *AA*, *AB*, and *BB*, respectively; μ, estimated mean age at onset; γ_{AA}, γ_{AB}, γ_{BB}, probability of developing AD persons with genotypes *AA*, *AB*, and *BB*, respectively; n_e, number of independently estimated parameters (no. estimated – no. dependent); brackets, parameter fixed.

[b]Models 2–5 compared against model 1; models 6–8 compared against model 5.

[c]Fixed at the boundary by the maximization function.

[d]$\tau_{AA} = \tau_{AB} = \tau_{BB}$.

[e]$\tau_{AA} = \tau_{AB} = \tau_{BB} = q_A$.

Source: Rao et al. (1994, Table 4).

(Schoenberg et al., 1987; Kokmen et al., 1988). Sex-and genotype-dependent susceptibilities were estimated under each model.

To interpret the results in Table 5.4, one must compare the models hierarchically starting with the most general (unrestricted) model. The test statistic, which is the large sample approximation of χ^2, is derived by computing the difference between the −2 ln likelihoods with degrees of freedom (df) equal to the difference in number of independent parameters of the two models in the comparison. The general model gave a significantly better fit than the no-major-gene, sporadic, and environmental models, suggesting that none of these adequately explain familial clustering of AD in these families. The arbitrary major gene model was not significantly different from the general model, suggesting that a single major gene model adequately fits the data. Because the arbitrary major gene model is not rejected, it is appropriate to compare it with the specific genetic models. Although dominant, recessive, and additive models cannot be rejected, the parameter estimates of the arbitrary major gene model are identical to those of the dominant model. Focusing on this model, the major gene for early-onset AD appears to be fully penetrant in both sexes by age 102 years and has a frequency of 1.5%. The dominant model also suggests that in these families genetically nonsusceptible males and females develop the disease at rates of 8.3 and 17.6%, respectively. Such persons may be phenocopies, a result that can be due to multifactorial inheritance, new mutations, inheritance of a second major gene, or environmental insult.

How *accurate* are these segregation analysis results? Similar to any studies that utilize amnestic diagnostic information, conclusions must be tempered by diagnostic uncertainty. Even with rigorous diagnostic evaluation, the accuracy of the clinical diagnosis, as determined by confirmation at autopsy, is only 85–95% (Joachim et al., 1988; Rao et al., 1994). This rate is certainly lower in unexamined relatives. Second, because conclusions from segregation analysis are achieved by comparing likelihoods, failure to reject some (wrong) models may be due to small sample size (e.g., recessive model), whereas failure to distinguish among certain models (e.g., dominant vs. additive) might reflect heterogeneity of AD in the sample or, on the other hand, might indicate that AD is better explained by a model more complex than the ones presented in Table 5.4 (e.g., an oligogenic model). Except by increasing sample size, these problems are not easily overcome with the current methodology. Third, segregation analysis, which is a genetic modeling exercise, cannot measure the effects of specific genes, per se. However, genotype information can be incorporated into segregation analysis by stratifying the data set or, in the case of regressive models, by treating the genotype as a covariate. In the MIRAGE study, for example, stratification of 636 AD families by *APOE* genotype of the proband revealed that there are likely other major genes for the common form of the disorder (Rao et al., 1996).

Twin Studies and Adoption Studies

Adoption studies have not been done in AD because of its very late age at onset, which also makes it impossible to trace the histories of most if not all biological relatives. In addition, twin studies have not greatly clarified the role of genes in AD. Concordance rates among co-twins of MZ probands with AD are 40–60%, suggesting a strong but not absolute genetic influence (Kallman, 1956; Cook et al., 1981; Nee et al., 1987; Rapoport et al., 1991; Breitner et al., 1995). In interpreting these data it is important to note that all studies have examined small numbers of twins, usually in a retrospective manner, have been without neuropathological confirmation of diagnosis in most cases, and tend to report a disproportionate number of MZ twin pairs compared to DZ twins pairs (St. George-Hyslop et al., 1989). Furthermore, few studies have followed the twin pairs for long enough periods of time to allow for varying age at onset.

EXAMPLE OF A QUANTITATIVE TRAIT: HIGH DENSITY LIPOPROTEIN CHOLESTEROL

Lipid levels are widely known to be associated with the development of coronary heart disease (CHD). High levels of low density lipoprotein cholesterol (LDL) increase one's risk of disease, while high levels of high density lipoprotein cholesterol (HDL) are protective. Among relatives of persons with CHD, an increased frequency of low HDL accompanied by high levels of triglyceride has been observed (Genest et al., 1992). A number of investigators have examined the familial aggregation and segregation of HDL. Studies of familial aggregation have suggested that the genetic effects upon HDL are greater than the environmental or cultural effects (Beaty et al., 1983; Namboodiri et al., 1985; Williams et al., 1990). Segregation analyses, on the other hand, have yielded conflicting results, with some finding no evidence of major genes for HDL (Hasstedt et al., 1984; Prenger et al., 1992) and others finding some evidence for a major gene, although the mode of inheritance remained unclear (Borecki et al., 1986; Amos et al., 1987).

In this example, we analyzed data collected by the Donner Laboratory Family Study, located in Berkeley, California. Most of the families were randomly ascertained without regard to history of CHD or lipid disorders and were of Mormon descent. Two families identified through a clinic population were not included in these analyses. These data were made available to the Genetic Analysis Workshop 8 in 1992 and were evaluated at that time (Cupples and Myers, 1993; Krauss et al., 1993).

Two-hundred seventy-seven subjects in 27 pedigrees of the 385 study subjects had recorded measurements of HDL (137 men and 140 women). The pedigrees ranged in size from 5 to 120. HDL levels were measured after precipitation with heparin-$MnCl_2$ following the Lipid Research Clinics guidelines (Krauss et al., 1993). The average HDL was 46.88 mg/dL, with a standard deviation of 11.4 (43.1±10.1 mg/dL for men and 50.6±11.5 for women mg/dL). To remove the effects of known environmental factors, HDL values were adjusted by linear regression techniques for age, age-squared, and gender. Few subjects in the sample reported smoking (5.8%) or consuming alcoholic beverages (13.5%), and we decided not to adjust for these variables. Other factors that may affect HDL levels are exercise, weight, and hormone treatment for women, but this information was not available in this study. Following the age and sex adjustment, we obtained the standardized residuals from the regression model, since many statistical genetic analysis computer packages assume that the measures being analyzed are normally distributed with a mean of zero and a standard deviation of one. These residuals were employed in subsequent analyses of familial aggregation and segregation. All analyses of these data used the SAGE analysis package.

Familial Aggregation and Heritability

To examine familial aggregation of HDL levels, we calculated correlations for relatives of different types. If there is familial aggregation, the correlations between blood relatives are expected to be greater than those of individuals who are not biologically related such as spouses. If a single major locus determines HDL levels, the correlations should decrease by 0.5 for each degree of relationship. Table 5.5 displays the correlations for the Donner Laboratory data. These data provide some evidence for familial aggregation, since the correlation for the spouses is considerably smaller than all the those for first-degree relatives (0.04 vs. 0.17). The fact that the parent–offspring correlation (0.166) is nearly identical to the sibling correlation (0.170) suggests that a Mendelian autosomal dominant model might be appropriate. However, the correlations do not fall off at a rate of 0.5 for each degree of relationship. The grandparent–grandchild correlation (second-degree relatives) is almost the same as those of the siblings and the parent–offspring. On the other hand, the avuncular correlation (also second-degree) is considerably weaker. On first glance, the correlation for cousins (third-degree relatives), which is the largest of all (0.25), is quite puzzling. However, the confidence limits on this estimate are large, meaning that this estimate may not be particularly accurate.

A simple estimate of the heritability can be obtained by doubling the parent–offspring correlation. This calculation yields an estimate of 34%. Quadrupling the estimate for second-degree relatives also provides an estimate, yielding from the grandparent–grandchild correlation an estimate of 60% or from the avuncular correlation,

TABLE 5.5 Familial Aggregation of HDL Cholesterol in the Donner Laboratory Family Study

Relationship	Correlation[a]	Confidence Interval
Spouse	0.036	−0.240, 0.307
Parent–offspring	0.166	0.034, 0.292
Sibling	0.170	0.003, 0.328
Grandparent–grandchild	0.150	−0.096, 0.378
Avuncular	0.047	−0.076, 0.168
Cousin	0.246	−0.265, 0.649

[a]Pearsonian correlation except for siblings and cousins, which are intraclass correlations; calculation of these correlations from FCOR in SAGE gave equal weight to pedigrees, which varied considerably in size in these data. The 95% confidence intervals are approximate, computed by estimating the number of effective pairs and applying Fisher's Z transformation for correlation coefficients.

20%. An estimate from the cousins would not be in the allowable range of 0–100%. Clearly, the estimate is quite variable from this analysis.

Several explanations may be offered for these observations, including a cohort or secular effect that may account for the relatively high correlation among cousins, who were probably born later than many of the other relatives in the study. It is possible that those born at the same time may have more similar levels owing to some undetermined factor. Also, these correlations may reflect a confounding environmental exposure or physical trait that was not accounted for in the regression models adjusting for gender and age. For example, since alcohol is highly associated with HDL levels, some of the associations may reflect differing levels of alcohol consumption and not shared genes. Another possibility is chance uncertainty in the estimates, as the standard errors might be large (and are, unfortunately, not available). There appears to be some familial aggregation, but no simple genetic model explains the expression of HDL levels. A more complicated model than that specified by a single major locus might be appropriate, such as a model postulating several major loci and an additional polygenic component.

These results are typical for complex traits which may be determined by several genes and/or environmental effects. For example, Namboodiri et al. (1985), using data collected in the Lipid Research Clinics Program Family Study, found correlations among first-degree relatives ranging from 0.17 to 0.29 and smaller correlations between spouses of 0.06 (suggesting a stronger influence of shared genes than shared environments). It is rare, however, that a quantitative trait is determined by a single major locus. If a single major locus did determine these levels, we would expect the correlations to fall off at a rate of 0.5 for each degree of relationship. For example, if the correlation among first-degree relatives were 0.2, then we would expect the correlation for second degree relatives to be approximately 0.1 and among third degree relatives, 0.05. Moreover, the correlation between spouses should be weak in comparison with that of first-degree relatives. In this example, we find that the correlations among first-degree relatives is indeed stronger than that for spouses, thus, suggesting some degree of heritability; but we do not find the correlations declining with the degree of relationship that would be expected if a single major locus determined HDL levels. Thus, for this complex trait a more complicated genetic etiology is likely.

Segregation Analysis

The residual HDL levels can also be evaluated by segregation analysis using the REGC procedure in SAGE for quantitative traits that implement the regressive models described by Bonney (1984). A variety of regressive models are possible:

class A models, which assume that the association among siblings in a nuclear family is due only to common parentage.

class B models, which assume that a subject's trait depends on the eldest sibling's trait in addition to the parents.

class C models, which assume, in contrast to class B, that a subject's trait depends on the immediately preceding sibling in addition to the parents.

class D models, which assume a common correlation among the siblings in addition to the correlation with the parents' traits. Class D models have been shown to be comparable to the mixed model (Demenais and Bonney, 1989; Demenais, 1991).

For this analysis, we used class A regressive models and assumed Hardy–Weinberg equilibrium. Since we considered only pedigrees in which both parents are in the study, these pedigrees approximate a random sample of subjects and thus no ascertainment correction was made. These data were transformed to obtain satisfactory bell–shaped distributions for both men and women.

Table 5.6 displays the results for several models fit to the data. The most general model (model 1) allowed for separate means and variances for both men and women, non-Mendelian transmission parameters, and different correlations for mothers and fathers with their offspring. By comparing reduced models, in which the means and/or variances were assumed to be the same for men and women, the comparison of the –2 ln likelihoods indicated no significant differences between men and women for either the variances (model 1 v 2: $\chi^2 = 2.93$, df = 3, $p > .40$) or the means (model 1 v 3: $\chi^2 = 9.81$, df = 6, $p > .13$). Hence, we used the reduced model 3 as a suitable general model to compare with further reduced models. Continuing on to examine the residual correlations assuming a major gene effect, significant sex-specific parent–offspring (model 3 v 4: $\chi^2 = 5.09$, df = 1, $p < .03$) and spouse (model 3 v 5: $\chi^2 = 6.14$, df = 1, $p < .02$) correlations were found. However, the father–offspring correlation (model 3 v 6: $\chi^2 = 0.06$, df = 1, $p > .80$) was not different from zero. The mother–offspring correlation was significant ($\chi^2 = 6.31$, df = 1, $p < .02$, data not presented). Hence, the codominant model with non-Mendelian t's (model 6) was the most parsimonious of the general models and it was used for further comparisons.

The environmental model with equal τ's (model 6 v 7: $\chi^2 = 27.79$, df = 3, $p < .0001$), the no-major-gene model (model 6 v 8: $\chi^2 = 37.80$, df = 8, $p < .0001$), and the no-transmission model (model 6 v 9: $\chi^2 = 42.45$, df = 10, $p < .0001$) are all rejected. The environmental model assumes that there may be familial aggregation, but it is not transmitted because all the τ's are equal. The difference between the no transmission and the no major gene models is that the former assumes no polygenic/multifactorial component as well as no major gene by setting the residual correlations to zero. Since all three are rejected in these data, the results suggest familial aggregation and transmission of HDL levels.

Among the Mendelian models, the additive major gene model provides the best fit for these data (models 10–13). However, the arbitrary Mendelian major gene model is rejected in favor of non-Mendelian τ's (model 6 v 10: $\chi^2 = 8.69$, df = 3, $p < .04$).

These results suggest that model 6, with the non-Mendelian τ's, common means and variances for men and women, and zero correlation for the residual father–offspring association is the most likely explanation of transmission of HDL

TABLE 5.6 Segregation Analysis of HDL Cholesterol in the Donner Laboratory Family Study[a]

					Means			Variances							
Models[b]	q_A	τ_{AA}	τ_{AB}	τ_{BB}	μ_{AA}	μ_{AB}	μ_{BB}	σ^2_{AA}	σ^2_{AB}	σ^2_{BB}	ρ_{sp}	ρ_{po}	$-2 \ln L$	n_e	p
Non-Mendelian															
1. General	0.49	0.72	0.47	0.0[c]	M: 1.38 W: 2.21	M: 0.11 W: −0.33	M: 0.11 W: −0.33	M: 0.83 W: 0.45	M: 0.53 W: 0.18	M: 0.34 W: 0.29	M: 0.29 W: −0.35	M: 0.04	693.13	19	
2. Common variances	0.48	0.76	0.49	0.0[c]	M: 1.71 W: 2.17	M: 0.85 W: 0.81	M: −0.02 W: −0.31	0.60	0.31	0.33	0.37	M: 0.04 W: −0.34	696.06	16	0.4025
3. Common means and variances	0.49	0.78	0.51	0.0[c]	1.91	0.80	−0.22	.68	0.30	0.33	0.42	M: 0.03 W: −0.29	702.94	13	0.1329
4. Common parent–offspring ρ	0.51	0.78	0.55	0.0[c]	1.78	0.74	−0.30	0.71	0.29	0.31	0.48	−0.21	708.03	12	0.0241
5. Spouse ρ = 0	0.49	0.79	0.52	0.005	1.84	0.78	−0.22	0.73	0.31	0.36	[0.0]	M: 0.14 W: −0.33	709.08	12	0.0132
6. Father–offspring ρ = 0	0.49	0.78	0.51	0.0[c]	1.90	0.80	−0.23	0.67	0.30	0.33	0.44	M:[0.0] W: −0.29	703.00	12	0.8065
7. Environmental	0.68	[0.68]	[0.68]	[0.68][d]	1.06	0.54	−0.78	0.97	0.27	0.21	0.04	M: [0.0] W: 0.27	730.79	9	0.0001

8. No major gene	0.65	[0.65]	[0.65]	0.86	[0.86]	[0.86]	0.10	M: [0.0]	740.80	4	0.0001	W: 0.11		
9. No transmission	0.64	[0.64]	[0.64]	0.86	[0.86]	[0.86]	[0.0]	M: [0.0]	745.45	2	0.0001	W: [0.0]		
Mendelian														
10. Arbitrary major gene	0.50	[1.0]	[0.5]	[0.0][e]	1.71	0.73	−0.25	0.73	0.38	0.33	0.41	M: [0.0] W: −0.25	711.69	9	0.0337
11. *A* dominant	0.22	[1.0]	[0.5]	[0.0]	1.36	[1.36]	0.25	0.65	0.70	0.54	0.20	M: [0.0] W: −0.11	721.92	8	0.0014
12. *B* dominant	0.77	[1.0]	[0.5]	[0.0]	1.10	[0.09]	0.09	0.71	0.53	0.32	0.18	M: [0.0] W: −0.06	725.25	8	0.0002
13. Additive	0.50	[1.0]	[0.5]	[0.0]	1.70	[0.73]	−0.25	0.74	0.38	0.33	0.41	M: [0.0]	711.69	8	1.0000

[a]Conventions: q_A, frequency of allele *A*; τ_{AA}, τ_{AB}, τ_{BB}, probability of transmitting allele *A* for persons with genotypes *AA*, *AB*, and *BB*, respectively; μ_{AA}, μ_{AB}, μ_{BB}, means of phenotype for person with genotypes *AA*, *AB*, and *BB*, respectively; σ^2_{AA}, σ^2_{AB}, and σ^2_{BB}, variances of the phenotype for persons with genotypes *AA*, *AB*, and *BB*, respectively; ρ_{sp}, ρ_{po} residual correlations between spouse pairs (sp) and between parents and offspring (ρ_0) after adjusting for major gene effects; n_e, number of independently estimated parameters (no. estimated – no. Dependent); M, men; W, women; brackets, parameter fixed.

[b]Models 4–6 compared against model 3; models 7–10 compared against model 6; models 11–13 compared against model 10.

[c]Fixed at the boundary by the maximization function.

[d]$\tau_{AA} = \tau_{AB} = \tau_{BB}$.

[e]$\tau_{AA} = \tau_{AB} = \tau_{BB} = q_A$.

levels. This model indicates that the frequency of the A allele is 0.49, representing individuals with high levels of HDL (m_{AA} = 1.9). This allele is nearly balanced with the B allele, representing individuals with low HDL levels (m_{BB} = –0.2), and heterozygotes fall at an intermediate level, suggesting a codominant model. Note that the means, being on a standardized residual scale, have little meaning as they stand. To make them more meaningful, one would have to multiply by a standard deviation and add some constant mean level. The spouse correlation is large, indicating significant residual environmental effects, not accounted for in the regression model. This may be due in part to residual confounding from environmental factors such as alcohol, weight, and physical activity, which were not adjusted for in the regression analysis. It is difficult to explain the negative correlation between mothers and their offspring, although it may be a reflection of a more complicated genetic model. The primary departure of this model from a Mendelian model is in τ_{AA}, estimated at 0.78 instead of 1. This suggests that parents with high levels are not transmitting these high levels to their children sufficiently often; hence these high levels may be phenocopies (nongenetic), or the children at low levels may be phenocopies. This estimate of τ_{AA} along with the large spouse correlation and the negative mother–offspring correlation reveal a complex genetic model, not explained by a single major gene or simple polygenic inheritance. A more complicated model, involving several major genes and/or gene–environment interaction, may provide a better fit.

The foregoing results, along with those for familial aggregation of HDL, are not unusual for a complex quantitative trait wherein a complicated mix of environmental and genetic factors may be determining the levels of the trait. Unfortunately, many of the tools currently available for segregation analyses permit only relatively simple models (e.g., a biallelic single major locus and perhaps a polygenic component). Models that specify several major genes and/or environmental factors interacting in the determination of the trait have not been well developed and implemented in computer programs. In addition, such models would require the estimation of many more parameters, and would consequently require even larger data sets to have sufficient power. It is important to note that regardless of the genetic model, these statistical approaches only examine patterns of levels within families; they do not implicate specific genes associated with a trait. Even if the results of a model demonstrate the involvement of a single major gene, the identity of this gene is still unknown. Further information is required to identify specific genes.

It is commonly believed that many quantitative traits that show strong familial resemblance are determined by a limited number of genes. Hence, investigators are frequently skipping over this step of evaluating the mode of transmission through segregation analyses and proceeding directly to linkage studies, employing the candidate gene or genomic search approache (Chapter 11). Under these circumstances (i.e., since the transmission model is unknown), it is necessary to perform the linkage analysis using nonparametric approaches, which do not require knowledge of the mode of transmission such as sib pair analysis (Haseman and Elston, 1972; Fulker and Cardon, 1994; Kruglyak and Lander, 1995; also see Chapters 12–15).

PROCEED WITH CAUTION

Before embarking on an ambitious project to collect and analyze clinical and family history information on a large population, an investigator should understand what can be accomplished from studies of these types and be cognizant of the limitations. The studies described in this chapter are a prerequisite to efforts for localizing genes for common and genetically complex diseases. Obviously, such efforts would be severely compromised, or perhaps doomed, if no major gene existed or if parameters of the genetic model were poorly estimated. The genetic epidemiological approaches described herein require at least three areas of expertise, namely, diagnosis, epidemiology, and statistical genetics, and are not mastered by simply learning how to plug data into computer programs.

Results of genetic risk estimation or modeling procedures are meaningful only when the assumptions underlying the distribution of the disease in the specified population are correct. Failure to correct adequately for ascertainment bias, age- and sex-adjusted rates of disease, and confounding among genetic and environmental risk factors may distort conclusions about recurrence in relatives, transmission properties, penetrance, gene frequency, mutation/phenocopy rate, and heterogeneity. Poor assumptions of the genetic model, in turn, can corrupt linkage studies, particularly those using parametric analytic approaches.

This chapter provides only a brief glimpse of the methods employed to evaluate the genetic contribution to a disease. In particular, we focused on methods of analysis that can help determine whether a linkage analysis will be profitable and provide reasonable estimates of genetic model parameters. A comprehensive or even a broad overview of all genetic analysis methods is beyond the scope of this review. For example, we have not described the application to familial data of path analysis (Li, 1975; Rao, 1985), which has been particularly helpful in twin studies in distinguishing shared environmental from inherited effects. The inability to identify a Mendelian model by segregation analysis may suggest that transmission of the disease is not adequately explained by a single locus model. Methods that permit a trait to be determined by the interaction of two genes, that is, two-locus models (Elandt-Johnson, 1970; Greenberg, 1984; Neuman and Rice, 1992), can be implemented in the PAP program. Greenberg demonstrated by simulation (Greenberg, 1984) and by analysis of empirical data for celiac disease (Greenberg and Lange, 1982) that two-locus models can often explain familial aggregation of disease better than single locus models. Investigators studying common and complex traits may also consider using bivariate segregation analysis for exploring the hypothesis that two traits have a common genetic mechanism (Blangero and Konigsberg, 1991).

ACKNOWLEDGMENTS

We thank Dr. Ron Krauss for permission to use the Donner Laboratory Family Study data used in the Genetic Analysis Workshop 8. This work was supported in part by National Institutes of Health grant AG09029 to LAF.

REFERENCES

Akaike H (1974): A new look at the statistical model identification. *IEEE Trans Autom Control*, AC-19:716–723.

Allen G, Harvald B, Shields J (1967): Measures of twin concordance. *Acta Genet*, 17:475–481.

Amos CI, Elston RC, Srinivasan SR, Wilson AF, Cresanta JL, Ward LJ, Berenson GS (1987): Linkage and segregation analyses of apolipoproteins A1 and B, and lipoprotein cholesterol levels in a large pedigree with excess coronary heart disease: The Bogalusa Heart Study. *Genet Epidemiol* 18:731–749.

Bailey-Wilson JE, Elston RC (1992): *Statistical Analysis for Genetic Epidemiology (SAGE)*. New Orleans: Louisiana State University Medical Center.

Baldwin CT, Weiss S, Farrer LA, DeStefano AL, Adair R, Franklyn B, Kidd KK, Korostishevsky M, Bonné-Tamir B (1995): Linkage of congenital, recessive deafness (*DFNB4*) to human chromosome 7q31 in the Middle Eastern Druze population and evidence for genetic heterogeneity. *Hum Mol Genet* 4:1637–1642.

Beaty TH, Self SG, Chase GA, Kwiterovish PO (1983): Assessment of variance components models on pedigrees using cholesterol, low-density, and high-density lipoprotein measurements. *Am J Med Genet* 16:117–129.

Bird TD, Nemens EJ, Kukull WA (1993): Conjugal Alzheimer's disease: Is there an increased risk in offspring? *Ann Neurol* 34:396–399.

Blangero J, Konigsberg LW (1991): Multivariate segregation analysis using the mixed model. *Genet Epidemiol* 8:299–316.

Boehnke M, Moll PP, Lange K, Weidman WH, Kottke BA (1986): Univariate and bivariate analyses of cholesterol and triglyceride levels in pedigrees. *Am J Med Genet* 23:775–792.

Bonney GE (1984): On the statistical determination of major gene mechanisms in continuous human traits: Regressive models. *Am J Med Genet* 18:731–749.

Bonney GE (1986): Regressive logistic models for familial disease and other binary traits. *Biometrics* 42:611–625.

Borecki IB, Rao DC, Third JLHC, Laskarzewski PM, Glueck CJ (1986): A major gene for primary hypoalphalipoproteinemia. *Am J Hum Genet* 38:373–381.

Breitner JCS, Folstein MF (1984): Familial Alzheimer dementia: A prevalent disorder with specific clinical features. *Psychol Med* 14:63–80.

Breitner JCS, Murphy EA, Folstein MF (1986): Familial aggregation in Alzheimer dementia. II. Clinical genetic implications of age dependent onset. *J Psychiatr Res* 20:45–55.

Breitner JCS, Silverman JM, Mohs RC, Davis KL (1988): Familial aggregation in Alzheimer's disease: Comparison of risk among relatives of early- and late-onset cases, and among male and female relatives in successive generations. *Neurology* 38:207–212.

Breitner JCS, Welsh KA, Gau BA, McDonald WM, Steffens DC, Saunders AM, Magruder KM, et al (1995): Alzheimer's disease in the National Academy of Sciences–National Research Council Registry of Aging Twin Veterans. *Arch Neurol* 52:763–771.

Cannings C, Thompson EA, Skolnick M (1976): The recursive derivation of likelihoods on complex pedigrees. *Adv Appl Prob* 8:622–625.

Chase GA, Folstein MF, Breitner JCS, Beaty TH, Self SG (1983): The use of life tables and survival analysis in testing genetic hypotheses, with an application to Alzheimer's disease. *Am J Epidemiol* 117:590–597.

Christensen K, Mitchell LE (1996): Familial recurrence-pattern analysis of nonsyndromic isolated cleft palate—A Danish Registry Study. *Am J Hum Genet* 58:182–190.

Conneally PM (1984): Huntington's disease: Genetics and epidemiology. *Am J Hum Genet* 35:506–526.

Cook RH, Schneck SA, Clark DB. Twins with Alzheimer's disease (1981): *Arch Neurol* 38:300–301.

Corder EH, Saunders AM, Risch NJ, et al (1993): Gene dose of apolipoprotein E type 4 allele and the risk of Alzheimer's disease in late onset families. *Science* 261:921–923.

Cupples LA, Myers RH (1993): Segregation analysis of high density lipoprotein in the Berkeley data. *Genet Epidemiol* 10:629–634.

Cupples LA, Terrin NC, Myers RH, D'Agostino RB (1989): Using survival methods to estimate age at onset distribution for genetic disease with an application to Huntington's disease. *Genet Epidemiol* 6:361–371.

Cupples LA, Risch N, Farrer LA, Myers RH (1991): Estimation of age at onset with missing information of onset. *Am J Hum Genet* 1991;49:76–87.

Cupples LA, Myers RH, Ordovas JM, Wilson PWF, Schaefer EJ (1995): Heritability of risk factors for coronary heart disease: The Framingham study. *Am J Hum Genet* 57(Suppl):A316.

Curnow RN, Smith C. (1975): Multifactorial models for familial disease in man. *J Stat Soc A* 138:131–169.

Demenais FM (1991): Regressive logistic models for familial diseases: A formulation assuming an underlying liability model. *Am J Hum Genet* 49:773–785.

Demenais FM, Bonney GE (1989): Equivalence of the mixed and regressive models for genetic analysis. I. Continuous traits. *Genet Epidemiol* 6:597–618.

Detera-Wadleigh S, Goldin L, Sherrington R (1989): Exclusion of linkage to 5q11-13 in families with schizophrenia and other psychiatric disorders. *Nature* 340:391–393.

Duyao M, Ambrose C, Myers RH, Novelletto A, Persichetti F, Frontali M, Folstein S, Ross C, Franz M, Abbott M, Gray J, Conneally P, Young A, Penny J, Hollingsworth Z, Shoulson I, Lazzarini A, Falek A, Koroshetz W, Sax D, Bird E, Vonsattel J, Bonilla E, Alvir J, Bickham Conde J, Cha J-H, Dure L, Gomez F, Ramos M, Sanchez-Ramos J, Snodgrass S, de Young M, Wexler N, Moscowitz C, Penchaszadeh G, MacFarlane H, Anderson M, Jenkins B, Srinidhi J, Barnes G, Gusella J, MacDonald M (1993): Trinucleotide repeat length instability and age at onset in Huntington's disease. *Nat Genet* 4:387–392.

Egland JA, Gerhard DS, Pauls DL, Sussex JN, Kidd KK, Allen CR, Hostetter AM, Housman DE. (1987): Bipolar affective disorder linkaed to DNA markers on chromosome 11. *Nature* 325: 783–787.

Elston RC, Stewart J (1971): A general model for the genetic analysis of pedigree data. *Hum Hered* 21:523–542.

Elandt-Johnson RC (1970): Segregation analysis for complex modes of inheritance. *Am J Med Genet* 22:129–194.

Emory AEH (1986): *Methodology in Medical Genetics: An Introduction to Statistical Methods*, 2nd ed. New York: Churchill Livingstone.

Evans DA, Funkenstein HH, Albert MS, et al (1989): Prevalence of Alzheimer's disease in an community population of older persons. *JAMA* 262:2551–2556.

Farrer LA, Conneally PM, Yu PL (1984): The natural history of Huntington's disease: Possible role of aging genes. *Am J Med Genet* 18:115–123.

Farrer LA, O'Sullivan DM, Cupples A, Growdon JH, Myers RH (1989): Assessment of genetic risk for Alzheimer's disease among first-degree relatives. *Ann Neurol* 25:485–493.

Farrer LA, Myers RH, Connor L, Cupples LA, Growdon JH (1991): Segregation analysis reveals evidence of a major gene for Alzheimer disease. *Am J Hum Genet* 48:1026–1033.

Falconer DS (1965): The inheritance of liability to certain disease estimated from the incidence among relatives. *Ann Hum Genet* 29:51–76.

Flugel FE (1929): Zür diagnostik der Alzheimershen Krankheit. *Z Ges Neurol Psychiat* 120:783–787.

Froguel P, Vaxillaire M, Sun F, Velho G, Zouoali H, Butel MO, Lesage S, Vionnet N, Clément K, Fougerousse F, Tanizawa Y, Weissenbach J, Beckmann JS, Lathrop GM, Passa P, Permutt MA, Cohen D (1992): Close linkage of glucokinase locus on chromosome 7p to early-onset non-insulin-dependent diabetes mellitus. *Nature* 356:162–164.

Fulker DW, Cardon LR (1994): A sib-pair approach to interval mapping of quantitative trait loci. *Am J Hum Genet* 54:1092–1103.

Genest JJ, Martin-Munley SS, McNamara JR, Ordovas JM, Jenner J, Myers RH, Silberman SR, Wilson PWF, Salem DN, Schaefer EJ (1992): Familial lipoprotein disorders in patients with premature coronary artery disease. *Circulation* 85:2025–2033.

Goate A, Chartier-Harlin M-C, Mullan M, Brown J, Crawford F, Fidani L, Giuffra L, Haynes A, Irving N, James L, Mant R, Newton P, Rooke K, Roques P, Talbot C, Pericak-Vance M, Roses A, Williamson R, Rossor M, Owen M, Hardy J (1991): Segregation of a missense mutation in the amyloid precursor protein gene with familial Alzheimer's disease. *Nature* 349:704–706.

Goradia TM, Lange K, Miller PL, Nadkarni PM (1992): Fast computation of genetic likelihoods on human pedigree data. *Hum Hered* 42:42–62.

Greenberg DA (1984): Simulation studies of segregation analysis: application of two-locus models. *Am J Hum Genet* 36:167–176.

Greenberg DA, Lange KL (1982): A maximum likelihood test of the two-locus model for coeliac disease. *Am J Med Genet* 12:75–82.

Haseman JK, Elston RC (1972): The investigation of linkage between a quantitative trait and a marker locus. *Behav Genet* 2:3–19.

Hasstedt SJ (1982): A mixed-model likelihood approximation on large pedigrees. *Comput Biomed Res* 15:295–307.

Hasstedt SJ (1993): PAP: Pedigree Analysis Package, revision 4. Salt Lake City: Department of Human Genetics, University of Utah Medical Center.

Hasstedt SJ, Albers JJ, Cheung MC, Jorde LB, Wilson DE, Edwards CQ, Cannon WN, Ash KO, Williams RR (1984): The inheritance of high density lipoprotein cholesterol and apolipoproteins A-I and A-II. *Atherosclerosis* 51:21–29.

Huff FJ, Auerbach J, Chakravarti A, Boller F (1988): Risk of dementia in relatives of patients with Alzheimer's disease. *Neurology* 38:786–790.

Huntington's Disease Collaborative Research Group (1993): A novel gene containing a trinucleotide repeat that is expanded and unstable on Huntington's disease chromosomes. *Cell* 72:971–983.

Joachim CL, Morris JH, Selkoe DJ (1988): Clinically diagnosed Alzheimer's disease: Autopsy results in 150 cases. *Ann Neurol* 24:50–56.

Kallman FJ (1956): Genetic aspects of mental disorders in later life. In: Kaplan OJ, ed. *Mental Disorders in Later Life*, 2nd ed. Stanford, CA: Stanford University Press, pp. 26–46.

Kaplan EL, Meier P (1958): Non-parametric estimation from incomplete observations. *J Am Stat Assoc* 53:457–481.

Kelsoe J, Ginns E, Egeland J, Gerhard DS, Goldstein AM, Bale SJ, Pauls DL, Long RT, Kidd KK, Conte G, Housman DE, Paul SM (1989): Re-evaluation of the linkage relationship between chromosome 11p loci and the gene for bipolar affective disorder in the old order Amish. *Nature* 342:238–243.

Kennedy J, Giuffra L, Moises H, Cavalli-Sforza LL, Pakstis AJ, Kidd JR, Castiglione CM, Sjogren B, Wetterberg L, Kidd KK (1988): Evidence against linkage of schizophrenia to markers on chromosome 5 in a northern Swedish pedigree. *Nature* 336:167–170.

Khoury MJ, Beaty TH, Cohen BH (1993): *Fundamentals of Genetic Epidemiology*. New York: Oxford University Press.

King MC, Lee GM, Spinner NB, Thomson G, Wrensch MR (1984): Genetic epidemiology. *Am Rev Pub Hlth* 5:1–52.

King RA, Rotter JI, Motolsky AG (1992): *The Genetic Basis of Common Diseases*. New York: Oxford University Press.

Kokmen E, Chandra V, Schoenberg BS (1988): Trends in incidence in dementing illness in Rochester, Minnesota, in three quinquennial periods, 1960–1974. *Neurology* 38:975–980.

Krauss RM, Williams PT, Blanche PJ, Cavanaugh A, Holl LG, Austin MA (1993): Lipoprotein subclasses in genetic studies: The Berkeley Data Set. *Genet Epidemiol* 10:523–528.

Krugylak L, Lander ES (1995): A nonparametric apporach for mapping quantitative trait loci. *Genetics* 139:1421–1428.

Lalouel J-M, Morton NE (1981): Complex segregation analysis with pointers. *Hum Hered* 31:312–321.

Lalouel J-M, Rao DC, Morton NE, Elston RC (1983): A unified model for complex segregation analysis. *Am J Hum Genet* 35:816–826.

Lange K, Boehnke M (1983a): Extensions to pedigree analysis. IV. Covariance components models for multivariate traits. *Am J Med Genet* 14:513–524.

Lange K, Boehnke M (1983b): Extensions to pedigree analysis. V. Optimal calculations of Mendelian likelihoods. *Hum Hered* 33:291–301.

Lautenschlager NT, Cupples LA, Rao VS, Auerbach SA, Becker R, Burke J, Chui H, Duara R, Foley EJ, Glatt S, Green RC, Jones R, Karlinsky H, Kukull WA, Kurz A, Larson EB, Martelli K, Sadovnick AD, Volicer L, Waring SC, Growdon JH, Farrer LA (1996): Risk of dementia among relatives of Alzheimer disease patients in the MIRAGE study: What's in store for the "oldest old"? *Neurology* 46:641–650.

Levy-Lahad E, Wasco W, Poorkaj P, et al. (1995): Candidate gene for the chromosome 1 familial Alzheimer's disease locus. *Science* 269:973–977.

Li CC (1975): *Path Analysis—A Primer*. Pacific Grove, CA: Boxwood Press .

Martin RL, Gerteis G, Gabrielli WF (1988): A family-genetic study of dementia of Alzheimer type. *Arch Gen Psychiatr* 45:894–900.

Merikangas KR, Risch NJ, Merikangas JR, Weissman MM, Kidd KK (1988): Migraine and depression: Association and familial transmission. *Psychiatr J Res* 22:119–129.

McKusick VA (1994): *Mendelian Inheritance in Man, 11th ed.* Baltimore: Johns Hopkins University Press.

Mohs CR, Breitner, JCS, Silverman JM, Davis KL (1987): Alzheimer's disease, morbid risk among first-degree relatives approximates 50% by 90 years of age. *Arch Gen Psychiatr* 44:405–408.

Morton NE, MacLean C (1974): Analysis of family resemblance. III. Complex segregation of quantitative traits. *Am J Hum Genet* 26:489–503.

Morton NE, Rao DC, Lalouel JM (1983): Vol 4: Methods in genetic epidemiology. In "Contributions to epidemiology and biostatistics: Klingberg MA (ed.). Karger, New York. Pp. 18–22.

Mullan M, Crawford F, Buchanan J (1984): Technical feasibility of genetic testing for Alzheimer's disease. *Alzheimer Dis Assoc Disord* 8:102–115.

Myers RH, Madden JJ, Teague JL, Falek A (1982): Factors related to age at onset of Huntington's disease. *Am J Hum Genet* 34:481–488.

Namboodiri KK, Kaplan EB, Heich I, Elston RC, Green PP, Rao DC, Laskarzewski P, Glueck CJ, Rifkind BM (1985): The collaborative Lipid Research Clinics Family Study: Biological and cultural determinants of familial resemblance for plasma lipids and lipoproteins. *Genet Epidemiol* 2:227–254.

Nee LE, Eldridge R, Sunderland T, Thomas CB, Katz D, Thompson KE, Weingartner H (1987): Dementia of the Alzheimer type: Clinical and family study of 22 pairs. *Neurology* 37:359–363.

Neuman RJ, Rice JP (1992): Two-locus models of disease. *Genet Epidemiol* 9:347–366.

Pericak-Vance MA, Elston RC, Conneally PM, Dawson DV (1983): Age-of-onset heterogeneity in Huntington's disease families. *Am J Med Genet* 14:49–59.

Pericak-Vance MA, Bebout JL, Gaskell PC, Yamaoka LH, Hung W-Y, Alberts MJ, Walker AP, Bartlett RJ, Haynes CA, Welsh KA, Earl NL, Heyman AL, Clark CM, Roses AD (1991): Linkage studies in familial Alzheimer disease: Evidence for chromosome 19 linkage. *Am J Hum Genet* 48:1034–1050.

Prenger VL, Beaty TH, Kwiterovich PO (1992): Genetic determination of high-density lipoprotein-cholesterol and apolipoprotein A-I plasma levels in a family study of cardiac catheterization patients. *Am J Hum Genet* 51:1047–1057.

Rao DC (1985): Application of path analysis in human genetics. In Krishnaiah PR, ed. *Multivariate Analysis* -IV North-Holland. Amsterdam pp. 467–484.

Rao VS, van Duijn CM, Connor-Lacke L, Cupples LA, Growdon JH, Farrer LA (1994): Multiple etiologies for Alzheimer disease are revealed by segregation analysis. *Am J Hum Genet* 55:991–1000.

Rao VS, Cupples LA, van Duijn CM, Kurz A, Green R, Chui H, Duara R, Auerbach S, Volicer L, Wells J, van Broeckhoven C, Growdon JH, Haines JL, Farrer LA (1996): Evidence for major gene inheritance of Alzheimer disease in families of patients with and without APOE e4. *Am J Hum Genet* 59: 664–675.

Rapoport SI, Pettigrew KD, Schapiro MB (1991): Discordance and concordance of dementia of the Alzheimer type (DAT) in monozygotic twins indicate heritable and sporadic forms of Alzheimer's disease. *Neurology* 41:1549–1533.

Risch N (1990): Linkage strategies for genetically complex traits. I. Multilocus models. *Am J Hum Genet* 46:222–228.

Rogaev E, Sherrington R, Rogaeva EA, Levesque G, Ikeda M, Liang G, Chi H, Lin C, Holman K, Tsuda T, Mar L, Sorbi S, Nacmias B, Placentini S, Amaducci L, Chumakov I, Cohen D, Lannfelt L, Fraser PE, Rommens JM and St George-Hyslop PH (1995): Familial Alzheimer's disease in kindreds with missense mutations in a gene on chromosome 1 related to the Alzheimer's disease type 3 gene. *Nature* 376:775–778.

Sadovnick AD, Irwin ME, Baird PA, Beattie BL (1989): Genetic studies on an Alzheimer clinic population. *Genet Epidemiol* 6:633–643.

Schoenberg BS, Kokmen E, Okazaki H (1987): Alzheimer's disease and other dementing illnesses in a defined United States population: Incidence rates and clinical features. *Ann Neurol* 22:724–729.

Sherrington R, Brynojolfsson J, Petursson H, Potter M, Dudleston K, Barraclough B, Wasmuth J, Dobbs M, Gurling H (1988): Localization of a susceptibility locus for schizophrenia on chromosome 5. *Nature* 336:164–166.

Sherrington R, Rogaev EI, Liang Y, Rogaeva EA, Levesque G, Ikeda M, Chi H, Lin C, Li G, Holman K, Tsuda T, Mar L, Foncin JF, Bruni AC, Montesi MP, Sorbi S, Rainero I, Pinessi L, Nee L, Chumakov I, Pollen D, Brookes A, Sanseau P, Polinsky RJ, Wasco W, Da Silva HAR, Haines JL, Pericak-Vance MA, Tanzi RE, Roses AD, Fraser PE, Rommens JM, St George-Hyslop PH (1995): Cloning of a gene bearing missense mutations in early-onset familial Alzheimer's disease. *Nature* 375:754–760.

Silverman JM, Breitner JC, Mohs RC, Davis KL (1986): Reliability of the family history method in genetic studies of Alzheimer's disease an related dementias. *Am J Psychiatr* 143:1279–1282.

Silverman JM, Li G, Zaccario ML, et al (1994): Patterns of risk in first degree relatives of Alzheimer's disease patients. *Arch Gen Psychiatr* 51:577–586.

Snell RG, MacMillan JC, Cheadle JP, Fenton I, Lazarou LP, Davies P, MacDonald ME, Gusella JF, Harper PS, Shaw DJ (1993): Relationship between trinucleotide repeat expansion and phenotypic variation in Huntington's disease. *Nat Genet* 4:393–397.

St George-Hyslop PH, Myers RH, Haines JL, Farrer LA, Tanzi RE, Abe K, James MF, Conneally PM, Polinsky RJ, Gusella JF (1989): Familial Alzheimer's disease: Progress and problems. *Neurobiol Aging* 10:417–425.

Strittmatter WJ, Saunders AM, Schmechel D, Pericak-Vance M, Enghild J, Salvesen GS, Roses AD (1993): Apolipoprotein E: High avidity binding to β-amyloid and increased frequency of type 4 allele in late-onset familial Alzheimer disease. *Proc Natl Acad Sci USA* 90:1977–1981.

van Duijn CM, Farrer LA, Cupples LA, Hofman A (1993): Genetic transmission for Alzheimer's disease among patients identified in a Dutch population based survey. *J Med Genet* 30:640–646.

Williams RR, Hopkins PN, Hunt SC, Wu LL, Hasstedt SJ, Lalouel JM, Ash KO, et al (1990): Population-based frequency of dyslipidemia syndromes in coronary-prone families in Utah. *Arch Intern Med 150:582–588.*

6

Patient and Family Participation in Genetic Research Studies

Pamela E. Cohen
Pacific Southwest Regional Genetics Network
California Department of Health Services
Berkeley, California

Chantelle Wolpert
Division of Neurology
Center for Human Genetics
Duke University Medical Center
Durham, North Carolina

INTRODUCTION

The genetic analysis methods described elsewhere in this book are extraordinary tools that have been used to identify many genes associated with different traits and disorders. As a result of the success of these gene mapping and positional cloning methods, genetic family studies, also called pedigree or linkage studies, are being done more often by clinical investigators new to this area of research. Although such studies offer researchers opportunities for identifying and studying disease genes, they also present specific challenges and quandaries. Therefore, it is neces-

Approaches to Gene Mapping in Complex Human Diseases, Edited by Jonathan L. Haines and Margaret A. Pericak-Vance. ISBN 0-471-17195-6

sary and beneficial for investigators to understand the particular issues related to family genetic research studies.

The distinction between human genetic research and traditional medical research arises primarily from the unique nature of personal genetic information. Unlike clinical trials used to evaluate medical treatment options of *individuals*, genetic studies involve the collection and analysis of DNA samples and medical information from entire *families*. All genetic studies, even those performed on individuals, can potentially generate genetic information that could have implications for the family as a whole. In addition, clinical trials typically offer participants more immediate benefits, such as improved therapies and treatments, whereas genetic studies rarely relate to direct clinical care and often offer no immediate or direct benefits to individuals. Furthermore, while clinical research may be associated with a risk of physical harm, the risks involved with participating in a genetic study are mainly psychological and social. Such risks are due both to the possible predictive nature of genetic information and to the conceivable use of personal genetic information as a basis for stigmatization and discrimination. Finally, while there are extensive regulations governing clinical research, codes of conduct for performing genetic family studies are still evolving.

In this chapter, we present a step-by-step approach for organizing family genetic research studies, hereafter referred to as "family studies." Practical methods are provided for maximizing family participation and increasing the efficiency of the research. In addition, recommendations are presented for providing high-quality family services while minimizing potential problems. Finally, suggested ethical and legal guidelines for conventional practices of genetic research are reviewed.

STEP ONE: PREPARING TO INITIATE A FAMILY STUDY

A Need for a Family Studies Director

The numerous tasks associated with conducting a family study include the administration of the study, family recruitment, obtaining informed consent from research participants, coordination of sample collection and field trip visits to families, and working with the institutional review board for study approval. Tasks may also include genetic counseling and other clinical activities. One professional should be designated as the family studies director to handle these diverse, but vital, responsibilities. (See Box 6.1 for a listing of duties of the family studies director.) Ideally, because of the diverse skills required, the family studies director should be a genetic counselor, social worker, psychologist, or other health professional with training in counseling and medical genetics. Some of the duties of the family studies director are discussed in greater detail elsewhere in this chapter.

Working with Families. The family studies director plays an important role in facilitating research while addressing the research participants' needs (McConkie-Rosell and Markel, 1995). Serving as the family's contact person with the research team, the family studies director works to establish rapport with families. In partic-

BOX 6.1: DUTIES OF THE FAMILY STUDIES DIRECTOR

Administrative

Submit IRB applications and renewals
Ascertain families for research studies
Serve as liaison between research team and families, referring professionals, and support organizations
Monitor and ensure compliance with genetic research regulations
Draft and implement protocols to ensure quality control measures
Assist with preparation of grant applications

Clinical Activities

Obtain informed consent from research participants
Elicit family histories
Phlebotomy*
Directed physical exams*
Genetic counseling*

Educational

Prepare presentations for professional conferences and disorder support groups
Prepare yearly research updates (newsletters) for families
Prepare educational materials as needed for research studies
Assist with preparation of abstracts, manuscripts, and other publications

Logistical

Arrange for sample collection
Arrange field study visits
Request medical records
Maintain databases

*Dependent on family studies director's background

ular, the family studies director helps foster realistic expectations about the study. In addition, individuals may have more questions related to the research study or to the genetic aspects of their condition or may need to be resampled or reexamined at a later date. By being available to families to answer questions and offer support, the family studies director helps meet the needs of participants, including the need for information, the need to be informed of the study progress, and the need to be understood. The family studies director is essential to providing referrals to families that would benefit from clinical services.

Working with Referring Professionals. In addition to working with families, the family studies director serves as the laboratory's representative to referring professionals and facilities. By being available to answer clinicians' questions about the research study, the study director builds credibility among clinical professionals who may serve as a referral source for families. Also, the director provides current and accurate information on the status of genetic research for a particular disease. For example, the director may provide referring professionals with regular bulletins or notifications of research discoveries with clinical implications. Thus, by maintaining a rapport with outside institutions, the family studies director or research coordinator enhances referrals, hence family participation.

The Researcher as the Family Studies Director. Those who do research in genetics are often among the few experts who can provide patients and professionals with accurate and comprehensive information about the disorder they are studying. While providing such information to research participants and referring professionals can be a time-consuming process, it is a vital part of conducting family studies. Many families with genetic disorders, especially rare disorders, may have little or only incomplete information about the condition. Additionally, newly diagnosed individuals may need more support and information about the disorder, its mode of inheritance, and genetics in general. Individuals who require formal counseling and genetic testing should be referred for these services when they are not provided as part of the research protocol.

A single investigator with limited funding may not be able to hire a professional for family ascertainment. One option in such instances is to consider collaborating with a research group that has a family studies director, who can assist with family ascertainment. If, alternatively, the investigator chooses to serve as the family studies director, he or she should allot time for communicating with the families, recognize the necessity of maintaining ongoing communication with families, and be aware of the families' needs and their reasons for participating in the study. Since the needs of a family may go beyond the expertise of the researcher, it is helpful for all researchers to have a resource list of genetic counselors and other health care professionals and disease support groups for family referral.

Working with Human Subjects

> The greatest challenge to securing continuing funding for the [Human] Genome Project does not originate from concerns about privacy, confidentiality, or coercive genetic testing. It is eugenics, manipulating the human genome to improve or enhance the human species, that is the real source of worry. This is reflected in the content of the futuristic horror scenarios spun by the Project's critics. It is also rooted in the historical reality of social policies based on eugenics that led to the deaths of millions in this century.
>
> —Arthur Caplan (Caplan, 1992a, p. 138)

In light of the history of abuse of genetic science, researchers should proceed with an awareness of the possible harms that can arise from the misuse of genetic information. Perhaps the most notorious example of abuse in the field of human genetics was the implementation of "racial hygiene" resulting in programs aimed at eliminating persons considered "genetically defective" by the Nazi regime in Germany (Proctor, 1988; Mueller-Hill, 1992). As in Europe, the United States had overt eugenics movements and proponents of compulsory sterilization of those with mental and physical disabilities as well as prison inmates and other "undesirables" (Reilly, 1991). In addition, over half of the states enacted laws barring marriages between whites and other ethnic groups and reinforcing existing stigmas of racism and segregation (Proctor, 1992). Several decades after the disintegration of the eugenics movement, critical accounts of modern genetic technologies continue to caution the scientific community and the public at large on the use and misuse of genetic information (Hubbard, 1986; Holtzman, 1989; King 1992; Paul 1994).

In addition to the unique concerns in genetic research, all types of research study involving human subjects may contain potential risks to participants. During the infamous Tuskegee syphilis study, vulnerable research subjects infected with syphilis were denied antibiotic therapy even though penicillin had become the treatment of choice for this condition (Caplan, 1992b). These subjects were also recruited with misleading promises of free treatment and were enrolled without their informed consent. The public disclosure of these abuses ultimately led to the National Research Act of 1974 (US OPRR, 1993), which mandates the approval of all federally funded proposed research with human subjects by an institutional review board (IRB).

The Institutional Review Board

In accordance with federal regulations, research studies involving human subjects, including noninvasive family studies, that are funded by federal agencies (e.g., National Institutes of Health, Department of Energy, Food and Drug Administration) must be reviewed by an institutional review board (US OPRR, 1993). IRB approval extends to every aspect of a pedigree study from family recruitment efforts to releasing research results. Typically, an IRB reviews the research protocol of the facility or medical center where the study will be performed. Research facilities without an existing IRB must either establish an IRB or submit their project proposals for review by an external IRB which satisfies federal requirements (US DOE brochure).

Although the majority of privately funded companies and institutions are connected to federal agencies via funding or facilities, research conducted strictly from private finances is not currently subject to federal regulations or IRB review (US DOE, 10CFR745; US DHHS, 1991). However, independent researchers may choose to contract with an external IRB for protocol review to ensure that participants' interests are protected.

Comprising of a multidisciplinary group of professionals, the IRB is defined as "an administrative body established to protect the rights and welfare of human research subjects to participate in research activities conducted under the auspices of the institution with which it is affiliated" (US OPRR, 1993, p. 1-1). The IRB review process is designed to be, as much as possible, beneficial to both the researcher and the study participant. In accordance with protecting the rights and welfare of participating families, IRB members offer additional insight into the study design and strengthen the research by assuring that family studies are conducted in accordance with federal and state regulations.

Human studies protocols are subject to a critical IRB review for scientific and ethical validity prior to initiation of a study. There are two types of review process: full board review and expedited review. While full board review involves all or most of the IRB committee members, expedited review is usually performed by a designated member (e.g., the committee chair) or a subset of members (US OPRR, 1993). Typically, the review process involves providing the IRB with a detailed description of the study protocol, a documentation of funding, and a model consent document for study participants. In addition, the researcher should present a rationale for maintaining identifying information, if any, on study participants and the methods, such as data coding, used to ensure participants' anonymity.

Identifiable Versus Anonymous Data

Full board review is generally required for all studies that maintain identifying information about research participants. Identifying information is any information that can be used to directly or indirectly link an individual with his or her sample (e.g., numerical code, name, address, pedigree information, medical records). The majority of genetic research studies necessitate the use of identifying information, such as pedigree information and documentation of diagnosis through medical records, to establish genetic relationships among family members and to perform genotype–phenotype analyses. In addition, some researchers maintain identifiable information so that participants can be recontacted for a variety of reasons as discussed later (see below, Maintaining Contact with Participants). It is therefore essential that all researchers using identifiable data maintain strict confidentiality practices (see below, Confidentiality).

IRB review is generally not required for studies using preexisting, anonymous data (Clayton et al., 1995). Anonymous data is that which cannot be linked to a participant, either directly (e.g., a name) or indirectly (e.g., a numerical code). In contrast, protocols utilizing data that has been "anonymized" through the irreversible removal of identifiers from samples are subject to IRB approval. In both cases, the use of anonymous or anonymized data offers the researcher the advantage of eliminating certain potential risks to individual study participants, such as genetic discrimination (see below, Genetic Discrimination), and social or psychological risks. Even with individually anonymous data, these risks may still exist for groups or populations if ethnic or social group information is retained (Clayton, 1995).

STEP TWO: ASCERTAINMENT OF FAMILIES FOR STUDIES

Family Recruitment

As part of the study protocol, family recruitment proposals must receive IRB approval before the initiation of the study and prior to beginning family ascertainment. As with all voluntary studies, no coercion should be used in ascertainment for family genetic studies. In addition, these recruitment methods must be conducted to ensure that potential participants are not contacted by the researcher until they have given their permission through a family member or through a referring professional. After permission to contact a potential research participant has been obtained, it is often appropriate to write a letter introducing the research team and the study and inviting the person to contact the family studies director for more information. Alternatively, the researcher may call the person directly and explain the study over the telephone.

Families can be ascertained using several methods, each with its own advantages and disadvantages. The difficulty in finding families interested in research studies will vary depending on the incidence of the disorder being investigated and the methods used to publicize the study. Four main sources of family referrals are disease support groups and similar organizations, health care providers, public databases and Web sites, and medical clinics (Box 6.2).

Support Groups and Organizations. Support groups for specific genetic disorders can provide general information about research studies to their membership and refer interested families to the researcher. Likewise, researchers can refer participants to support groups as a valuable resource for families seeking further information and services. Among the many genetic support groups in (and out of) the United States, there are three umbrella organizations that can facilitate finding a specific disorder support group: the Alliance of Genetic Support Groups, the National Organization of Rare Disorders (NORD), and the Self-Help Clearinghouse (a federally funded program that creates an information office in each U.S. state).

Referrals from Health Care Providers. Advertising in medical journals, announcements at medical conferences, or word-of-mouth referrals among professional colleagues are additional methods for reaching families. When referral is through a health professional, the family should be given the researcher's name and contact information or, alternatively, the referring professional should obtain the family's permission to be contacted by the researcher directly. It is reemphasized that families should never be contacted directly unless permission to do so has been obtained through a referring professional or support group.

Databases and the Internet. There are several databases with which researches can register and list their study descriptions and contact information directly. These databases are usually accessed by health care professionals seeking out researchers to whom they can refer their patients. With access to the World Wide Web via the Internet now becoming an integral part of communication, it is feasible and, perhaps, preferable to consider ascertaining families through Internet connections. This can be done by posting an announcement online with a disorder support group or by developing a home page where families can contact the researcher (Biesecker and Derenzo, 1995). Researchers should prepare for a high volume of inquiries in response to Internet advertising.

BOX 6.2: SOURCES OF FAMILY REFERRAL

Support groups
Medical clinics
Helix directory of clinical and research DNA laboratories
Genetics professional societies
Internet

Family Ascertainment Through Medical Clinics. Researchers who ascertain families through medical clinics must be especially careful to clearly define the difference between the patient's clinical care and his or her participation in a study protocol. Patients may mistakenly believe that the research study is part of their medical care. Alternatively, patients may have the impression that their medical care is provided contingent upon participation in the proposed research study (which contradicts the principle of a voluntary study). In addition, potential study participants may confuse the research study with a clinical service and may anticipate test results or treatment. Therefore, the researcher should clarify the difference between the clinical and research settings and emphasize their independence (e.g., a research lab does not provide clinical services).

In addition to clearly delineating the line between research and clinical services, clinician researchers who also provide medical care to research participants should maintain study files separate from medical records (Earley and Strong, 1995). Furthermore, medical practitioners should avoid referencing the research study in the patient's medical chart. In this way, information gathered by the study is not subject to disclosure requests by insurance companies, except in legal proceedings. As described later, however, researchers may apply for a certificate of confidentiality that grants research data immunity from such subpoenas.

Recruitment by Family Members. In most genetic studies, especially linkage studies, samples from entire families are necessary for analysis. As has been pointed out, studies that involve individual participants may provide "[genetic] information about relatives who are not in the study and who therefore did not have the opportunity to give or withhold informed consent to their participation" (Frankel and Teich, 1993, p. 31). All family members who participate in a genetic research study must do so voluntarily. This is the most important issue in work with families (Parker and Lidz. 1994). In some instances, people are tempted to coerce their own family members to join a research study and may in fact apply such pressure. A researcher should guard against coercion, albeit well-intentioned, by speaking with all family members individually and eliciting their opinions about participating in the research study. Logistically, it may be necessary for one family member to inform all the other family members about the study, but the researcher should ask for permission to communicate individually with each potential participant.

Informed Consent and Family Participation

Participation in a genetic research study involves the comprehension of complex information. This information must be presented to each family member participating in the study in a clear and concise manner (Box 6.3). The process of communicating this information to potential study participants is the basis on which that individual can give his or her informed consent. The researcher must provide the potential participant with an appropriate explanation of the study, for only in this way can informed consent assure voluntary participation and informed decision making on behalf of the individual. When families understand the research protocol, they will be able to set realistic expectations about the benefits and the limitations of participating in a genetic research study.

In response to documented human rights violations in the area of human studies research, regulations have evolved to ensure the protection of research participants. The importance of "voluntary consent" in research did not emerge until the end of World War II in accordance with the Nuremberg Code. The required elements of the informed consent process are set forth by federal regulations (US OPRR, 1993). In addition to federal regulations, state laws and case law precedents may apply to research involving human subjects.

Potential research participants should understand that their participation is optional and that they may withdraw their consent at any time after enrolling in the study without harm or penalty to themselves or to their family members. Participants should be told the purpose of the study, as well as the risks, benefits, and alternatives to participation, the policy on disclosure of results (if any), and the financial and time commitment involved (Alliance of Genetic Support Groups brochure; US DOE, brochure; Weir and Horton 1995a, 1995b).

BOX 6.3: ELEMENTS OF INFORMED CONSENT

Emphasize the voluntary aspect of the study
Answer all questions thoroughly
Offer a clear and concise explanation of the study including information on:
- Risks (e.g., medical, nonmedical/psychosocial, future insurability)
- Benefits
- Type and timing of information disclosure (if any)
- Limits of confidentiality
- Time/financial commitment
- Right to withdraw
- Right not to be recontacted or to know research results
- Ownership of DNA sample

In accordance with the Clinical Laboratories Improvement Amendments (CLIA) of 1988, laboratories that disclose results to participants must obtain approval to do so as described later (see below, CLIA-Approved Research Laboratories: Releasing Diagnostic Results) and should set a policy on the results to be disclosed and the circumstances of disclosure. This policy should be clearly explained to potential research participants as part of the informed consent process. In addition to disclosure of results, genetic family studies raise unique issues with respect to confidentiality and the protection of genetic information, genetic discrimination, psychological risks, and DNA banking and ownership. These issues require further discussion and are addressed later in this chapter.

The informed consent process should be documented, typically with a signed consent form (see Appendix 6.1). Participants should be given a copy of the consent form for their files. While standard research practice requires a signed consent form before enrolling a research participant in a protocol, the existence of such a document does not guarantee that a participant fully understands all the ramifications of the research. To comply with the IRB approval process, consent forms and procedures must meet the facility's standards as well as federal and state regulations.

Vulnerable Populations. According to federal regulations, individuals belonging to vulnerable populations are afforded special protection with regard to their participation in research protocols. Such populations include children, individuals with a mental disability, prisoners, pregnant women, fetuses, and economically or educationally disadvantaged persons (US OPRR, 1993). In addition to federal laws, studies involving research participants who receive services from state-funded agencies or institutions may require additional approval by those state agencies (e.g., state department of mental retardation). Therefore, as part of the IRB review process, researchers must define the populations to be included and provide a rationale for their inclusion in genetic research protocols.

Minors. By nature of their design, genetic studies, such as linkage analyses, typically involve sampling entire families. While a parent or legal guardian ultimately gives consent for a child's participation in a research protocol, federal regulations suggest that children should receive an explanation of the study at an age-appropriate level and, whenever possible, should give their assent in which they affirm their agreement to participate in the study, especially if the study has no direct medical benefit to the child (US OPRR, 1993). While the validity of assent obtained from children continues to be debated, some researchers feel that children should have the opportunity to refuse to participate in a study (Frankel and Teich, 1993). In addition, according to the specific protocol and the associated risks, limits may be placed on the amount of blood, if any, that may be drawn from a child for a research study (US OPRR, 1993).

The two most likely scenarios involving the sampling of minors for genetic studies involve the sampling of an affected child for linkage and/or mutation studies and the sampling of a child (regardless of affected status) to enable the reconstruction of parental genotypes. However, as with all research results, results should be disclosed only by CLIA-approved laboratories within the context of genetic counseling as discussed later (see below, CLIA-Approved Research Laboratories: Releasing Diagnostic Results).

Persons with Cognitive Impairment. Many family studies involve researching the genetic basis of disorders that affect mental and cognitive abilities in a congenital, progressive, or late-onset manner. Persons lacking mental capacity (e.g., persons with mental retardation or dementia) who are, thus, unable to give informed consent for participation in a study, are afforded special consideration. As discussed above with regard to children, in conjunction with obtaining consent from the legal guardian, assent should be sought from the potential participant whenever possible. Researchers should check with their IRB about federal and state guidelines for including persons with mental impairment in their research protocols.

STEP THREE: SAMPLE COLLECTION

As required when designing a study protocol, researchers should define their methods for obtaining DNA samples prior to initiating sample collection. The study requirements will determine which samples should be collected. (See Chapter 8.) Depending on the requirements of the study, DNA samples can be collected by the researcher at a medical clinic or obtained via field studies. Alternatively, sample collection kits can be mailed to research participants. In these instances the researcher or the participant must arrange for the collection of blood or other tissue samples. The researcher should clarify in the protocol whether the participant will be reimbursed for any expenses associated with sample collection, such as blood-drawing fees, parking, and food. In addition, researchers should check with their IRB to ensure that packaging of biological samples sent through the mail or delivery service meets Federal Occupational Safety and Health Administration (OSHA) requirements, especially when samples are shipped between countries.

Confirmation of Diagnosis

A good study design requires the confirmation of family history information and medical diagnoses. An accurate pedigree is critical to establishing the genetic relationships among family members necessary for genetic research. Likewise, diagnostic information on affected individuals is essential for proper analysis. Diagnostic confirmation can be obtained in several ways depending on many factors including the type of disorder being investigated, proximity of the patients to the researcher, research funding for travel, and previous medical workup of research participants. Participants can be examined in a clinic or during a field study (see below, The Art of Field Studies). Alternatively, researchers may confirm diagnoses or traits by reviewing medical records.

For some genetic studies, especially linkage studies, it is as important to rule out a diagnosis in an apparently unaffected family member as it is to confirm the diagnosis in an affected individual. This rigorous clinical evaluation may lead to the diagnosis of a genetic disorder in an unsuspecting family member who did not know that he or she was affected. Before participating, research participants should be informed that there is a possibility that they will be found to have the disorder being investigated. The participants should be given the option of deciding whether to be advised if such information develops and should be reminded that their participation in the study is voluntary. If the researcher feels it is appropriate, patients interested in learning whether they have a disorder can be referred to a health care professional for clinical and follow-up care (see below, Need for Additional Medical Services). It should be decided in advance by the research team and the IRB how potential dilemmas should be handled. As with all studies, unanticipated problems arise and should be dealt with on a case-by-case basis in conjunction with the researcher's IRB.

Accurate clinical diagnoses of affected individuals are crucial to performing genetic studies aimed at identifying disease genes. Occasionally, a diagnosis cannot be confirmed unless additional clinical studies are done. In such instances, participants should be informed of the rationale for requesting additional tests and told how the costs of further clinical evaluations will be covered (e.g., by the researcher, by the participant, or by the participant's insurer). Again, people should be reminded of their right to refuse testing. Furthermore, participants should be told that, as with all medical information, results obtained from clinical evaluations or testing will be placed in their medical records. As discussed previously (under Family Ascertainment Through Medical Clinics), it is important for the researcher to clearly delineate the distinction between research participation and clinical care to ensure the family's continued informed and voluntary participation in the protocol. Any clinical diagnostic procedure performed specifically as part of the research protocol must be approved by the IRB.

The Art of Field Studies

The field study has a long tradition in genetic research. While no formal assessment exists, many researchers attest to the success of field studies as the single most valuable practice to ensure success in family ascertainment for genetic studies. Typically, researchers who do field studies travel to meet with study volunteers in their homes during times convenient for the different families. This increases the efficiency of the research study by minimizing the toll on participating family members of arranging travel for physical examinations and tissue sampling.

Field Study Preparation. A field study takes place only after a family has met the criteria for participation in the study, the diagnosis in the family has been confirmed by reviewing the available medical records of the individuals with the disorder, and all family members who are interested in participating in the research have received documentation of the purpose of the study and other detailed information. Detailed information about what a field study involves needs to be provided to each individual in the family who may participate. This information includes a meeting location (e.g., the research subject's home or another site, such as a clinical facility) and an estimate of how much time will be required for each individual and for the group.

A research team should consist of a clinician who can perform the directed physical examinations to confirm the diagnosis of the disorder being studied, an experienced phlebotomist, and a researcher who can coordinate the field visit, elicit a family history, explain the purpose of research, answer participants' questions, and obtain informed consent. Typically, a research team consists of a physician or physician assistant and a genetic counselor. For field visits that involve working with a large family, more professionals may be needed to facilitate the examination, sampling, and informed consent processes.

In conducting a field study, it is customary to introduce the entire research team to the family and to explain what each team member will be doing and to whom questions should be directed. The research team should try to complete the exams and sample collection in a timely fashion and allow time at the end of the visit for closing questions and comments by family members. Before leaving the family, one member of the research team (usually the family studies director) should define how contact and follow-up will be maintained with the family and research center. Finally, it is courteous to follow up a visit to a family with a thank-you letter. (See Boxes 6.4–6.6)

Special Issues in Family Studies. Although all studies involving human subjects entail such issues, there is heightened concern that home studies somehow

BOX 6.4: SUGGESTED GUIDELINES FOR FIELD STUDIES

Do:
- Introduce the field team to everyone in the family
- Explain the purpose of the research to each family member (see Box 6.3)
- Travel with two researchers
- Have complete supplies available (see Box 6.6)
- Tell individuals approximately how long a visit will take
- Be respectful about being in someone's home
- Refer family to a physician or genetic counselor, when indicated

Do not:
- Put a pedigree on a table for everyone to see
- Reveal information about one family member to another family member
- Offer diagnostic information (unless part of an IRB-approved clinical protocol)

BOX 6.5: FIELD STUDY PREPARATION

Pre–Field Trip Checklist

Obtain medical records
Review pedigree and medical records and approve field visit to family
Contact all relevant family members and confirm attendance
Send confirmation letter to all relevant family members
Confirm travel arrangements
Notify laboratory of the field trip date and expected number of samples
Pack field trip bag (see Box 6.6)

Post–Field Trip Checklist

Update pedigree (including affected status changes, age of onset, physical exam findings, dates of birth)
Match clinical evaluation forms with blood samples and consent forms
Take blood samples to laboratory
Write post–field trip note synopsis of what took place, significant clinical findings, and follow-up plan (e.g., if other family members need to be seen in the future)
Send letter of appreciation to family

might be viewed as coercive. Therefore, it is especially important to remind research participants prior to and during the visit that their participation is voluntary. To minimize the possibility of family coercion to participate in the study (Parker and Lidz, 1994), it is a good idea whenever possible to obtain informed consent from individual family members before visiting the home. It is vital that the informed consent process include an explanation of what should and should not be expected during the field visit. For instance, while a directed physical examination may be performed to confirm the diagnosis, clinical services, like medical treatment, are usually not provided by researchers in the field. In addition, individual decisions about whether to participate in the study should not be discussed with other family members unless permission has been given to do so. A written summary of these discussions needs to be sent to all the members of the family who will participate.

Interacting with families is qualitatively different from interacting with individuals and presents unique circumstances for both the families and the researcher. Visiting families in their homes allows for a more detailed discussion of family history and the family folklore surrounding a disorder. However, while some families are very open and accepting of their disease, others view it more as a stigma and treat the disorder as a taboo within the family. Thus, in addition to observing normal social conventions and being courteous when in someone's home, it is important for

BOX 6.6: FIELD STUDY SUPPLY CHECKLIST FOR FIELD TRIPS

Clinical Supplies

8.5 mL yellow stopper tubes (containing solution ACD solution)*
Other tubes for special studies (e.g., chromosomes, CPK)
21 × 3/4, 12′ Butterfly tubing (purple)
23 × 3/4, 12′ Butterfly tubing (orange)
Alcohol preps
Gauze pads
Band-Aids
Luer adapters
Vac holders
10/mL Syringes
Tourniquets (latex-free tourniquets)
Dirty needle box
Marker pens and 2 regular pens
Wire test tube for stopper tubes
Buccal swabs
Reflex hammer
Tuning fork
Other medical equipment (as needed)
Gloves (latex-free)
4 Mailing kits

Paperwork

Summary of the study*
Consent forms*
Pedigree forms*
Clinical evaluation forms
Medical record release forms*
Post–field trip checklist (see Box 6.5)
Educational materials (e.g., fact sheet, support group information)*

*These materials as well as packaging material for return shipment with instructions, should be included in all kits mailed to families.

the researcher to anticipate and be sensitive to different attitudes of family members toward the disease in question.

Confidentiality

Appropriate measures should be taken to secure all data obtained or generated by the research study (Box 6.7). Data should be coded to maintain individual confiden-

BOX 6.7: CONFIDENTIALITY

Maintain research information separately from medical records
Secure research files and computer databases
Restrict access to identifying information
Code samples and data
Obtain certificate of confidentiality for data protection from third parties

tiality. Likewise, databases should be secured (e.g., password protection), and access to databases and research files should be restricted to a designated individual(s), such as the family studies director. This person should maintain a list and family contact information (names, addresses, etc.) in a secure and safe place. Finally, continuing IRB review ensures that study protocols adhere to the strict enforcement of confidentiality as described during the informed consent process (Mac Kay, 1993).

Certificate of Confidentiality. Most researchers agree that they have an ethical duty to protect personal genetic information. Although no law exists that expressly guarantees genetic data protection, data is protected from third parties (e.g., insurers and employers) when accompanied by a certificate of confidentiality. This protection includes protection from subpoenas by third parties and from court orders. However, the certificate pertains only to federally funded research studies and does not protect information that is voluntarily provided by research participants to third parties (Earley and Strong, 1995). Furthermore, the certificate of confidentiality does not offer protection to a participant who is asked directly, for example, by an employer or insurer, whether he or she has participated in a study and/or has received any information as a result of that participation (Philip Reilly, personal communication). Researchers conducting federally funded studies may apply for a certificate of confidentiality through the Department of Health and Human Services.

Most studies maintain confidentiality and protection of information from third parties. However, despite efforts to uphold confidentiality such as data coding, securing files, and using certificates of confidentiality, there exists a possibility that information will be inadvertently disclosed to third parties (e.g., indirectly, through a notation in a medical record alluding to the study). Such information could preclude a person's chance for obtaining health or life insurance coverage (Wertz, 1992) (see below, Genetic Discrimination). Therefore, some professionals recommend including a discussion of the risk of future insurability, however small, in the informed consent process and document (Kass, 1993). Furthermore, researchers should be aware that disclosure to third parties of personal information without a participant's consent may be subject to legal action (Andrews, 1990).

STEP FOUR: FAMILY FOLLOW-UP

Need for Additional Medical Services

When appropriate, the family studies director should refer individuals to other health care professionals or to clinics for additional medical services. Research participants with a family history of a genetic disorder may have specific questions about their personal reproductive risk and should be referred for genetic counseling. In addition, families with an affected child may seek community and federal services such as early intervention programs and financial compensation for medical care. The family studies director can refer these families to a local genetic counselor, social worker, or support group to serve as a family advocate by exploring available programs. Finally, family members needing supportive counseling or additional medical or psychological care should be referred to the appropriate clinician when indicated.

Duty to Recontact Research Participants

In addition to the hope of learning research results (see below, Releasing Diagnostic Results), many families participate in genetic studies because they want to contribute to the research effort and are interested in following the progress of the research. These families often express a need to be given information about the study following their participation and may periodically contact the laboratory to inquire about the status of the project.

While the release of individual research results is regulated by federal legislation (Andrews et al., 1994), the legal duty of a research laboratory to recontact study participants to prevent potential harm—the "duty to warn"—is ambiguous and controversial. According to one report, "The fact that the health care provider or the laboratory had only brief contact with the person providing the DNA or that much time has passed since the DNA was originally deposited *does not provide a defense for not recontacting the patient*" (Andrews, 1990, p. 225, italics added). Nevertheless, unlike the case of clinical practice (*Tarasoff* v. *Regents of University of California*, 1976; Macklin, 1992; Wertz et al., 1995), the researcher's duty to warn participants of potential reproductive or other medical risks based on their genetic status or to provide them with general information (e.g., availability of commercial DNA testing) has yet to be decided in court. Although there is presently no legal precedent binding researchers to recontact participants, researchers may follow some general guidelines by sorting research results into categories as described (US OPRR 1993).

In the absence of a law or a court precedent creating a duty to recontact participants in the research setting, the question of whether researchers have a moral duty to recontact participants is being debated among geneticists, ethicists, lawyers, public health professionals, and consumers. Some professionals believe that genetic researchers have an obligation to inform participants of the progress of the research study (MacKay, 1993), especially if that information may impact participants' treatment options and reproductive and medical decision making. Furthermore, it has been suggested that families be sent an abstract or reprint of any published studies that resulted from their participation (de Leon and Lustenberger, 1990) and be given a summary of the data gathered during a study (Kodish et al., 1994).

Regardless of the research team's view of the ethical duty to recontact, participants should be told, as part of the informed consent process, whether they will receive any information about the status of the research following their participation in the study. Facilities that do maintain contact with study participants should clarify how the information will be relayed, as discussed below.

Maintaining Contact with Participants. Maintaining contact with study participants provides both the researchers and the participants with several advantages. The participant is informed of pertinent information relating to a particular genetic disorder, including any breakthroughs that have resulted from the research study. The researcher maintains contact with the family, providing the potential for recontact, reexamination, and/or resampling in the future. Ideally, given adequate funding and staffing, the family studies director provides participants with yearly updates on the progress of the study as well as bulletins when research breakthroughs occur.

Contact can be maintained through a variety of ways: a follow-up letter to the families, periodic updates on the status of the research (e.g., a newsletter), and telephone calls. It is helpful to designate a contact person in each family and to direct all correspondence to that person. In addition, each family contact person should receive the telephone number of the researcher (e.g., family studies director), in the event that future questions arise. It is beneficial to inform families who wish to be recontacted about their responsibility to notify the research team of any change in address or phone number so that they may receive timely reports on research progress and breakthroughs. In case contact is lost with a family, mailing a certified or registered ("return receipt requested") letter to the family at their last known address is suggested.

Guidelines for Releasing Genetic Information

> [I]f researchers are aware of identities of the individuals whose samples they have received, do they have a moral or legal obligation to notify the participants of information that may save, or at least change, [the participants'] lives?
>
> —Researcher of Huntington disease, quoted in Frankel and Teich, (1993, p. 16)

The nature of genetic information is unique in comparison to other types of personal information for three key reasons as delineated by Annas (1995, p. 1196): "it can predict an individual's likely medical future; it divulges personal information about one's parents, siblings, and children; and it has a history of being used to stigmatize and victimize individuals." Thus the disclosure of genetic information must occur in accordance with federal laws and ethical guidelines.

CLIA Regulations: Separation of Research and Clinical Laboratories. The Clinical Laboratories Improvement Amendments of 1988 (CLIA88) is federal legislation enacted to regulate the quality assurance of testing and reporting of results by clinical laboratories. In accordance with CLIA88 and other federal regulations, genetic testing protocols differ between research and clinical laboratories (Andrews et al., 1994). Under CLIA88, laboratories that disclose test results to patients for the purposes of the diagnosis, prevention, or treatment of a disease are subject to regulation. Research laboratories that release results to families for clinical decision making (e.g., laboratories that study rare diseases for which commercial testing is not available) are included under this provision. In addition to legal considerations, study participants should understand this separation of research and clinical laboratories as part of their decision to participate in the research (Miller, in preparation).

It is essential that research laboratories define the extent of information disclosure to determine whether the CLIA88 regulations are applicable and to develop a policy about releasing results to review with research as part of the informed consent process. Many research laboratories follow a protocol of complete nondisclosure of information to families (i.e., they do not provide research results to individuals). However, some research laboratories may desire to be authorized to release results under CLIA regulations. Laboratories that disclose results or other information to participants must consider additional issues as discussed below.

Although most research labs do not release genetic results, researchers should draft their study protocol in anticipation of scenarios that could necessitate the unforeseen disclosure of genetic information to participants, such as incidental findings with medical implications. Reilly has suggested that researchers define three categories of disclosure: "1) findings that are of such potential importance to the subject that they *must* be disclosed immediately; 2) [findings] that are of importance to subjects . . . , but about which [the researcher] should exercise judgment about the decision to disclose; 3) [findings] that do *not* require special disclosure" (Reilly, 1980, pp. 5, 12, original italics).

Incidental Findings and Special Disclosure. Incidental findings, such as nonpaternity or adoption status, reveal information other than that sought by the researchers. Such information, while important in establishing genetic relationships, is both potentially harmful when revealed to unsuspecting family members (Mac Kay, 1993) and potentially helpful when used in conjunction with a clinical protocol. For example, if nonpaternity were uncovered in a family in which the father had Huntington disease, such information could at once be devastating to and welcomed by the children, who were assumed to carry a 50% risk for developing this neurodegenerative disease. In addition, a genetic disorder other than the one being studied may be revealed in an individual's sample. Cases that are not clearly defined under the study protocol or involve serious or immediate medical implications should be referred to an ethics committee or an IRB on an individual basis to determine appropriate action.

CLIA-Approved Research Laboratories: Releasing Diagnostic Results. Research laboratories that are CLIA-approved for disclosing genetic data for diagnostic and other medical purposes must define a policy detailing which results will be disclosed, as well as the manner of their disclosure, and they must explain this disclosure policy during the consent process. In addition, laboratories should provide participants with a genetic counseling service (US OPRR, 1993), either directly or, more typically, through a referral to an appropriate center. As part of the counseling process, the problematic issues associated with disclosing genetic information, such as the potential personal and social ramifications of learning one's genetic status, are addressed (MacKay, 1993; Earley and Strong, 1995). In this way, test results are tactfully disclosed to the family and interpreted in a meaningful manner.

Psychological Risks. The potential harm to an individual of learning his or her genetic status—for example, by eliciting feelings of guilt, low self-esteem, or similar responses—is well recognized (US OPRR, 1993; Clayton et al., 1995). Such psychological and social risks should be discussed with participants as part of the informed consent process and prior to disclosing genetic results. This discussion should enable each individual to reach a decision about whether to receive test results. In this manner, an individual's "right not to know" is regarded equally with the right to know his or her genetic status.

Genetic Testing of Children. Genetic testing of an apparently unaffected child in the absence of any direct medical benefit is a controversial practice (Wertz, et a., 1994) and should be done only within the context of genetic counseling and according to specific guidelines (ASHG/ACMG Report, 1995; Council on Ethical and Judicial Affairs, 1995). Concern arises over the potential risk of stigmatization of a child known by family members and future employers and insurers to carry a genetic mutation (US OPRR, 1993). In addition, questions about the child's autonomy and his or her right not to know personal genetic information may conflict with parental autonomy. Parents wishing to have their children tested for genetic disorders should be referred to a genetic counselor to discuss the potential risks and benefits of testing.

Genetic Discrimination

The stigmatizing nature of genetic information and its history of abuse necessitate special provisions for protection (Holtzman and Rothstein, 1992a). Although a basic right to privacy has been established by the U.S. Supreme Court (*Roe* v. *Wade*, 1973), the nonspecific nature of this privacy protection fails to adequately guard against unauthorized disclosure of personal genetic information. Reports show that genetic discrimination is practiced by employers and insurers wishing to avoid potential medical expenses of individuals with a family history of a genetic disease, including asymptomatic individuals (Gostin, 1991; Billings et al., 1992; Natowicz et al., 1992a; Wertz, 1992; Billings, 1993).

The Americans with Disabilities Act of 1990. The Rehabilitation Act of 1973, revised in 1992, was the first law to prohibit employment discrimination by federally funded agencies and institutions based on physical disability. This legislation defines an "individual with a disability" as any person who "is regarded as having [a physical or mental] impairment [1992 Reauthorization of the Rehabilitation Act of 1973 , Section 7(8-B)]. Such a definition could include asymptomatic gene carriers perceived as being "sick" by their employer. The 1990 Americans with Disabilities Act (ADA) significantly broadened the scope of the 1973 legislation by prohibiting discrimination against disabled individuals in most areas of employment and public transportation. While the ADA's definition of disability excludes persons with a "characteristic predisposition to illness or disease," some professionals believe that the legislation allots sufficient protection to persons against genetic discrimination in the workplace (cited in Natowicz et al., 1992a). Although there is disagreement about whether persons with a genetic predisposition to disease are adequately protected under the ADA (Holtzman and Rothstein 1992a; 1992b; Natowicz et al., 1992b), a recent ruling by the Equal Employment Opportunity Commission (EEOC) has interpreted the ADA as affording protection to individuals from employer discrimination based on genetic test results (Leary, 1995). Nevertheless, the judicial system may ultimately be called upon to define the extent of the ADA legislation within the context of case precedent.

Although the ADA prohibits employer discrimination, this law, along with the majority of state laws, does not adequately protect those with genetic disorders—both symptomatic and asymptomatic—against discrimination by insurers (Natowicz et al., 1992a). Thus, one could argue that employment opportunities may be limited, especially by smaller employers, who may refuse to offer employment because of a desire to avoid paying the higher premiums demanded by insurers. Equally disconcerting is the potential for employers to institute genetic screening programs prior to offering employment (Gostin, 1991). Given the legal latitude afforded to insurance companies, genetic information could become part of the standard application process for insurance policies (McEwen et al., 1992).

There have been reports of denial of insurance coverage based on participation in genetic studies (Earley and Strong, 1995; Hudson et al., 1995). Therefore, it is essential that all information collected and generated as part of the family study be kept confidential. Because of recognition of the potential harm of disclosure of a person's genetic status to a third party, the need for federal legislation guaranteeing a right to genetic privacy is gaining support (Reilly, 1992; Annas, 1995). In addition, working groups have been formed to evaluate the impact of genetic information on individual insurability and on the insurance industry (US NIH-DOE Working Group on Ethical, Legal, and Social Implications of Human Genome Project, 1993; (ASHG Ad Hoc Committee on Genetic Testing/Insurance Issues, 1995).

DNA Banking

Genetic research laboratories are inherently DNA banking facilities. Therefore, research laboratories should develop policies that address banking issues, including length of DNA storage and accessibility to banked samples. These policies should be incorporated into the informed consent process (Hall et al., 1991; Weir and Horton, 1995b).

The use of previously collected DNA samples for a new research study presents a problematic scenario in terms of informed consent. Suggestions for developing consent forms that address the issues of DNA banking and informed decision making have been presented (ASHG Ad Hoc Committee on DNA Technology, 1988; Knoppers and Laberge, 1989; Gold et al., 1993; ACMG Storage of Genetics Materials Committee, 1995; Weir and Horton, 1995b). Research facilities that bank DNA samples as a service to families in order to maintain a sample for future diagnostic purposes should clarify the differences between the banking and use of research versus clinical DNA samples.

Some practical guidelines for informing the participants about DNA banking options include the following.

1. Incorporation of the research center's policy on DNA banking into the informed consent process. This document should state clearly whether DNA samples collected for research purposes will be available to the research participants or their surviving family members.
2. Advising research participants about the availability of commercial DNA banking facilities. This advice may be particularly relevant for affected elderly family members or when a sole individual in a family has an unknown or terminal disorder. Commercial DNA banking of a DNA sample enables a family or individual to access a DNA sample and pursue diagnostic genetic testing when it becomes available.

It is advantageous for both the family and the research team to have interested family members deposit DNA samples with a commercial bank. This allows family members, not researchers, to designate which family members may or may not have access to the banked DNA sample. Commercial DNA banks usually have procedures that allow donors to designate those who can access the DNA sample.

DNA Ownership

In a landmark decision in 1980, the U.S. Supreme Court ruled that biological materials could be patented, opening the door for the patenting of genes and gene products (*Diamond* v. *Chakrabarty*, cited in Annas and Elias, 1992, p. 235). Ten years later, the California Supreme Court ruled that a person has no ownership rights to his or her cells once they have been removed from his or her body (*Moore* v. *Regents of University of California*). Therefore, although the legal status of DNA ownership has yet to be completely resolved, one could argue that the DNA derived from discarded tissue from research participants is the property of the researcher.

The right of a research participant to control the fate of his or her sample remains a sensitive and ambiguous issue. Several reports suggest that in addition to DNA storage and accessibility issues, participants should be offered a choice in determining the fate of their samples in terms of research and commercial uses (Clayton et al., 1995). Guidelines for DNA banking as well as legislation establishing an individual's ownership rights to his or her DNA have been proposed (Annas, 1993, 1995). Although the dispute over DNA ownership is an important one, it appears that unless a sample proves to have commercial value (*Moore* v. *Regents of University of California*), research participants are generally more concerned that the information derived from genetic studies be secure than about the fate of their individual sample material.

Future Considerations

As mentioned at the beginning of this chapter, many of the issues in genetic research studies discussed here are not completely resolved, and the procedures currently being used are constantly evolving. Some of the strongest debates center around such questions as what constitutes sufficient informed consent (and for which age groups), what is the status of DNA ownership, and how DNA banking should be handled. Two other issues still being hotly debated are the status of samples collected under earlier (and generally less detailed) informed consent procedures, and laboratory legal liabilities and responsibilities to the participants and their families. The resolution of these debates will have a major and lasting impact on the research community.

REFERENCES

ACMG Storage of Genetics Materials Committee (1995): ACMG statement: Statement on storage and use of genetic materials. *Am J Hum Genet* 57:1499–1500.

Alliance of Genetic Support Groups (brochure): Informed consent: Participation in genetic research studies. (Available by calling 1-800-336-GENE).

Andrews LB (1990): DNA testing, banking, and individual rights. In: Knoppers BM, Laberge CM, eds *Genetic Screening: From Newborns to DNA Typing*. London: Elsevier Science Publishers.

Andrews LB, Fullarton JE, Holtzman NA, Motulsky AG (1994): *Assessing Genetic Risks: Implications for Health and Social Policy*. Washington, DC: National Academy Press, Institute of Medicine

Annas GJ (1993): Privacy rules for DNA databanks: Protecting coded "future diaries." *JAMA* 270:2346–2350.

Annas GJ (1995): Editorial: Genetic prophecy and genetic privacy—Can we prevent the dream from becoming a nightmare? *Am J Public Health* 85:1196–1197.

Annas GJ, Elias S (1992): *Gene Mapping: Using Law and Ethics as Guides*. New York: Oxford University Press.

ASHG/ACMG Report (1995): Points to consider: Ethical, legal, and psychosocial implications of genetic testing in children and adolescents. *Am J Hum Genet* 57:1233–1241.

ASHG Ad Hoc Committee on DNA Technology (1988): DNA banking and DNA analysis: Points to consider. *Am J Hum Genet* 42:781–783.

ASHG Ad Hoc Committee on Genetic Testing/Insurance Issues (1995): Background statement: Genetic testing and insurance. *Am J Hum Genet* 56:327–331.

Biesecker LG, DeRenzo EG (1995): Internet solicitation of research subjects for genetic studies. *Am J Hum Genet* 57:1255–1256.

Biesecker LG, Collins FS, DeRenzo EG, Grady C, MacKay CR (1995): Case: Responding to a request for genetic testing that is still in the lab. *Cambridge Q Healthcare Ethics* 4:387–400.

Billings PR (1993): Genetic discrimination. *Healthcare Forum J* Sept–Oct:35–37.

Billings PR, Kohn MA, de Cuevas M, Beckwith J, Alper JS, Natowicz MR (1992): Discrimination as a consequence of genetic testing. *Am J Hum Genet* 50:472–482.

Caplan AL (1992a): If gene therapy is the cure, what is the disease? In: Annas GJ, Elias S, eds. *Gene Mapping: Using Law and Ethics as Guides*. New York: Oxford University Press, pp.128–141.

Caplan AL (1992b): Twenty years after: The legacy of the Tuskegee syphilis study. *Hastings Center Report* 6:29–32.

Clayton EW (1995): Panel comment: Why the use of anonymous samples for research matters. *J Law Med Ethics* 23:375–377.

Clayton EW, Steinberg KK, Khoury MJ, Thomson E, Andrews L, Ellis Kahn MJ, Kopelman LM, Weiss JO (1995): Informed consent for genetic research on stored tissue samples. *JAMA* 274:1786–1792.

Council on Ethical and Judicial Affairs (1995): Testing children for genetic status. *Code of Medical Ethics Reports, American Medical Association* 6:47–58.

de Leon D, Lustenberger A (1990): Issues raised in gene linkage studies. *Birth Defects* (original article series) 26:1391–1394.

Earley CL, Strong LC (1995): Certificates of confidentiality: A valuable tool for protecting genetic data. *Am J Hum Genet* 57:727–731.

Frankel MS, Teich AH (1993): *Ethical and Legal Issues in Pedigree Research*. Washington, D.C: Directorate for Science and Policy Programs, American Association for the Advancement of Science.

Gold RL, Lebel RR, Mearns EA, Dworkin RB, Hadro T, Burns JK (1993): Model consent forms for DNA linkage analysis and storage. *Am J Med Genet* 47:1223–1224.

Gostin L (1991): Genetic discrimination: The use of genetically based diagnostic and prognostic tests by employers and insurers. *Am J Law Med* 17:109–144.

Hall J, Hamerton J, Hoar D, Korneluk R, Ray P, Rosenblatt D, Wood S (1991): Policy statement concerning DNA banking and molecular genetic diagnosis. *Clin Invest Med* 14:363–365.

Holtzman NA (1989): *Proceed with Caution: Predicting Genetic Risks in the Recombinant DNA Era*. Baltimore: Johns Hopkins University Press.

Holtzman NA, Rothstein MA (1992a): Invited editorial: Eugenics and genetic discrimination. *Am J Hum Genet* 50:457–459.

Holtzman NA, Rothstein MA (1992b): Reply to Natowicz et al. *Am J Hum Genet* 51:897.

Hubbard R (1986): Eugenics and prenatal testing. *Int J Health Serv* 6:227–242.

Hudson KL, Rothenberg KH, Andrews LB, Ellis Kahn MJ, Collins FS (1995): Genetic discrimination and health insurance: An urgent need for reform. *Science* 270:391–393.

Jordan E (1992): Invited editorial: The Human Genome Project: Where did it come from, where is it going? *Am J Hum Genet* 51:1–6.

Kass NE (1993): Participation in pedigree studies and the risk of impeded access to health insurance. *IRB* 15(September–October):7–10.

King PA (1992): The past as prologue: Race, class, and gene discrimination. In: Annas GJ, Elias S, eds. *Gene Mapping: Using Law and Ethics as Guides*. New York: Oxford University Press, pp. 94–111.

Knoppers BM, Laberge C (1989): DNA sampling and informed consent. *Can Med Assoc J* 140:1023–1028.

Kodish E, Murray TH, Shurin S (1994): Cancer risk research: what should we tell subjects? *Clin Res* 42: 396–402.

Leary WE (1995): "Using gene tests to deny jobs is ruled illegal." *New York Times*, April 8.

MacKay CR (1993): Discussion points to consider in research related to the human genome. *Hum Gen Ther* 4:477–495.

Macklin R (1992): Privacy and control of genetic information. In: Annas GJ, Elias S, eds. *Gene Mapping: Using Law and Ethics as Guides*. New York: Oxford University Press, pp. 157–172.

McConkie-Rosell A, Markel D (1995): Facilitating research: The many roles of the genetic counselor. *Perspectives in Genetic Counseling* (newsletter, National Society of Genetic Counselors) 17:1, 4.

McEwen JE, McCarty K, Reilly PR (1992): A survey of state insurance commissioners concerning genetic testing and life insurance. *Am J Hum Genet* 51:785–792.

Miller JL (in preparation): Needs and rights of participants in human genetic research: Perspective of the subjects. Evanston, IL: Northwestern University, Department of Obstetrics and Gynecology

Moore v. *Regents of University of California*, 793 P2d 479 (Sup. Ct. CA 1990).

Mueller-Hill B (1992): Eugenics: The science and religion of the Nazis. In: Caplan AL, ed. *When Medicine Went Mad: Bioethics and the Holocaust*. Totowa, NJ: Humana Press, pp. 43–52.

Natowicz MR, Alper JK, Alper JS (1992a): Genetic discrimination and the law. *Am J Hum Genet* 50:465–475.

Natowicz MR, Alper JK, Alper JS (1992b): Genetic discrimination and the Americans with Disabilities Act. *Am J Hum Genet* 51:895–897.

Parker LS, Lidz CW (1994): Familial coercion to participate in genetic family studies: Is there cause for IRB intervention? *IRB* 16(January–April):6–12.

Paul DB (1994): Is human genetics disguised eugenics? In: Weir RF, Lawrence SC, Fales E., eds. *Genes and Human Self-Knowledge: Historical and Philosophical Reflections on Modern Genetics*. Iowa City: University of Iowa Press, pp. 67–83.

Pelias MZ (1991): Duty to disclose in medical genetics: A legal perspective. *Am J Med Genet* 39:347–354.

Proctor R (1988): *Racial Hygiene: Medicine Under the Nazis*. Cambridge, MA: Harvard University Press.

Proctor RN (1992): Genomics and eugenics: How fair is the comparison? In: Annas GJ, Elias S, eds. *Gene Mapping: Using Law and Ethics as Guides*. New York: Oxford University Press, pp. 57–93.

Reilly P (1980): When should an investigator share raw data with the subjects? *IRB* 2(November):4–5, 12.

Reilly PR (1991): *The Surgical Solution: A History of Involuntary Sterilization in the United States*. Baltimore: Johns Hopkins University Press.

Reilly PR (1992): ASHG statement on genetics and privacy: Testimony to United States Congress. *Am J Hum Genet* 50:640–642.

Roe v. *Wade* 410 U.S. Sup. Ct. 113 (1973).

Shore D, Berg K, Wynne D, Folstein MF (1993): Legal and ethical issues in psychiatric genetic research. *Am J Med Genet* 48:17–21.

Tarasoff v. *Regents of University of California*, 551 P2d 334 Sup. Ct. CA (1976).

US Department of Energy (DOE) Title 10 Code of Federal Regulations Part 745 (10CFR745): Federal Policy for the Protection of Human Subjects–Notices and Rules.

US Department of Energy (DOE) (brochure): Protecting Human Research Subjects within the U.S. Department of Energy. Brochure available by calling 301-903-5037.

US Department of Health and Human Services (DHHS) (1991): Title 45, Code of Federal Regulations, Part 46 (45CFR46), Protection of Human Subjects.

US NIH-DOE Working Group on Ethical, Legal, and Social Implications of Human Genome Research (1993): Genetic information and health insurance: Report of the Task Force on Genetic Information and Insurance. NIH publication 93-3686. Bethesda, MD: National Institutes of Health.

US Office for Protection from Research Risks (OPRR) (1993): *Protecting Human Research Subjects: Institutional Review Board Guidebook*. Washington, DC: Department of Health and Human Services, National Institutes of Health.

Weir RF, and Horton JR (1995a): DNA banking and informed consent—Part 1. *IRB* 17(July–August):1–4.

Weir RF, and Horton JR (1995b): DNA banking and informed consent—Part 2. *IRB* 17(September–December):1–8.

Wertz DC (1992): Ethical and legal implications of the new genetics: Issues for discussion. *Soc Sci Med* 35:495–505.

Wertz DC, Fanos JH, Reilly PR (1994): Genetic testing for children and adolescents: Who decides? *JAMA* 272:875–881.

Wertz DC, Fletcher JC, Berg K (1995): Guidelines on ethical issues in medical genetics and the provision of genetic services, WHO/HDP/GL/ETH/95.1 Geneva: World Health Organization: Hereditary Diseases Programme.

APPENDIX 6.1

Example of Standard Consent Form

PURPOSE

The purpose of this study is to locate and characterize the gene or genes associated with *disorder*. This disorder is characterized by one or more of the following: *disorder symptoms*. I have been asked to participate in this study because I have been diagnosed with *disorder* or I am a spouse or family member of a person diagnosed with this disorder.

PROCEDURES

Sample Collection

If I agree to participate, a blood sample (about 4 tablespoons) will be drawn from my arm. In rare circumstances, the inside of my cheek will be swabbed in lieu of a blood sample. I understand that discarded tissue and tumor tissue removed at surgery may be obtained and studied when it is no longer needed to establish diagnosis.

I understand that I may be recontacted if another sample is necessary for my continued participation in this study. I may refuse to be contacted in the future or to submit another sample and may withdraw from this study at any time.

Medical Records

I understand that if I agree to participate in this study, I will be asked questions regarding my medical and family history. My medical records will be reviewed and results of tests associated with the diagnosis of *disorder* will be recorded for the study.

Confirmation of Diagnosis

I understand that the researcher may request that I undergo further medical testing as part of the standard diagnostic workup for *disorder* to confirm or exclude the diagnosis of *disorder* and that I may refuse to undergo such tests. If I wish to be examined, I will be referred to my primary care physician to discuss the purposes, risks, and benefits of this workup. I understand that, as with any medical care, such tests will only be performed with my consent and only if they are determined necessary by my physician as part of my medical care and that information generated from such tests will become part of my medical record.

Family Members

I may be asked if I am willing to inform other family members about possible participation in this study so that more samples can be obtained for analysis. I am not obligated in any way to recruit family members. If I agree to contact other family members, I will inform them to call the genetic counselor for more information. The research team will not discuss with me or my family members which family members do or do not decide to participate in this study.

Disclosure of Results

The specific research results from this study will Wnever be shared with me. In addition, these results will not be disclosed to my family members, to my physician, or to other third parties. In some cases, samples collected for this study may yield information which could be used in a clinical setting.

I do _________ do not __________ wish to be recontacted in the event that clinical genetic testing for *disorder* becomes available to my family. If I wish to be recontacted, I will be offered a referral to a clinical service and have the opportunity to discuss the implications of this testing with a health care professional.

Storage of Information and Confidentiality

There are many safeguards in place to prevent unintentional disclosure of information obtained for or produced by this study. All information and samples obtained for this study will be assigned a code. A key to the code will be kept in a separate locked file in the investigators office.

Coded research results (including the data collected from the medical charts) will be entered into a research database. No names are entered into this database, only the codes assigned to the research record. A Certificate of Confidentiality will also been obtained to protect data from third party subpoena or other legal action.

Uses of Samples and Information

My sample may be used to establish an immortal cell line. This means that an inexhaustible supply of my DNA (genetic material) will be available for genetic research. Samples and information may be shared with the researcher's collaborators for research purposes but all information will be coded to maintain my confidentiality.

I authorize the researchers and *Institution* to use my cells, tissue, blood, or other specimens for scientific, research, or teaching purposes, or for the development of a new product, which may be distributed commercially. I understand that, as a participant in this study, I relinquish all ownership and rights to tissue samples and DNA extracted from those samples.

COSTS

No charges will be billed to me to participate in this study. If I choose to have my blood drawn by someone other than the investigator, a prepaid blood kit and mailing envelope will be sent to me. If I am billed for the blood draw, I may submit this bill to the investigator for reimbursement.

If, at the investigator's suggestion, I decide to undergo further diagnostic testing for *disorder*, a referral will be made to my primary care physician. Any medical care I receive will be billed to my insurance carrier. If I do not have medical insurance, the investigator may reimburse me for my medical expenses on a case-by-case basis.

RISKS AND DISCOMFORTS

There are minor risks and discomforts associated with blood sampling. This includes a brief amount of pain and possibly a small bruise at the needle site. Occasionally a person feels faint when blood is drawn. Rarely an infection develops, which can be treated. There are no risks to a cheek swab. There may a very minor discomfort when the cheek is swabbed, but this passes immediately when the swab is removed.

Due to the maintenance of strict confidentiality measures and the policy of this laboratory that research results and other genetic and clinical information are not disclosed, any potential psychological and/or social harm to me that is associated with genetic information is minimized. As with any study, the voluntary disclosure by me about my participation in this study could influence my future insurability or employability. I understand that not sharing information about my participation in this study with third parties will minimize these risks.

BENEFITS

There are no direct benefits to me from participation in this study. It is possible that in the future, the genetic cause(s) of *disorder* may be learned. It is possible that, in future years, a diagnostic genetic test for *disorder* will become commercially available. This information, as well as updates on research progress, will be available to me through yearly newsletter from the investigator. If I choose to receive the newsletter, my name and address will be entered into a computer database mailing list.

I do ______ do not ______ wish to receive a yearly newsletter updating me on the research and clinical status of *disorder*.

ALTERNATIVES

I understand that my participation is voluntary and that I may refuse to participate or may withdraw consent and discontinue participation in the study at any time without prejudice to my present or future care at *Institution*.

REQUEST FOR MORE INFORMATION

I understand that I may ask more questions about the study at any time. *Principal Investigator* and *Genetic Counselor* at ###-###-#### are available to answer my questions or concerns.

If during the study, or later, I wish to discuss my participation in or concerns regarding it with a person not directly involved, I am aware that the Patient Care Representative of *Institution* at ###-###-#### is available to talk with me. A copy of this consent form will be offered to me to keep for careful re-reading.

INJURY STATEMENT

In the unlikely event that I am injured as a direct result of participation in this study, I understand that, if I report the injury promptly to *Principal Investigator* at ###-###-####, *Institution* will provide free medical treatment to me for that injury.

I confirm that ______________________________ has explained to me, the participant/legal guardian of ___________________________, the purpose of the research, the study procedures that I will undergo and the possible risks and discomforts as well as potential benefits that I may experience. Alternatives to my participation in the study have also been discussed. I have read and I understand this consent form. Therefore, I agree to give my consent to participate as a subject in this research project.

Participant/Legal Guardian ______________________ Date __________

Witness to Signature ______________________ Date __________

7

Sample Size and Power

Marcy C. Speer
Section of Medical Genetics
Center for Human Genetics
Duke University Medical Center
Durham, North Carolina

INTRODUCTION

Before undertaking a linkage analysis, it is critical to know whether the available pedigree information is sufficient to allow detection of the gene(s) underlying the trait of interest. In certain circumstances, estimates can be obtained prior to the collection of samples to develop sampling strategies that will ensure adequate power to detect genetic effects. Such information can be useful in assessing the costs of a study and in determining optimal study design.

The interpretation of power studies is based on fundamental statistical concepts. In general, the investigator wants to know whether two loci are linked to one another. In this case, the null hypothesis is that the two loci are not linked and the alternate hypothesis is that the two loci are linked. In genetically complex diseases this scheme may be modified so that the statistical testing may be applied to search for an association as opposed to (or in addition to) a linkage. Here, the question is whether the trait allele and an allele at a marker locus occur together more frequently than expected by chance. Following a statistical test, the null hypothesis is either rejected or not. In the latter case, insufficient data are available to reject the null hypothesis. In statistical testing, the alternate hypothesis is never "accepted"; evidence can only be accumulated in support of the alternate hypothesis.

Four possible outcomes of an experiment exist. If the null hypothesis is rejected,

Approaches to Gene Mapping in Complex Human Diseases, Edited by Jonathan L. Haines and Margaret A. Pericak-Vance. ISBN 0-471-17195-6

the investigator may be correct or incorrect. Similarly, if the null hypothesis is not rejected, the investigator may be correct or incorrect. These four outcomes are shown in Table 7.1. A type I error, quantified as α, occurs when the null hypothesis is incorrectly rejected. This occurrence is known as a false positive outcome. A type II error, quantified by β, occurs when a false null hypothesis is not rejected; this occurrence is known as a false negative outcome. The power of the study, or the detection of a correct positive result, is quantified by $1 - \beta$ and is the probability of correctly detecting a linkage.

Studies of power in linkage analysis are frequently divided into two general categories, depending on whether the underlying genetic model is known or unknown. In most diseases with a simple, Mendelian inheritance pattern, the components of the genetic model are known with enough confidence to allow straightforward power calculations that predict well the sample sizes requirements, type I and type II errors, and power. However, in most genetically complex diseases where the mode of inheritance is uncertain, these straightforward approaches can be misleading and are often not the best choice. Although the "answers" about sample size, error frequencies, and power usually would be very desirable to have, they are frequently based on assuming about the disease information that is not known with certainty.

An important consideration in determining required sample sizes in complex disease involves the ability to assess whether two genes are identical by descent (IBD) or identical by state (IBS) (Chapter 13). Other important variables in the analysis of complex disease and the requisite sample size required to detect linkage include the relative risk to relatives (λ) value of the disorder and the expected proportion of alleles shared IBD for any particular type of relative pairs.

As defined in Chapter 5, the λ value is a ratio of two risks which describes the increase in risk to a relative of an affected individual compared to the prevalence of the trait in the general population. Most frequently, a λ_s value is reported, which is:

$$\frac{\text{trait risk for siblings of an affected individual}}{\text{general population trait prevalence}}$$

As the λ increases, so does the presumptive genetic contribution to the trait. It should be noted, however, that as the trait incidence in the population increases, the overall λ values will generally decrease.

Throughout this chapter, different strategies for collecting affected relative pairs are discussed. Figure 7.1 shows relative pairs of different types and the average pro-

TABLE 7.1 The Relationship between Type I and Type II Error

	True State of Nature	
Decision	Linkage	No Linkage
Linkage	$1 - \beta$	α
No linkage	β	$1 - \alpha$

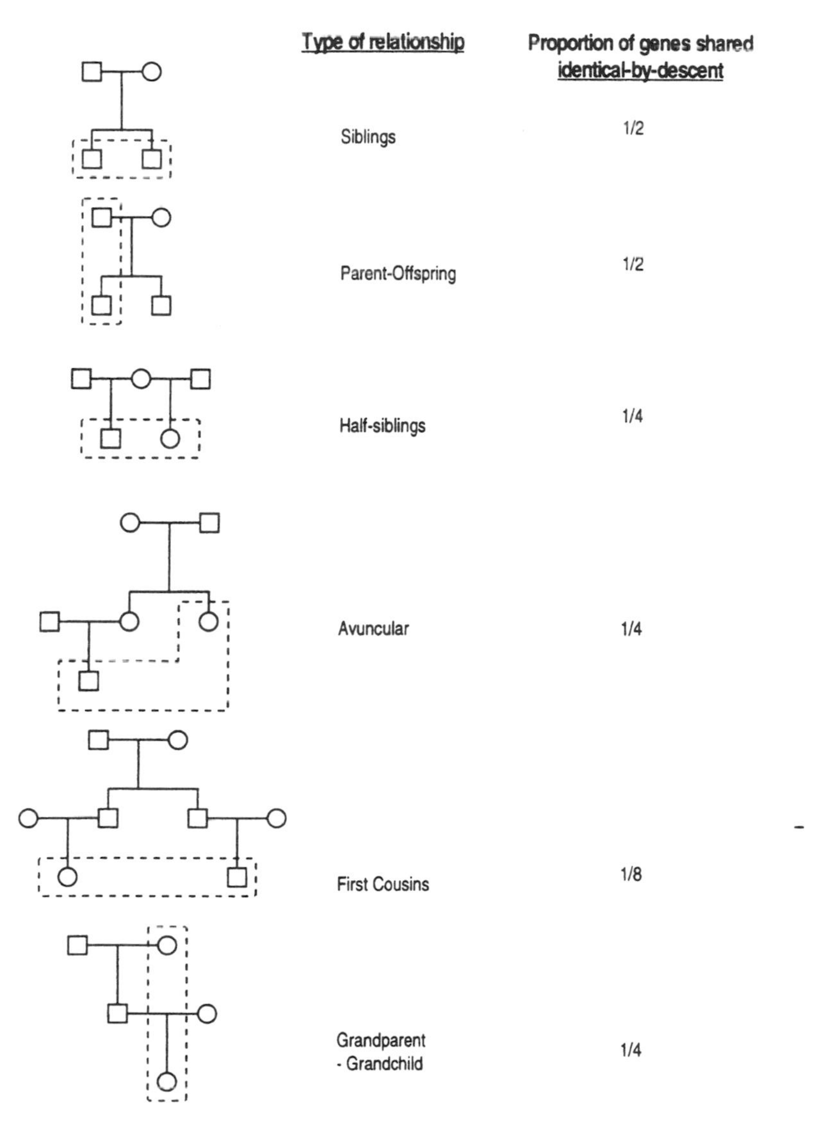

Figure 7.1. *Allele sharing between relatives of different types. Note that although sibling pairs and parent–child pairs each share 50% of their genetic material in common, the type of sharing is different: a parent transmits exactly 50% of his or her genetic material to each offspring, while on average a pair of siblings will share 50% of their genetic material in common—some portion of the 50% from the mother and the remainder from the father.*

portion of genes shared between them. A general rule of thumb is that the proportion of genes shared in common between a particular type of relative pair is equal to $1/2^n$, where n is the degree of relationship.

Several factors are known to affect the power to detect linkage or association, regardless of the type of analysis. These include locus heterogeneity, frequency of phenocopies, marker polymorphism, recombination between disease and marker locus, and the correct specification of marker allele frequencies. General principles regarding these factors are outlined in Table 7.2.

THE INFORMATION CONTENT OF PEDIGREES FOR MENDELIAN TRAITS

In small pedigrees with all critical individuals available for both phenotyping and genotyping, it is sometimes possible to assess the available information in the pedigree by visual inspection. Guidelines for the contribution of particular offspring to the lod score under linkage phase known conditions are shown in Table 7.3. In an autosomal dominant disease when linkage phase is known and there is no recombination between the disease and marker locus, affected individuals contribute 0.30 to the lod score. The contribution of unaffected individuals is dependent on the penetrance of the disease allele. As penetrance decreases, so does the contribution of an unaffected individual to the lod score, because the ability to accurately score these unaffected individuals (who may or may not be gene carriers) as recombinants or nonrecombinants is reduced. Thus, at a minimum in a disorder that is autosomal dominant and fully penetrant and linkage phase is known, 10 meiotic events (children) are necessary to generate a lod score of 3.0 when the recombination fraction between the trait and marker locus is 0.0. Under similar conditions for autosomal recessive disorders, affected individuals contribute 0.60 to the lod score. In recessive disease, gametes from both parents contribute linkage information, whereas in dominant disease only the affected parent contributes linkage information. The contribution of unaffected individuals is minimal, regardless of the penetrance. Thus

TABLE 7.2 Factors Affecting Power to Detect Linkage and/or Association

Factor	Effect
Heterogeneity	When more than one locus leads to the trait phenotype, more families are needed to detect a significant result.
Recombination fraction between the trait and marker locus result.	The further the marker locus is from the trait locus, the more families required to detect a significant.
Marker allele frequency specification	Incorrect specification of marker allele frequencies can lead to an increase in either type I or type II errors.
Phenocopies and diagnostic misclassification	The higher the frequency of phenocopies or misdiagnosis, the more families required to detect a significant result.
Marker polymorphism	The lower the heterozygosity value of a marker locus, the more families (or genotyped markers) required to detect a significant result.

TABLE 7.3 The Approximate Contribution of Affected and Unaffected Offspring to the Lod Score Under Dominant and Recessive Inheritance, Assuming Known Linkage Phase, No Recombination Between the Trait and Marker Locus, and No Phenocopies

	Penetrance		
	1.00	0.90	0.80
Dominant			
Affected	0.30	0.30	0.30
Unaffected	0.30	0.26	0.22
Recessive			
Affected	0.60	0.60	0.60
Unaffected	0.12	0.11	0.10

for a recessive disorder, at a minimum, five fully informative nuclear pedigrees with two affected offspring and a marker that demonstrates no recombination with the disease gene are required to obtain a lod score of 3.0.

In common and genetically complex disorders, the assumption of a rare disease allele does not hold, making the assumption that the spouse does not carry the disease allele invalid in the autosomal dominant case. Thus, in a complex disease the determination of whether a parent is heterozygous at the trait locus is not straightforward, especially if penetrance of the disease allele is incomplete. Furthermore, given the suspected underlying heterogeneity of most complex diseases (e.g., genetic and/or environmental contributions), estimating the underlying genetic model in advance is virtually impossible.

COMPUTER SIMULATION METHODS IN MENDELIAN DISEASE

Although small pedigrees are amenable to visual inspection and casual assessment of information content, for most pedigrees obtained in a research setting, such an approach is not feasible and assessments of power are frequently done with the assistance of computer programs.

For Mendelian diseases, studies of power are frequently done after an investigator has talked to families willing to participate in the study and obtained the relevant pedigree information including pedigree structure, which members are affected and unaffected, and which members are willing to participate in the study. Available computer programs for assessing power in Mendelian disease are based on Monte Carlo methods, which utilize pseudorandom numbers to assign values for analysis to particular situations.

To initiate a simulation process, the investigator must provide to the computer program information about the genetic model of the trait under study, including the degree of dominance for the trait, the penetrance and the frequency of the trait allele characteristics of the genetic marker (number and frequency of alleles and the recombination fraction between the trait and marker loci) the pedigree structure and disease phenotypes for family members and their availability for the study. The investigator must also provide a value for the extent of heterogeneity. The computer in turn will simulate genetic marker data for the number of replicates specified by the investigator. Each replicate is analyzed and the results compiled and summarized.

Three computer programs are generally available for power studies in Mendelian diseases, SIMLINK (Boehnke, 1986; Ploughman and Boehnke, 1989); SLINK (Ott, 1989; Weeks et al., 1990b) and the companion analysis programs MSIM, LSIM, and ISIM; and SIMULATE (Ott and Terwilliger, 1992). All three programs allow simulation of a marker locus unlinked to a disease locus. SLINK and SIMLINK, however, allow the simulation of genetic marker data for a locus linked to a disease gene. These two programs are very similar in their workings. One of the major algorithmic differences between the two programs is in the way they assign genetic marker genotypes given information on the trait phenotypes of members in the pedigree ("conditional" on disease status): SLINK uses a series of risk calculations and SIMLINK first traverses the entire pedigree, assigning disease genotypes conditional on disease phenotypes. The most critical differences from a practical perspective are that SIMLINK allows a thorough evaluation of the power and type I error rates in one analysis, while multiple analyses must be performed to generate the same information with SLINK. However, SLINK allows the simulated data to be analyzed under a different model than that used to generate the data, which SIMLINK cannot. These differences and others are summarized in Table 7.4.

Examples of the process for both conditional and unconditional Monte Carlo simulation are presented in the subsections that follow.

Practical Example of Unconditional Simulation in Mendelian Disease: Trait and Marker Locus Unlinked

There are numerous uses for simulation of genetic marker data unlinked to a trait marker locus. For instance, investigators may wish to know about the frequency of false positive results, or they may require assistance in developing genomic screening strategies based on average exclusion from available pedigree material. For this first example, the disease locus was assumed to demonstrate 50% recombination (the "true" recombination fraction) with a four-allele marker with equally frequent alleles; in other words, the disease and marker locus are unlinked to each other. The disease is segregating as an autosomal dominant disorder with complete penetrance of the disease allele, which has frequency of 0.0001 in the general population. Appendix 7.1 shows the process for generating the simulated genetic marker genotypes utilizing Family 1713 (Fig. 7.2). Note that although the investigator knows the trait phenotypes of members of the pedigree and this information is provided to the computer, it is not used in the calculations since the trait and marker loci are not linked to one another.

Appendix 7.2A gives the lod scores associated with the simulated genetic marker genotypes as described in Appendix 7.1, analyzed under the identical genetic model as was input into the simulation (i.e., dominant, fully penetrant disease, dis-

TABLE 7.4 Practical Differences Between Computer Simulation Programs SIMLINK and SLINK

Property	SIMLINK	SLINK
Ability to handle penetrance functions	Often cannot duplicate penetrance function the user would choose to utilize in linkage analysis, especially when using the LINKAGE programs. Program simulates and analyzes data under the identical generating model.	Since LINKAGE and its variants are the most frequently utilized linkage analysis packages, penetrance functions can be identical or different between simulation studies and actual data analysis.
Output	Output is thorough and detailed; calculates results assuming both a linked and an unlinked marker, so the investigator can obtain both power and exclusion information from a single run.	Output is less detailed than SIMLINK; additionally, because analysis modules MSIM, LSIM, and ISIM are separate from simulation module SLINK, two separate runs, one for data simulation and then a second for analysis, must be performed to generate results. To obtain both results under the null hypothesis of linkage and nonlinkage, two separate sets of simulations and analyses must be performed. At the end of the analysis runs, a count of the total number of replicates that exceeded specified threshold values is output, providing an intuitive summary of results.
Flexibility	Program does not routinely output replicate pedigree data, but can be modified to do so.	Program outputs a very large file with the pedigree replicates in them. These data can be extremely useful if additional studies, (e.g., for investigating the robustness of conclusions to model misspecification) need to be performed.
		Program can simulate genetic marker data given that some individuals have already been genotyped at the marker locus.

ease allele frequency = 0.0001, tetraallelic marker with equally frequent alleles). Note that for this particular replicate, the pedigree allows the exclusion of approximately 12 cM on either side of the marker for the location of the disease gene using the traditional –2.0 lod score for exclusion.

Although in this example only one replicate was performed, in practice several hundred—or in many cases up to 1000 replicates—usually are performed for such power studies, to obtain estimates of the results that are as accurate as possible. By averaging the exclusion region obtained in these series of replicates, the investigator can develop a rational plan for screening the genome using the available pedigree material. For instance, if after performing 1000 replicates of family 1713 as de-

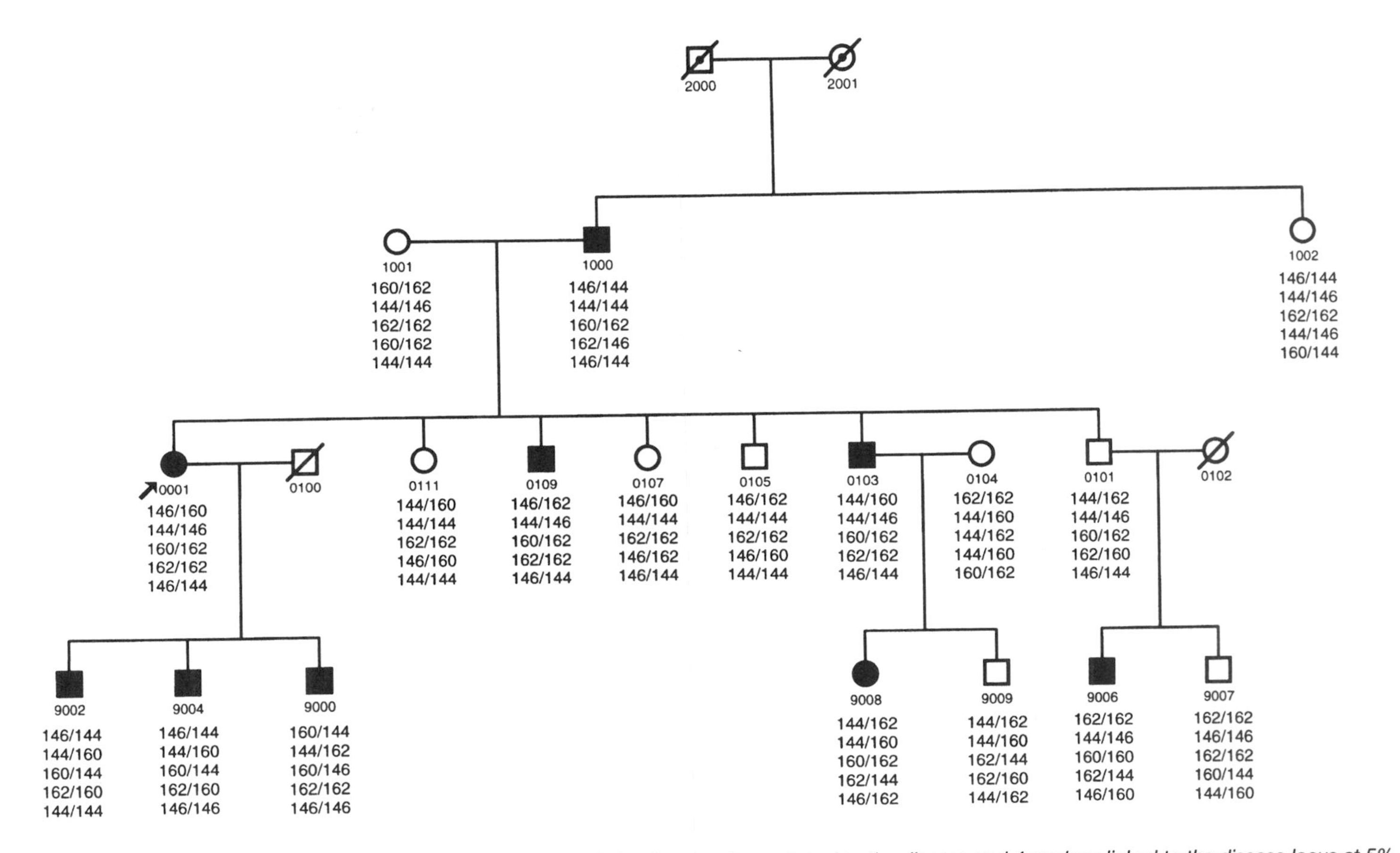

Figure 7.2. *Autosomal dominant pedigree used in examples of simulation for a marker unlinked to the disease and 4 markers linked to the disease locus at 5% recombination. The first set of genotypes under the pedigree individual numbers are those generated from the unconditional simulation in Appendix 7.1. The next four sets of genotype results beneath each individual in the pedigree are the genotypes generated for each of four separate replicates from the simulation which assumes linkage between the disease and marker locus at $\theta = 0.05$. Genotypes are given only for sampled individuals.*

scribed above, the investigator found that the average exclusion interval for this pedigree was 10 cM, he could choose to perform a genomic screen using markers spaced at 20 cM intervals.

Practical Example of Conditional Simulation in Mendelian Disease: Trait and Marker Locus Linked

Usually of more interest is the ability to detect a true linkage, or the power of the study. In this case, the true recombination fraction between the disease and marker locus would be set at less than 0.50. Simulating genetic marker data conditional on disease status also allows the investigation of the rate of false negatives. Appendix 7.3 shows an example of the process of conditional simulation, first assuming complete penetrance at the disease locus and then allowing for reduced penetrance. Note that there are several algorithms for performing this process; this very simple example simply illustrates the process.

For this example, in which complete penetrance for the trait locus has been assumed, four separate sets of genotype results, or replicates, were performed assuming a true recombination fraction between the disease and marker locus of 0.05, and these are shown on the pedigree in Figure 7.2. The results are summarized in Appendix 7.2B. Note by looking at the results in the pedigree that in replicate 1, by chance, a key individual in the pedigree (individual 1000) is uninformative because he is homozygous at the marker locus. In replicate 2, the pedigree is fully informative with no recombination events between the disease and marker locus, even though the "true" recombination fraction between the disease and marker loci is 5%. In replicate 2, this series of chance events provides the highest attainable lod score for this particular pedigree. If, by chance, in the entire series of replicates performed this particular scenario had not occurred, the simulated maximum lod score would be lower than this one. In replicate 3, there is no recombination between the disease and marker loci, although the pedigree is not fully informative. In this replicate, by chance individuals 0001 and 0103 are homozygous at the marker locus, thereby reducing the attained lod score even though the highest lod score still occurs when $\theta = 0.00$. In replicate 4, the pedigree is fully informative but two recombination events have occurred between the disease and marker locus, so that the best estimate of the recombination fraction in this replicate is 0.15 even though the "true" recombination fraction is 0.05. These results can be used to summarize the information content of pedigree data as described next.

IMPORTANT DEFINITIONS IN ASSESSING THE POWER OF LINKAGE STUDIES

The *average lod score* is calculated as the sum of the lod scores at a selected value of the recombination fraction divided by the number of replicates. This result is most usefully interpreted in conjunction with its standard deviation, which provides an assessment of the variability of the lod score. For instance, the average lod score for family 1713 at θ = 0.05 is calculated as (0.66 + 3.86 + 2.46 + 1.30)/4 = 2.07. This value provides the investigator with an estimate of how much information, on average, should be expected from the available pedigree information at a particular recombination fraction.

The *maximum maximum lod score* is the highest lod score obtained at any value of the recombination fraction in the entire set of replicates. In this example, the maximum maximum lod score is 4.21, which is attained in replicate 2 at θ = 0.0. This value is very useful in that it allows the investigator to gain insight into the full information content of the available pedigree material, since the true maximum maximum lod score is the highest potential lod score when all meioses are fully informative under the assumed model and allele frequencies. **Note that the true maximum maximum lod score may not be attained even in a series of multiple replicates if, by chance, not all the meioses are informative.** In considering the significance of a particular lod score, it is useful to know whether all the available linkage information has been captured. The observed maximum lod score of a pedigree may not match the attainable maximum maximum lod score for a pedigree for several reasons, including lack of informativeness of key individuals and differences in allele frequencies from simulated results and true results, differences between true and simulated penetrances, among others.

The *average maximum lod score* is the sum of each of the highest lod scores obtained in each replicate, regardless of where it occurs on the lod score curve, divided by the number of replicates. In this example, the average maximum lod score is calculated as (0.84 + 4.21 + 2.71 + 1.59)/4 = 2.34. Note that the highest lod score for any replicate need not, and usually will not, occur at the true recombination fraction. There are many caveats associated with the use of this particular measure of information content of a pedigree, the most compelling of which is that it tends to paint too rosy a picture of the true expected outcomes. Note that while the average lod score is additive across pedigrees, the average maximum lod score is not, which is an additional disadvantage to using the average maximum lod score as a measure of pedigree information within a study.

Available pedigree simulation packages will also directly compute the type I and type II error rates and the power of the available material for specific constant values of the lod score. For instance, the investigator may need to know the probability of obtaining a false positive (a lod score of 3.0 or greater when the data was simulated assuming free recombination between the disease and marker locus), the probability of obtaining a false negative (a lod score of –2 or less, or some other specified constant value when the disease and marker loci were fixed as linked to one another), and the probability of correctly obtaining a lod score of 3.0 or more when the disease and marker loci are set as being linked to one another. In the example using the data presented in Appendix 7.2, the type I error rate cannot be calculated because only one replicate was performed; the type II error rate is calculated as 0/4 or 0%, and the power (to obtain a lod score exceeding 3.0, which is representative of a true linkage) is calculated as 1/4 = 25%. Here it is important to note that in most cases, the investigator will choose to simulate hundreds or even thousands of replicates to obtain estimates of the necessary values (and their standard deviations when appropriate) that are as precise as possible.

Another useful application of computer simulation methods is in determining the empiric *P* value associated with a study outcome. By generating genetic marker genotypes under the assumption of free recombination between the disease and marker locus for many replicates of the pedigree structures, the investigator can generate a series of lod score results that could be obtained from a random genomic screen. By counting all simulated lod scores exceeding *x* and dividing by the total number of replicates, the investigator can find the empiric *P* value associated with the observed result *x* (Weeks et al., 1990a). For this application, since the observation of the outcome is so rare, the investigator will typically perform tens of thousands of replicates to obtain a reasonable estimate of the empiric *P* value.

Another application of computer simulation involves assessing the effect of maximization over models, which may be a more powerful approach than assuming a genetic model when in truth the model is unknown (e.g., Hodge, et al., 1997). If the true underlying genetic model is unknown, an investigator may choose to analyze several different genetic models (e.g., different degrees of dominance, disease allele penetrance, and frequency) with the same marker data to see which model gives the highest lod score (Elston, 1989; Greenberg, 1989). In these cases, some correction for multiple comparisons must be performed. This procedure was illustrated with an example in which chromosome 5 markers were studied in schizophrenia pedigrees (Sherrington et al., 1988). Using three diagnostic schemes that differed in their classification of affected individuals, Sherrington and coworkers found that one of the three models generated a lod score of 6.5 versus chromosome 5 markers but failed to correct for the fact that three different diagnostic schemes had been used. Weeks et al. (1990a) used computer simulation methods and an approximate reconstruction of the original linkage report to show that this lod score of 6.5 was inflated by 0.7–1.5 lod score units.

Hodge et al. (1997) performed a simulation study to assess the magnitude of the increase in type I error when the investigator has maximized the linkage analysis over penetrance and whether the disease is dominant or recessive. They developed guidelines for increasing the test criterion under either or both these scenarios. As a general rule of thumb, the critical value of the test statistic should be increased by a lod value of 0.3 when one of the foregoing parameters is maximized and by 0.6 when both penetrance and dominance/recessiveness are maximized.

POWER STUDIES IN COMPLEX DISEASE

In complex diseases, power studies are not as straightforward as in Mendelian diseases but can nonetheless help to guide an investigator in determining whether a particular disorder is amenable to gene mapping. For complex diseases, discussions about power are generally limited to approximate estimates of sample size required to detect a linkage or association *making various assumptions about the underlying genetic model*. Guidelines for sample size depend on how closely related the sampled individuals (usually affected) are, how tightly linked the disease and marker locus are, how polymorphic the marker system is, and the mode of inheritance of the trait. These issues are frequently interrelated, and reliable estimates for these parameters are often unknown; consequently, developing standard guidelines is difficult. Furthermore, current technology limits our gene identification efforts to one locus at a time, so most studies of power in complex disease are based on this premise. These issues are discussed shortly. However, it is critically important to recognize that such studies are entirely dependent on *assumptions* underlying a genetic model, the validity of which is entirely unknown.

Two of the primary assumptions that must be made about an underlying genetic model involve the frequency of the trait allele and the mode of inheritance of the disease-predisposing locus. The frequency of the trait allele affects power and sample size calculations as well: if the disease allele is rare, then the probability that two affected individuals within a pedigree are identical by descent (IBD) at the trait locus is high, whereas if the trait allele is common, the IBD probability decreases. Considered heuristically, the contribution of a distantly related affected relative pair (ARP) such as an affected cousin pair should be greater than the contribution of an affected sibling pair when the number and frequency of contributing loci is small. It is a priori more striking if a pair of affected cousins (who share only 1/8 of their genetic material in common by chance alone) share alleles in common than if a pair of affected siblings (who share half of their genetic material in common by chance alone) share alleles in common.

While the true underlying mode of inheritance is rarely known in a genetically complex disease, the investigator can often develop a "gestalt" for the potential underlying genetic factors based on visual inspection of pedigrees. For instance, if a significant proportion of the pedigrees show transmission from affected parent to affected child, these data suggest that an underlying genetic effect may be dominant. Similarly, if carefully screened pedigrees fail to show affected relative pairs other than siblings, a dominance component is unlikely, unless the investigator is willing to assume a dominant allele with low penetrance. As a general rule of thumb, if the underlying trait is recessive, then affected sibling pairs are the preferred sampling unit. If the trait is dominant, more distantly related ARPs usually represent the most efficient approach. However, for a dominant trait, the most powerful type of affected relative pair is dependent on the λ value and on the frequency of the trait. In general, as the frequency of the trait allele decreases, the choice of ARPs should tend toward more distantly affected relatives. For example, for a rare trait allele, the most powerful affected relative pair sampling unit is grandparent/grandchild. However, for a common trait allele, the most powerful affected relative pair sampling unit is affected siblings. For candidate genes, when the recombination fraction between the trait and candidate gene is 0.0, the power of these different types of ARP is virtually the same regardless of the underlying mode of inheritance.

Since currently available genetic analysis methods are relatively intractable for the simultaneous consideration of two or more trait loci, the following discussions assume that the investigator is searching for an individual gene that contributes to the underlying phenotype.

Discrete Traits

Affected Sibling Pairs. Much of the work in assessing power and sample size in complex disease has focused on the use of affected sibling pairs. For example, Risch (1990) showed that the power of a sample consisting of affected sibling pairs is dependent on λ_R, the ratio of the recurrence risk to a relative of type R of a proband to the prevalence of the disease in the general population, and the recombination fraction between the trait and marker locus. He showed, for instance, that approximately 60 affected sibling pairs would provide 80% power to detect linkage in a disorder in which the λ_S (where s indicates that the affected relative pair is siblings) = 5.0 when the recombination fraction between the trait and marker locus is 0.0. As expected, a larger sample size is required to detect linkage as the recombination fraction between the trait and marker locus increases. Some estimates of power, sample size, and effects of λ are shown in Table 7.5 (see also Chapter 13).

This work was recently extended by Hauser et al. (1996) for multipoint analysis. As in multipoint linkage analysis for Mendelian disease, the ability to flank a particular interval with markers increases the information in a data set and the potential to exclude an area of interest. When searching for a locus that confers a λ_s of 2.0, a 10 cM interval flanked with markers with heterozygosity value of 0.75 and a sample size of 200 affected sibling pairs can generate a lod score of 3.0 with a power of 0.51. As the number of affected sibling pairs or the λ_s increases, so does the power. Similarly, ability to exclude an interval is dependent on the available sample size, the magnitude of the genetic effect, and the inter-marker genetic distance. Table 7.6 summarizes data about sample size used to detect and exclude linkage in a study designed around a multipoint affected sibling pair approach.

The Inclusion of Unaffected Siblings. The decision about whether to include unaffected siblings in a study is usually based on availability and cost of recruitment, and the cost associated with the additional genotyping of these individuals. The advantages of collecting unaffected siblings are many. For instance, the inclusion of the genotypic marker data on unaffected siblings may increase the ability to determine the IBD status of genetic markers in affected sibling pairs, which is especially important in late-onset disorders, where parents are unavailable for sampling. Additionally, as pointed out by Elston et al. (1996), performing a linkage study utilizing only affected sibling pairs is akin to performing a study without controls. Without the unaffected individuals, it is virtually impossible to determine whether a significant result is due to a true linkage or to some other factor, such as meiotic drive at a marker locus. The collection and availability of unaffected siblings also allows for the eventual assessment of penetrance and expression, and for considering genes that may potentially modify the phenotype. Although some available methods allow for the inclusion of unaffected relatives, they may make the assumption that unaffected relatives are non–gene carriers. If the underlying disease gene is of low penetrance, such an assumption may be unrealistic. Most current applications rely exclusively on allele sharing in affected relatives only.

TABLE 7.5 Guidelines on Sample Size for Affected Relative Pairs

A. Approximate minimum λ_s for power = 0.80 and 0.90 for specified number of affected sibling pairs when $\theta = 0.0$ for a fully informative marker

Number of Affected Sibling Pairs	Power to Detect Linkage	
	0.80	0.90
60	5.0	7.0
80	4.0	5.0
100	3.0	4.0
200	2.0	2.3

B. Approximate power to detect linkage for specified number of affected sibling pairs and values for λ_S and θ for a fully informative marker

Number of Affected Sibling pairs	θ					
	0.0	0.05	0.10	0.0	0.05	0.10
	$\lambda_S = 2.0$			$\lambda_S = 5.0$		
40	0.05	0.01	0.01	0.40	0.10	0.05
100	0.05	0.18	0.35	0.28	0.60	0.95
300	0.95	0.90	0.40	> 0.95	> 0.95	> 0.95

Source: Risch (1990).

Special Problems with Sib Pair Data: Considerations of Independence. Pedigree material, particularly with respect to sibling pairs, is usually summarized by simple counts. Some families, however, will have more than one set of affected siblings. The formula for calculating the number of affected sibling pairs is: $n = m(m - 1)/2$, where n is the total number of affected sibling pairs and m is the number of affected siblings. For example, in Figure 7.3*A*, the parents have three affected children and the number of affected sibling pairs that can be formed are $3 \times 2/2 = 3$ (pairs = individuals [3,4], [3,5], [4,5]). Similarly, the parents in the pedigree Figure 7.3*B* have 5 affected children so $5 \times 4/2 = 10$ affected sibling pairs (pairs = individuals [3,4], [3,5], [3,6], [3,7], [4,5], [4,6], [4,7], [5,6], [5,7], [6,7]) can be created. When multiple affected sibling pairs from the same nuclear family are used, the issue of independence is important. Hodge (1984) showed that the pairs formed from more than 2 affected siblings within a sibship are not independent, but the consequences of this nonindependence have been subject to considerable debate in the literature. The pedigree in Figure 7.3*C* illustrates the nonindependence of the affected sibling pairs. For instance, once it has been determined that individuals 1 and 2 share one allele identical by descent (IBD) and individuals 2 and 3 share two alleles IBD, then the number of alleles (one) that individuals 1 and 3 share is also known. The best mechanism for dealing with this nonindependence is still a matter of intense debate; however, it seems fair to say that it is less problematic, the larger the overall sample size. Therefore, for general screening in complex disease, it is generally accepted practice to assume that the sibling pairs are in fact independent. How-

TABLE 7.6 The Power to Detect or Exclude Linkage for Various Sample Sizes *N* of Affected Sibling Pairs and Contributions of the Locus to the Genetic Effect (λ_s)

	Number of Affected Sibling Pairs[b]								
	100			200			400		
λ_S	P(lod) > 3.0	% Excl	% All	P(lod) > 3.0	% Excl	%All	P(lod) > 3.0	% Excl	% All
1.4	0.02	1.9	1.5	0.06	13.8	10.7	0.23	49.0	43.9
1.6	0.05	10.9	7.9	0.19	43.9	38.1	0.59	82.1	78.5
2.0	0.15	33.6	27.6	0.51	73.1	68.4	0.94	95.7	94.5
3.0	0.45	69.9	64.0	0.90	94.9	93.1	1.00	99.8	99.7

[a]An intermarker distance of 10 cM is assumed between flanking markers, each of which has four equally frequent alleles (heterozygosity = 0.75).

[b]P(lod) > 3.0, the probability of obtaining a lod score > 3.0 given that the interval contains a disease locus with λ_S as specified; %Excl, the average percentage of a 10 cM interval excluded when no disease locus is present; %All, percentage of intervals in 1000 replicates excluded in their entirely when no disease locus is present.

Source: Hauser et al. (1996).

ever, once "interesting" areas of the genome have been identified, it is critically important to assess the significance of any nonindependent data, since ignoring it entirely can lead to falsely high significance levels. Daly and Lander (1996) have suggested a simulation approach to determining the correct empiric p value (see above) of such an "interesting" result for a study in which some families have more than two affected siblings, given information on the sibship sizes and frequencies of the marker alleles.

Affected Relative Pairs of Other Types. In many complex diseases, other types of affected relative pairs will be identified; indeed, trends for different types in a disease may shed some light on the underlying genetic models. For instance, consider a trait in which two different loci are involved. If the disease alleles at the loci are rare and act epistatically, it is unlikely that affected relative pairs more distantly related than siblings will be observed. However, if the disease alleles at the loci are relatively common and have low penetrance, the pedigrees may demonstrate affected avuncular or cousin pairs. Because it is unusual for a whole study to be designed around affected relative pairs other than siblings, estimates of sample size for studies specifically geared at other types of ARPs are rare. When other types of ARP are frequent, the investigator will frequently collect all affected patients and their connecting relatives and utilize an analysis approach other than one dependent on a sib pair design (Chapter 14).

Other Considerations. Some difficulties in the estimates of sample size or power of available pedigree material in the investigation of complex traits warrant consideration. For instance, the presence of phenocopies or genetic heterogeneity will increase the amount of material necessary to correctly identify a linkage (Bishop and Williamson, 1990). Additionally, the lower the heterozygosity of the

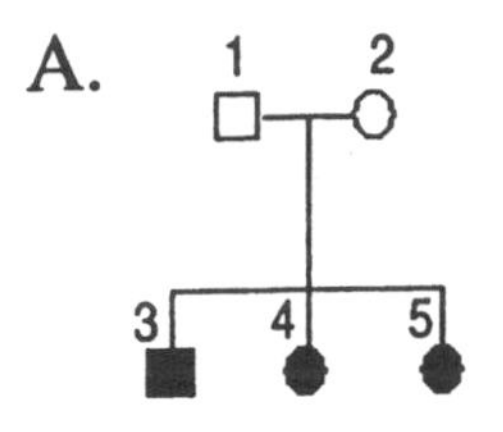

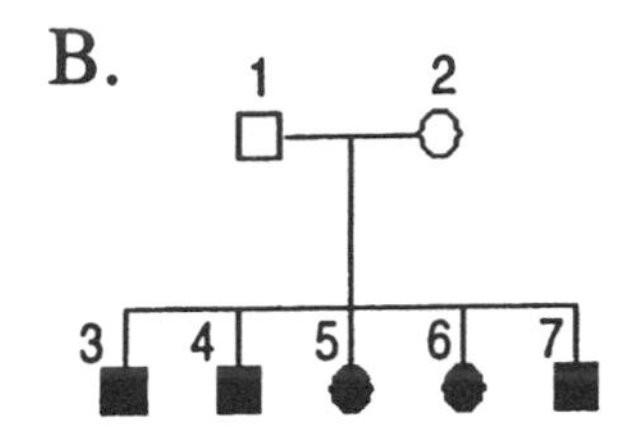

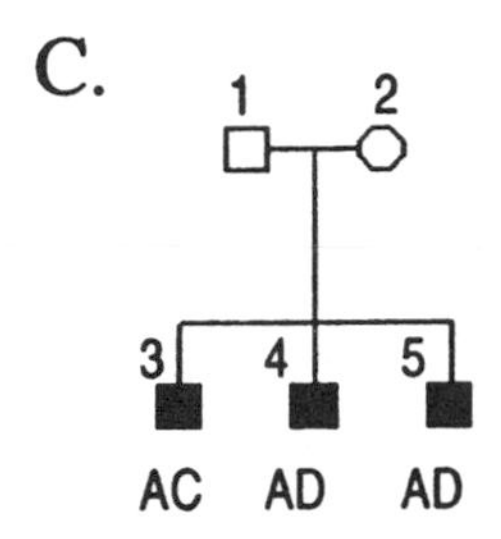

Figure 7.3. (A) *Three affected siblings yield three nonindependent affected sibling pairs.* (B) *Five affected siblings yield 10 nonindependent affected sibling pairs.* (C) *When the number of alleles shared identical by descent (IBD) is known for any two of the three pairs of affected siblings, the IBD sharing is known for the third affected sibling pair.*

marker to be tested, the more material will be needed on average to detect a significant effect.

Empiric Examples. Associations for common and genetically complex disorders have been identified in many disorders using the APM (affected-pedigree member) method of linkage analysis. Examples include late-onset Alzheimer's disease (Pericak-Vance et al., 1991), which utilized 87 ARPs (including those generated by inference of genotypes in deceased or unavailable individuals utilizing their offspring, both affected and unaffected), early-onset familial breast cancer (Hall et al., 1990), which utilized 69 ARPs, and familial melanoma (Cannon-Albright et al., 1992), which utilized 185 ARPs. In melanoma and breast cancer, these associations were observed in the less frequent Mendelian forms of these common disorders and actually represented true linkages. Since the pedigrees were large enough to permit the identification of the underlying linkage, an empiric assessment of the actual number of affected relative pairs in each of the studies is less meaningful. The linkages were also identified with standard, parametric linkage analysis. The meaning of the significant association results in late-onset Alzheimer disease were more difficult to unravel. These initial studies were performed using the original version of the APM (Weeks and Lange, 1988) and the resulting effect of the *APOE* gene was found because it was so large. Interestingly, recent extensions to the APM method allow the inference of missing genotypes when possible, and these recent extensions approximate the more powerful identity-by-descent methods (Davis et al., 1996). In the late-onset Alzheimer disease data set (Pericak-Vance et al., 1991), when ARPs generated by inference of missing genotypes in affecteds are included, the available numbers of ARPs are increased by 29. Clearly, the addition of this much data just by changing the computing algorithm represents a considerable increase in power to detect genetic effects. This scenario—a data set comprising varying numbers of ARPs of various types—is consistent with many realistic data sets, particularly late-onset disorders.

Genomic Screening Strategies: One-Stage Versus Two-Stage Designs. Interestingly, although much literature is focused on the use of affected sibling pairs to study complex disease, it frequently is not the most powerful approach. Elston (1992) and Elston et al. (1996) consider the issue of study design for complex genetic disease in terms of "cost" of the study. Their computing algorithm DESPAIR is available through the SAGE package. These authors (Elston, 1992; Elston et al., 1996) point out that for a wide range of scenarios in which the genomic screening is done via a two-stage process (sometimes called a grid-tightening approach), the optimal sampling strategy for ARPs is affected grandparent–grandchild pairs because of the limited opportunity for recombination. A two-stage genomic screen involves genotyping a certain number of genetic markers and then following up regions of interest defined by reaching or exceeding a certain predefined significance level. This is done by genotyping markers on either side of the "interesting" markers. Grandparent–grandchild pairs are difficult to find, particularly if the disorder has onset in adulthood. After grandparent–grandchild pairs, the optimal sampling strategies in terms of study cost usually follow the sequence of half-siblings followed by avuncular pairs followed by sibling pairs. Affected cousin pairs have varying degrees of utility depending on several considerations. In general, as λ increases so does the power of affected cousin pairs. Ease of sampling is almost the reverse, however (sibling pairs are generally the easiest to collect).

DESPAIR is valuable for designing the optimal two-stage genomic screening strategy and also for estimating the number of pairs of affected relatives required to attain a certain power. Elston et al. (1996) show that for a disease gene with a λ_s of 2.0 and a cost ratio R of 100 (where cost is defined as the ratio of ascertainment costs for a person to the cost of performing one genotype), the least expensive two-stage genomic screening strategy is to collect 350 affected grandparent–grandchild pairs and genotype 85 markers spaced at 38.8 cM throughout the genome; markers generating a significance level of 0.045 should then be "followed up" by testing two additional markers on either side of the "interesting" marker. Table 7.7 shows additional examples of optimal two-stage designs using different types of affected relative pairs.

Holmans and Craddock (1997) recently reported the results of simulation studies for genomic screening strategies in a complex disease for which nuclear families with one affected sibling pair were available. The researchers compared two strategies for maximizing the power of the study and the efficiency of the screen, measured in total number of genotypes performed. The strategies were grid tightening and sample splitting (which involves genotyping only a portion of the total sample in the initial screening). Decisions regarding sample-splitting strategies involved whether to genotype the parents or just the affected sibling pairs and whether to genotype the entire sample or half the sample. Overall the investigators concluded that a strategy that uses both a grid-tightening and a sample-splitting approach is typically most efficient. From a practical viewpoint, decisions regarding whether to genotype parents will depend on the relative costs of genotyping in a first stage or second stage and on investigator's knowledge about underlying sample homogeneity (and how important the availability of complete IBD information will be relative to the required genotyping—in other words, will allele frequencies make a big difference in the outcome?). As always, decisions about strategies will depend on the interplay between correctly detecting a linkage and following up false positives.

Recently, several genomic screens for complex disease have been published. The designs generally consisted of affected sibling pairs, but in some cases unaffected siblings and other affected relative types were included. A summary of sampling strategies in selected genomic screens is shown in Table 7.8.

Quantitative Traits

Methods of mapping human quantitative trait loci (QTLs) can be applied to phenotypes ranging from height and weight to obesity, blood pressure, "intelligence," and any other trait that is measurable on a continuous scale. Until recently, the prospect of looking for loci predisposing to human traits of a quantitative nature had been daunting because the sample size necessary for adequate power had been estimated to be exceptionally large unless the trait had an underlying major locus accounting for a significant amount of the phenotypic variance (Blackwelder and Elston, 1985). However, the ascertainment scheme has been shown to be critically important in minimizing the necessary sample (Boehnke and Moll, 1989; Risch and Zhang, 1995; 1996). By way of example, consider a Mendelian disease: an investigator maximizes his ability to detect linkage by sampling as many affected individuals as possible. Similarly, this scenario translates directly into the investigation of complex discrete traits—the optimal sampling scheme usually involves the sampling of pairs of affected relatives. The translation of this approach into the study of human QTLs is to sample pairs of relatives that are either very different from (extremely discordant) or very similar to (extremely concordant) one another. This approach ensures, to the maximum extent possible, that the family is "seg-

TABLE 7.7 Optimal Two-Stage Genomic Screening Strategies for Independent Pairs of Affected Relatives of Various Types, Utilizing a Fully Informative Marker and a Value of 100 for the Ratio of the Cost of Ascertaining a Patient to the Cost of Performing a Genetic Marker Genotype[a]

	$\lambda_i = 2.0$					$\lambda_i = 10.0$				
Affected Relative Pair Type:	Full Sib	Half-sib	Avunc	Cousins	GP-GC	Full sib	Half-sib	Avunc	Cousins	GP-GC
Number of stage 1 markers	168	147	175	166	85	317	224	279	208	131
Marker spacing, cM	19.6	22.4	18.9	19.9	38.8	10.4	14.7	11.8	15.9	25.2
Number of pairs of ARPs	320	359	362	477	350	26	25	25	32	24
Number of additional markers[b]	2	2	2	2	2	2	2	2	2	2
α^*[c]	0.060	0.056	0.059	0.055	0.045	0.096	0.083	0.078	0.07	0.066
Estimated cost[d], $	199,979	203,826	232,093	292,216	143,035	28,231	20,096	23,503	23,695	12,933

[a]Analysis assumes that a single locus accounts for λ_i (a locus-specific λ), where *i* refers to the type of affected relative pair, although the disease may have an overall λ larger than the locus-specific λ (and this will always be the case for a complex disease in which two or more loci act to define phenotype). Analysis was performed using DESPAIR.

[b]For stage 2 of the genomic screen, this number represents the number of additional markers to genotype *on each side* of a marker where $P \leq \alpha^*$. Note that the total number of markers to genotype for any marker with $P \leq \alpha^*$ will be twice the number in this column.

[c]Markers with P values $\leq \alpha^*$ are followed up in stage 2 of genomic screen by genotyping additional markers.

[d]Assuming that each marker genotype costs $1.00.

Source: Elston et al. (1996).

TABLE 7.8 Characteristics of Study Design for Selected Published Genomic Screens

Disease	Design	Sample Characteristics	Ref.
Type 1 (insulin-dependent) diabetes mellitus	Affected sibling pairs (ASPs)	Genomic screening in 96 British ASPs Follow-up of interesting regions in a second set of 102 British ASPs and 84 American ASPs	Davies et al. (1994)
Type 1 (insulin-dependent) diabetes mellitus	Affected sibling pairs	Genomic screening in 61 French and North American Caucasian nonindependent ASPs. Follow-up of interesting regions in additional 253 ASPs of French, North American, and North African Caucasian descent *Note*: Data stratified based on previously identified HLA-DR3 and DR4 associations	Hashimoto et al. (1994)
Multiple sclerosis	Affected sibling pairs	Genomic screen in 100 nonindependent ASPs of Canadian descent Follow-up of interesting regions in additional 122 ASPs. Unaffected siblings sampled and genotyped to allow inference of missing parental genotypes	Ebers et al. (1996)
Multiple sclerosis	Affected sibling pairs	Genomic screening in 143 non-independent ASPs of British descent Follow-up of interesting regions in additional 108 ASPs	Sawcer et al. (1996)
Multiple sclerosis	Affected sibling pairs and extended pedigrees with affected relatives of various other types	Genomic screening in 52 families including 81 nonindependent ASPS and 58 pairs of nonsibling affected relatives Follow-up of interesting regions in additional affected relatives Follow-up of 23 families, including 45 ASPs and 30 pairs of nonsibling affected relatives	The Multiple Sclerosis Genetics Group (1996)

regating" the trait and genes of interest. In other words, if one samples individuals regardless of their trait status (e.g., high value or low value), they may include numerous pedigrees for which the predisposing trait allele is not present in either of the parents.

Extreme Discordant Pairs. The sampling of extremely discordant relative pairs is the most powerful approach to identifying genes that predispose to human QTLs. Specifically, such relative pairs (usually siblings) should share very few genes IBD at the trait locus (Risch and Zhang, 1995, 1996). Risch and Zhang

(1996) have shown that the smallest genetic effect that can be identified using this approach with a reasonable sample size (i.e., < 1500) is one that accounts for as little as 10% of the variance. Suppose, for instance, a researcher wants to detect a gene that accounts for 10% of the underlying phenotypic variance, has a trait allele frequency of 0.30, and has no residual correlation by employing a sample scheme that includes pairs of relatives in the top 10 and bottom 10% of the overall distribution. Approximately 1482 sibling pairs would be required to detect linkage allowing for a type I error rate of 0.0001 and power of 0.80. Table 7.9 summarizes additional data on required sample size to detect QTLs. In general, the presence of residual correlation between siblings, in which siblings appear to be more phenotypically similar than expected by the presence of a major genetic locus, will tend to decrease the necessary sample size, especially as the heritability decreases. As explained by Risch and Zhang (1996), this reduction in estimated sample size is because

> phenotypically discordant sib pairs will be more likely to be genetically discordant at the locus of interest. This is important because it is for loci with low heritability that there is likely to be residual sib correlation due to other genetic effects. At high heritability, most of the sib correlation is probably due to that locus, and hence there is unlikely to be a large residual correlation.

Extreme Concordant Pairs. Sampling extremely concordant relative pairs is usually a less effective though often more practical ascertainment scheme than sampling extreme discordant pairs. The specific decisions about sampling are more complicated and require the researcher to have at hand information about trait allele frequency, the parental mating types, and the degree of dominance of the trait (Zhang and Risch, 1996). As a general rule of thumb, however, extremely concordant pairs are more powerful than extremely discordant pairs only when the underlying gene is a rare recessive.

Although in retrospect the collection of extremely discordant or extremely concordant relative pairs as compared to pairs with less extreme phenotypes seems intuitive, this idea has served to revolutionize the consideration of human QTLs. There are some potential drawbacks with either approach. To ascertain the extremes of the distribution, hundreds or thousands of families may need to be screened. If the screening tests are expensive, time-consuming, or logistically difficult, such an approach may not be feasible. However, if such screening can be undertaken with relative ease, a much smaller data set can be subjected to the rigors of a genomic screen and ultimate fine-mapping of susceptibility genes.

POWER IN ASSOCIATION STUDIES

Transmission Disequilibrium Test (TDT)

The TDT tests for evidence of both linkage and association (linkage disequilibrium) in triplets of father–mother–affected child (Chapter 15). Here, linkage disequilibri-

TABLE 7.9 Estimated Number of Sibling Pairs Required to Detect a QTL Under Various Genetic Models[a]

Allele Frequency	Heritability = 0.10[b]				Heritability = 0.30[b]			
	$\rho = 0.0$	N	$\rho = 0.4$	N	$\rho = 0.0$	N	$\rho = 0.4$	N
				Additive Model				
0.10	1647	19120	342	21441	155	2356	52	2833
0.30	1482	17357	346	22336	120	1958	42	2829
				Dominant Model				
0.10	1567	18153	359	22320	143	2171	59	3119
0.30	1454	16777	384	23957	127	1920	65	3915
				Recessive Model				
0.10	19984	205218	849	41016	18996	195204	753	35514
0.30	2049	22615	398	21777	276	3531	185	2943

[a]Assuming a power of 0.80 and a type I error rate of 0.0001, allowing for screening from the top 10% of the distribution and the bottom 10% of the distribution.

[b]N, estimated number of siblings to be screened to obtain the necessary number of sibling pairs fitting the ascertainment criterion; ρ, the residual correlation.

Source: Risch and Zhang (1996).

um is defined as an increase in frequency of a particular disease–marker haplotype over that expected based on Hardy–Weinberg equilibrium, and it requires both linkage of the disease and marker loci and association between one or more alleles at the locus and the disease allele.

The TDT has recently been extended to allow for multiallelic marker systems (Spielman and Ewens, 1996). Formulas for investigating samples sizes necessary to obtain a specified power and type I error rate are derived in Kaplan et al. (1997). Table 7.10 shows estimated sample sizes under dominant and recessive models for a variety of disease allele frequencies and penetrances. These estimates assume that only one locus contributes to the phenotype so that the attributable risk for that locus is 1.0. In the presence of heterogeneity, the required samples sizes will be increased. For instance, consider a trait that is autosomal dominant in action with disease allele frequency of 0.01 and penetrance of 0.80 (equivalent to a prevalence of 0.0159) and confers a relative risk of 2.9 for disease in disequilibrium with an allele at a marker locus that demonstrates 0% recombination. To attain a type I error rate of 5% and power of 80%, for this trait, a researcher would need a sample comprising 194 TDT families. Note that as the relative risk for disease conferred by the disease-associated allele increases from 2.9 to 10.3, the number of families required is reduced dramatically to 40.

It is also interesting to note that when the disease is recessive, the sample sizes required are generally smaller than when the disease is dominant. This reduction occurs because for recessive diseases, both parents can potentially contribute to the overall statistic (for a child to be affected, since each parent must carry and transmit a disease allele to that child). In a dominant disease, however, only one parent needs to transmit a disease allele for the offspring to be affected. Although the transmitted allele from each parent is scored in the $2 \times n$ contingency table (where n is the number of alleles), the only transmission that provides support for a disease–marker allele association comes from the carrier parent, and this status cannot be specified in advance.

Genomic Screening Strategies for Association Studies. Risch and Merikangas (1996) suggest that "the future of the genetics of complex diseases is likely to require large-scale testing by association analysis" via the TDT. They argue that the sample sizes required to identify genes that make a minor to modest contribution to the disease phenotype under an affected sibling pair design are virtually unattainable, but the collection of mother–father–affected child triplets and the application of the TDT across the genome is a more realistic approach. Currently, such an approach is not technically feasible, since the resolution of polymorphic markers required for such a fine-scale effort has not been attained.

As pointed out by Terwillinger (personal communication), the assumptions underlying Risch and Merikangas's discussion merit careful consideration. For linkage disequilibrium to be detectable, the disease phenotype in apparently unrelated individuals must have arisen from a single mutation in a common ancestor many generations ago. Any different mutation in the same gene will likely not demonstrate the same associated marker allele and will tend to dilute the effects of any single major

TABLE 7.10 Estimated Number of Families (Mother–Father–Affected Child) Necessary to Attain Desired Power and Type I Error Rates for Various Disease Models at Recombination Fractions Between the Disease and Marker Loci = 0.00 or 0.01[a,b]

	Autosomal Dominant: $d = 0.01$, $f = 0.80$, $k = 0.0159$							
	$\theta = 0.00$				$\theta = 0.02$			
	Power = 0.80		Power = 0.95		Power = 0.80		Power = 0.95	
	$\alpha = 0.05$	$\alpha = 0.0001$	$\alpha = 0.05$	$\alpha = 0.0001$	$\alpha = 0.05$	$\alpha = 0.0001$	$\alpha = 0.05$	$\alpha = 0.0001$
$RR = 2.9$	194	473	302	628	211	514	327	682
$RR = 10.3$	40	99	63	131	44	107	68	142
	Autosomal Dominant: $d = 0.05$, $f = 0.80$, $k = 0.078$							
	$\theta = 0.00$				$\theta = 0.02$			
	Power = 0.80		Power = 0.95		Power = 0.80		Power = 0.95	
	$\alpha = 0.05$	$\alpha = 0.0001$	$\alpha = 0.05$	$\alpha = 0.0001$	$\alpha = 0.05$	$\alpha = 0.0001$	$\alpha = 0.05$	$\alpha = 0.0001$
$RR = 2.9$	223	544	346	721	242	590	376	783 $RR =$
10.3	47	114	73	151	51	124	79	164

Autosomal Recessive: $d = 0.20$, $f = 0.50$, $k = 0.02$

	$\theta = 0.00$				$\theta = 0.02$			
	Power = 0.80		Power = 0.95		Power = 0.80		Power = 0.95	
	$\alpha = 0.05$	$\alpha = 0.0001$	$\alpha = 0.05$	$\alpha = 0.0001$	$\alpha = 0.05$	$\alpha = 0.0001$	$\alpha = 0.05$	$\alpha = 0.0001$
RR = 2.9	86	209	133	277	93	226	144	308 *RR* =
10.3	17	42	27	56	19	46	29	61

Autosomal Recessive: $d = 0.05$, $f = 0.20$, $k = 0.0005$

	$\theta = 0.00$				$\theta = 0.02$			
	Power = 0.80		Power = 0.95		Power = 0.80		Power = 0.95	
	$\alpha = 0.05$	$\alpha = 0.0001$	$\alpha = 0.05$	$\alpha = 0.0001$	$\alpha = 0.05$	$\alpha = 0.0001$	$\alpha = 0.05$	$\alpha = 0.0001$
RR = 2.9	59	144	92	191	64	156	99	208 *RR* =
10.3	12	30	19	40	13	32	21	43

[a]Model assumes availability of penta allelic marker with equal allele frequencies (heterozygosity = 0.80) in the control population.

[b]Symbols: *d*, disease allele frequency; *f*, penetrance; *k*, disease prevalence; θ, recombination fraction between marker and disease loci; α, type I error rate; *RR*, relative risk for disease given presence of associated allele.

mutation, if one exists. Since several disease genes have been shown to have numerous mutations (e.g., cystic fibrosis, breast cancer, neurofibromatosis), the TDT may not be a universally applicable approach. If the same mutation arises multiple times, the TDT may still work if the gene mutation itself is tested. In addition, for heterogeneous populations with diverse genetic backgrounds such as in the United States, the chances of successfully identifying evidence for linkage disequilibrium are reduced.

If application of the TDT demonstrates evidence for linkage disequilibrium, these data are extremely useful in narrowing the disease gene interval. For example, evidence for linkage disequilibrium in cystic fibrosis (Kerem et al., 1989) was critically important in identifying the cystic fibrosis gene (although this evidence came from a statistical approach other than the TDT).

Risch and Merikangas's article has frequently been misinterpreted as suggesting that traditional linkage analysis (including both parametric, model-free, and nonparametric methods) is no longer useful. However, Scott et al. (1997) point out that Risch and Merikangas parameterize their argument in terms of λ, the relative risk for disease in the heterozygote, instead of λ_s, which is the more familiar means of considering risk in relatives. Scott et al. (1997) allow λ_s to be partitioned into contributions from many loci, or gene-specific contributions to phenotype (λ_{gs}), and they demonstrate that genes with moderate effect (e.g., $\lambda_{gs} < 2$) can produce γ values that are detectable by means of reasonable samples of affected sibling pairs. Given currently available maps and molecular technology, the TDT may best be applied in diseases in which a plethora of candidate genes have been identified (e.g., the neural tube defects), in relatively isolated populations, or in attempts to narrow a previously identified candidate region rather than in large-scale genomic screening efforts.

SUMMARY

Power studies in genetic linkage analysis are invaluable in helping to develop a strategy for designing studies to identify disease genes. In Mendelian disease, the "strategy" often involves confirming that the material collected is sufficient to detect evidence for linkage with reasonable power. In complex genetic diseases, however, the "strategy" is more difficult to define and depends critically on factors inherent in the underlying genetic model. Since the underlying genetic model for the disease is virtually unknown in almost all cases, traditional applications of power studies in Mendelian disease are unable to be applied directly to complex disease.

For Mendelian disease, computer simulation programs are available to determine the power of an available sample to detect linkage, the type I error rate, and the type II error rate. The Monte Carlo methods employed in these programs can be used for a wide breadth of applications, including assessing potential increases in error associated with maximization of lod scores over disease model parameters and estimating empiric P values.

These approaches can also be utilized to provide some broad generalizations about power and sample size in complex disease, as long as investigators understand that all the results are dependent on the assumptions they are willing to make about how the disease works. Programs like DESPAIR are valuable in providing estimates of necessary sample size under an assumed genetic model and also for evaluating various genomic screening strategies.

One of the most valuable outcomes anticipated from the plethora of genomic screens currently under development for various complex diseases will be the ability to compare different genomic screening strategies, sample designs, and type I and type II errors. The availability of lod scores, *P* values, or other statistics from these genomic screens will allow important cross-study comparisons for a variety of diseases whose underlying models are different. The compilation of these empiric experiences will provide important insights into design and analysis of genetic studies in human complex disease.

REFERENCES

Bishop DT, Williamson JA (1990): The power of identity-by-state methods for linkage analysis. *Am J Hum Genet* 46:254–265.

Blackwelder WC, Elston RC (1985): A comparison of sib-pair linkage tests for disease susceptibility loci. *Genet Epidemiol* 2: 85–97.

Boehnke M (1986): Estimating the power of a proposed linkage study: A practical computer simulation approach. *Am J Hum Genet* 39:513–527.

Boehnke M, Moll PP (1989): Identifying pedigrees segregating at a major locus for a quantitative trait: An efficient strategy for linkage analysis. *Am J Hum Genet* 44:216–224.

Cannon-Albright LA, Goldgar DE, Meyer LJ, Lewis CM, Anderson DE, Fountain JW, Hegi ME, Wiseman RW, Petty EM, Bale AE, Olopade OI, Diaz MO, Kwiatkowski DJ, Piepkorn MW, Zone JJ, Skolnick MH (1992): Assignment of a locus for familial melanoma, *MLM,* to chromosome 9p13-p22. *Science* 258:1148–1152.

Daly MJ, Lander ES (1996): The importance of being independent: Sib pair analysis in diabetes. *Nat Genet* 14:131–132.

Davies JL, Kawaguchi Y, Bennett ST, Copeman JB, Cordell HJ, Pritchard LE, Reed PW, Gough SCL, Jenkins SC, Palmer SM, Balfour KM, Rowe BR, Farral M, Barnett AH, Bain SC, Todd JA (1994): Genome-wide search for human type 1 diabetes susceptibility genes. *Nature* 371:130–136.

Davies JL, Kawaguchi Y, Bennett ST, Copenan JB, Cordell HJ, Pritchard LE, Reed PW, Goun SCL, Jenkins SC, Palmer SM, Balfour KM, Rowe BR, Farrall M, Barnett AM, Bain SC, Todd J (1994): A genome-wide search for human type I diabetes susceptibility genes. *Nature* 371:130–136.

Davis S, Schroeder M, Goldin LR, Weeks DE (1996): Nonparametric simulation-based statistics for detecting linkage in general pedigrees. *Am J Hum Genet* 58:867–880.

Ebers GC, Kukay K, Bulman DE, Sadovnick AD, Rice G, Anderson C, Armstrong II, Cousin K, Bell RB, Hader W, Paty DW, Hashimoto S, Oger J, Dupuette P, Warren S, Gray T, O'Connor P, Nath A, Auty A, Metz L, Francis G, Paulseth JE, Murray TJ, Pryse-Phillips W, Nelson R, Freedman M, Brunet D, Bouchard JP, Hinds D, Risch N (1996): A full genome search in multiple sclerosis. *Nat Genet* 13:472–476.

Elston RC (1989): Man bites dog? The validity of maximizing lod scores to determine mode of inheritance. *Am J Med Genet* 34:487–488.

Elston RC (1992): Designs for the global search of the human genome by linkage analysis. In: *Proceedings of the XVIth International Biometric Conference, Hamilton, New Zealand*, December 7–11, pp. 39–51.

Elston RC, Guo X, Williams LV (1996): Two-stage global search designs for linkage analysis using pairs of affected relatives. *Genet Epidemiol* 13:535–558.

Greenberg DA (1989): Inferring mode of inheritance by comparison of LOD scores. *Am J Med Genet* 34:480–486.

Hall JM, Lee MK, Newman B, Morrow JE, Anderson LA, Huey B, King MC (1990): Linkage of an early onset familial breast cancer gene to chromosome 17q21. *Science* 250:1684–1689.

Haseman JK, Elston RC (1972): The investigation of linkage between a quantitative trait and a marker locus. *Behav Genet* 2:3–19.

Hashimoto L, Habita C, Beressi JP, Delepine M, Besse C, Cambon-Thomsen A, Deschampes I, Rotter JL (1994): Genetic mapping of a susceptibility locus for insulin-dependent diabetes mellitus on chromosome 11q. *Nature* 371:161–164.

Hauser ER, Boehnke M, Guo SW, Risch N (1996): Affected-sib-pair interval mapping and exclusion for complex genetic traits—Sampling considerations. *Genet Epidemiol* 13:117–137.

Hodge SE (1984): The information contained in multiple sibling pairs. *Genet Epidmiol* 1:109–122.

Hodge SE, Abreu PC, Greenberg DA (1997): Magnitude of type I error when a single-locus linkage analysis is maximized over models: a simulation study. *Am J Hum Genet* 60:217–227.

Holmans P, Craddock N (1997): Efficient strategies for genome scanning using maximum-likelihood affected-sib-pair analysis. *Am J Hum Genet* 60:657–666.

Kaplan NL, Martin ER, Weir BS (1997): Power studies for transmission/disequilibrium tests with multiple alleles. *Am J Hum Genet* 60:691–702.

Kerem B-S, Rommens JM, Buchanan JA, Markiewicz D, Cox TK, Chakravarti A, Buchwald M, Tsui L-C (1989): Identification of the cystic fibrosis gene: Genetic analysis. *Science* 245:1073–1080.

Kruglyak L, Daly MJ, Reeve-Daly MP, Lander ES (1996): Parametric and nonparametric linkage analysis: A unified multipoint approach. *Am J Hum Genet* 58:1347–1363.

Multiple Sclerosis Genetics Group (1996): A complete genomic screen for multiple sclerosis underscores a role for the major histocompatibility complex. *Nat Genet* 13:469–476.

Ott J (1989): Computer simulation methods in human linkage analysis. *Proc Natl Acad Sci USA* 86:4175–4178.

Ott J, Terwilliger JD (1992): Assessing the evidence for linkage in psychiatric genetics. In: Mendlewicz J, Hippius H, eds. *Genetic Research in Psychiatry*. Berlin: Springer-Verlag, pp. 245–249.

Pericak-Vance MA, Bebout JL, Gaskell PC, Yamaoka LH, Hung W-Y, Alberts MJ, Walker AP, Bartlett RJ, Haynes CS, Welsh KA, Earl NL, Heyman A, Clark CM, Roses AD (1991): Linkage studies in families with Alzheimer's disease: Evidence for chromosome 19 linkage. *Am J Hum Genet* 48:1034–1050.

Ploughman LM, Boehnke M (1989): Estimating the power of a proposed linkage study for a complex genetic trait. *Am J Hum Genet* 44:543–551.

Risch N (1990): Linkage strategies or genetically complex traits. II. The power of affected relative pairs. *Am J Hum Genet* 46:229–241.

Risch N, Zhang H (1995): Extreme discordant sib pairs for mapping quantitative trait loci in humans. *Science* 268:1584–1589.

Risch N, Zhang H (1996): Mapping quantitative trait loci with extreme discordant sib pairs: Sample size considerations. *Am J Hum Genet* 58:836–843.

Risch N, Merikangas K (1996): The future of genetic studies of complex human disorders. *Science* 273:1516–1617.

Sawcer S, Jones HB, Feakes R, Gray J, Smaldon N, Chataway J, Robertson N, Clayton D, Goodfellow PN, Compston A. A genome screen in multiple sclerosis reveals susceptibility loci on chromosome 6p21 and 17q22 (1996): *Nat Genet* 13:464–476.

Scott WK, Pericak-Vance MA, Haines JL. (1997): Genetic analysis of complex diseases. *Science* 275:1327.

Sherrington R, Brynjolfsson J, Petursson H, Potter M, Dudleston K, Barraclough B, Wasmuth J, Dobbs M, Gurling H (1988): Localization of a susceptibility locus for schizophrenia on chromosome 5. *Nature* 336:164–167.

Spielman RS, Ewens WJ (1996): The TDT and other family-based tests for linkage disequilibrium and association. *Am J Hum Genet* 59:983–989.

Weeks DE, Harby LD (1995): The affected-pedigree member method: Power to detect linkage. *Hum Hered* 45:13–24.

Weeks DE, Lange K (1988): The affected-pedigree member method of linkage analysis. *Am J Hum Genet* 42:315–326.

Weeks DE, Lehner T, Squires-Wheeler E, Kaufmann C, Ott J (1990a): Measuring the inflation of the LOD score due to its maximization over model parameter values in human linkage analysis. *Genet Epidmiol* 7:237–243.

Weeks DE, Ott J, Lathrop GM (1990b): SLINK: A general simulation program for linkage analysis. *Am J Hum Genet* 47:A204.

Zhang H, Risch N (1996): Mapping quantitative-trait loci in humans by use of extreme concordant sib pairs: Selected sampling by parental phenotypes. Published erratum in *Am J Hum Genet* 60:748–750 (1997). [cf *Am J Hum Genet* 59:951–957 (1996)].

APPENDIX 7.1

Example of Monte Carlo Simulation Assuming that the Trait and Marker Loci Are Unlinked to Each Other

Step 1: Assign genotypes to the founders

Decision rules:

If $0.00 \leq$ random number < 0.25, assign allele 144.
If $0.25 \leq$ random number < 0.50, assign allele 146.

If 0.50 ≤ random number < 0.75, assign allele 160.
If 0.75 ≤ random number ≤ 1.00, assign allele 162.

Process: A random number is selected to generate each of the two alleles of the founder. The paternally derived allele is arbitrarily assigned to be the "left" allele and the maternally derived allele is therefore the "right" allele. For instance, for individual 2000 a random number of 0.24 is selected (Table A7.1.1). Since this value is less than 0.25, assign the left allele as 144. Next, a random number of 0.34 is selected. Since this value is equal to or more than 0.25 and less than 0.50, assign the right allele as 146. The process continues for each of the founders.

Note that genotypes are simulated regardless of whether the individual is available for study. In this case, individuals 2000, 2001, 0100, and 0102 are deceased and unavailable for genotyping. Although their genotypes are simulated for use by the computer program, the simulated genotypes are not printed into the output file. These genotypes are listed as the first set of marker genotypes in Figure 7.2.

Step 2: Transmit alleles throughout the family according to mendel's first law of segregation

Decision rules:

If 0 ≤ random number < 0.50, then transmit paternally derived ("left") allele.
If 0.50 ≤ random number < 1.0, then transmit maternally derived ("right") allele.

Because the simulation here is performed without regard to whether the individual is affected with the disease, the selection of the marker allele which is transmitted is independent of the disease status of the parents.

Process: For example, in individual 1000 the first random number selected is 0.53, which is utilized to select which of the two alleles was transmitted from individual 2000, his father. Since 0.53 exceeds 0.50, the "right" allele (146) is selected. The second random number selected is 0.87, which is utilized to select which of the two alleles was transmitted from his mother, individual 2001. Since 0.87 is exceeds

TABLE A7.1.1 Example of Genotype Assignment for Founders

Individual	Random Number for Allele 1	Random Number for Allele 2	Assigned Genotype
2000	0.24	0.34	144/146
2001	0.33	0.04	146/144
1001	0.55	0.78	160/162
0100	0.03	0.19	144/144
0104	0.82	0.99	162/162
0102	0.87	0.47	162/146

0.50, again the "right" allele (144) is transmitted from the mother. The assignment of genotypes for the remainder of the nonfounders in the pedigree is demonstrated in Table A7.1.2.

Note that the nonrandom transmission of alleles (meiotic drive) can be simulated here if the decision rules are changed. For instance, if any random number ≤ 0.70 causes the program to select the paternally derived allele, then there is a bias in transmission in favor of the paternally derived allele.

APPENDIX 7.2

Example Lod Score Results for Pedigree in Figure 7.2

The simulation model is described in the text. Note that a lod score of –99.99 at $\theta = 0.0$ represents an impossible likelihood (the occurrence of a recombination event when the recombination fraction between the disease and marker locus is 0.0) and is often represented in text as $-\infty$.

A. True Recombination Fraction Between the Disease and Marker Locus Is 0.50

	θ			
Replicate	0.00	0.05	0.10	0.15
1	–99.99	–4.30	–2.59	–1.66

B. True Recombination Fraction Between the Disease and Marker Locus is 0.05; Italicized Lod Scores Are the Highest Lod Score for the Calculated θ Values.

TABLE A7.1.2 Example of Genotype Assignment for Non-Founders

	Random Number to Assign Allele Transmitted from:		
Individual Number	Father	Mother	Genotype
1000	0.53	0.87	146/144
1002	0.57	0.59	146/144
0001	0.01	0.19	146/160
0111	0.73	0.18	144/160
0109	0.11	0.83	146/162
0107	0.12	0.33	146/160
0105	0.27	0.90	146/162
0103	0.67	0.44	144/160
0101	0.93	0.99	144/162
9002	0.12	0.37	146/144
9004	0.32	0.91	146/144
9000	0.48	0.42	160/144
9008	0.16	0.01	144/162
9009	0.40	0.83	144/162
9006	0.71	0.18	162/162
9007	0.64	0.29	162/162

	θ			
Replicate	0.00	0.05	0.10	0.15
1	−99.99	0.66	0.81	*0.84*
2	*4.21*	3.86	3.49	3.10
3	*2.71*	2.47	2.21	1.95
4	−99.99	1.30	1.58	*1.59*

APPENDIX 7.3

Example of Simulation of Genetic Marker Genotypes Conditional on Trait Phenotypes Allowing for Complete and Reduced Penetrance

Complete Penetrance

For this example, a small nuclear pedigree is utilized to illustrate the concepts (Fig. 7.4); this simple example underscores the complexity of these conditional simulation problems. For this example, the trait locus is transmitted as an autosomal dominant with complete penetrance. The frequency of the trait allele d is 0.001 and the frequency of the normal allele D is 0.999. In this model, there are no phenocopies. Note that to simplify the example, the consideration of unknown phenotypes is not discussed. The marker locus has three alleles with frequency 0.40, 0.30, and 0.30, and it demonstrates 5% recombination with the trait locus.

Step 1*: *Calculate genotypic probabilities for founders, and calculate the probabilities of genotypes given known phenotype of a founder. The frequencies of the four possible genotypes at the trait locus are calculated assuming Hardy–Weinberg equilibrium as follows:

$$p(DD) = 0.999 \times 0.999 = 0.998001$$

$$p(Dd) = 0.999 \times 0.001 = 0.000999$$

$$p(dD) = 0.001 \times 0.999 = 0.000999$$

$$p(dd) = 0.001 \times 0.001 = 0.000001$$

Because the penetrance of the trait allele has been specified as 1.00, the following conditional probabilities for phenotype given genotype are known:

$p(\text{affected} \mid DD) = 0.00$	$p(\text{unaffected} \mid DD) = 1.00$
$p(\text{affected} \mid Dd) = 1.00$	$p(\text{unaffected} \mid Dd) = 0.00$
$p(\text{affected} \mid dD) = 1.00$	$p(\text{unaffected} \mid dD) = 0.00$
$p(\text{affected} \mid dd) = 1.00$	$p(\text{unaffected} \mid dd) = 0.00.$

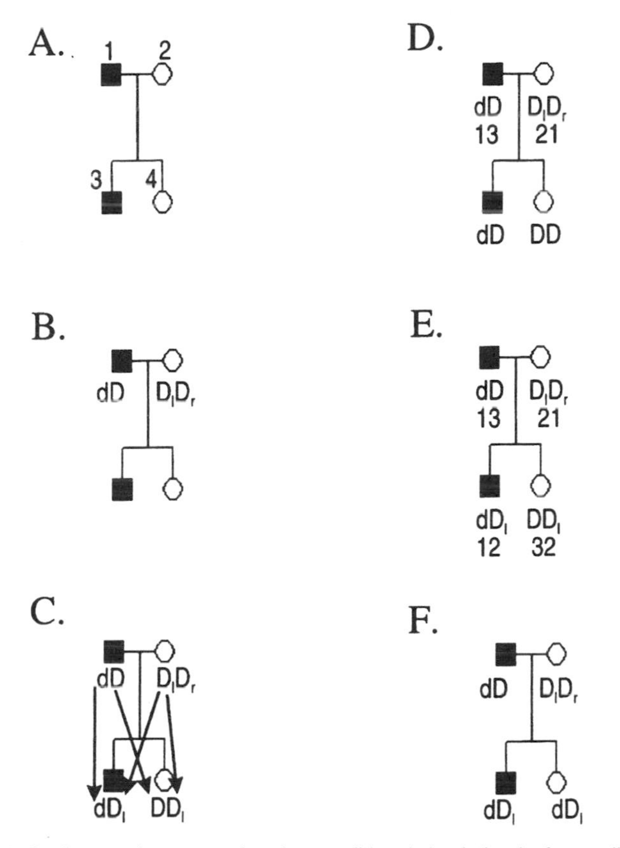

Figure 7.4. *Pedigrees demonstrating the conditional simulation in Appendix 7.3.*

The conditional probabilities for genotype given phenotype must be calculated, and these calculations can be done utilizing the Bayes rule.

So, for example:

$$\frac{p(DD \mid \text{unaffected}) = (0.998001) \times (1.0)}{(0.998001 \times 1.0) + (0.000999 \times 0.0) + (0.000999 \times 0.0) + (0.000001 \times 0.0)} = 1.0$$

In summary (see also Fig. 7.5):

	Conditional Probability	Cumulative Probability
$p(DD \mid \text{unaffected})$	1.000000	1.000000
$p(Dd \mid \text{unaffected})$	0.000000	1.000000
$p(dD \mid \text{unaffected})$	0.000000	1.000000
$p(dd \mid \text{unaffected})$	0.000000	1.000000

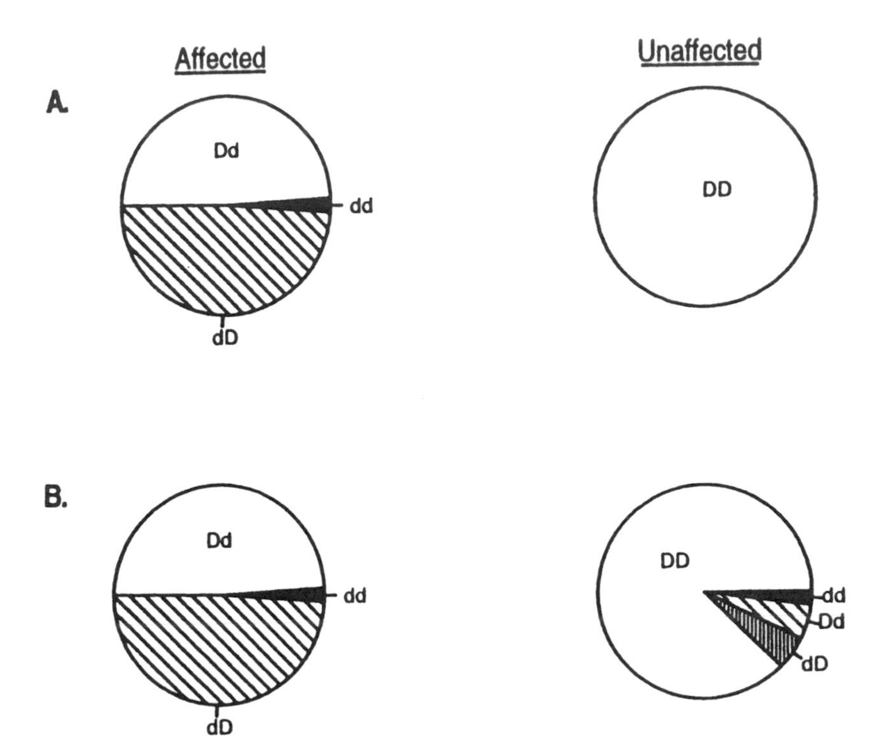

Figure 7.5. *Diagrams showing the different proportions of possible genotypes underlying the disease phenotype.* (A) *Complete penetrance. The most likely disease genotypes for an affected individual are heterozygous* (dD, Dd). *However, an affected individual can be homozygous at the disease locus* (dd). *The probability for a founding affected individual to be of genotype* dd *is calculated from disease allele frequencies provided by the investigator. Since penetrance is complete, an unaffected individual can only be of disease genotype* DD. (B) *Reduced penetrance. Note that the probabilities for the underlying genotypes at the disease locus remain unchanged for an affected individual regardless of the penetrance. However, the genotypes for unaffected individuals at the disease locus when reduced penetrance is allowed are different from those when penetrance is complete. Specifically, now an unaffected individual has a defined probability of being a disease gene carrier in the heterozygous state* (dD, Dd) *or in the homozygous state* (dd).

$p(DD \mid \text{affected})$	0.000000	0.000000
$p(Dd \mid \text{affected})$	0.499750	0.499750
$p(dD \mid \text{affected})$	0.499750	0.999500
$p(dd \mid \text{affected})$	0.000500	1.000000

And the decision rules for assigning genotypes given phenotypes are as follows.

If the founder's phenotype is affected, then:
If 0.0 ≤ random number < 0.000000, then genotype => *DD*.
If 0.000000 ≤ random number < 0.499750, then genotype => *Dd*.
If 0.499750 ≤ random number < 0.999500, then genotype => *dD*.
If 0.999500 ≤ random number ≤ 1.000000, then genotype => *dd*.

Note here that if an individual's phenotype is affected, then the genotype *DD* is impossible.

If the founder's phenotype is unaffected, then:
If 0.0 ≤ random number ≤ 1.00, then genotype => *DD*.

Note that because penetrance is complete, all phenotypically asymptomatic founders must not carry the disease allele and are therefore always of genotype *DD*.

Next, select random numbers for use in assigning trait genotypes to founders. For individual 1, random number 0.8731 is selected so trait genotype is *dD*; for individual 2, random number 0.3215 is selected so trait genotype is *DD* (Figure 7.4*B*).

Step 2: Assign trait locus genotypes to offspring given their phenotypes and the parental trait locus genotypes. Since the parental mating type is *dD* by *DD*, there are two possible genotypes for offspring which occur with equal frequency: *dD* and *DD*. This probability is designated as *P*(genotype of offspring | parents). For the sake of simplicity, we will assume that the unaffected mother always transmits her "left" allele (designated as D_l) to offspring. Since she is homozygous for the normal allele, which allele at the trait locus the mother transmits to her children is irrelevant to their trait phenotype and trait genotype. In practice, this process is also accomplished via selection of random numbers.

Decision rules for assignment of genotypes are dependent upon these Mendelian transmission rules and on the correspondence between the phenotype and genotype (i.e., the penetrance).
If the offspring phenotype is affected, then:

Possible Offspring Genotypes (*g*)	*P*(g | parents)	Conditional Probability	Cumulative Distribution
DD_l	½	*P*(*DD* | affected) = 0.0	0.0
dD_l	½	*P*(*Dd* | affected) – 1.00	1.0

If the offspring phenotype is unaffected, then:

Possible Offspring Genotypes (g)	P(g | parents)	Conditional Probability	Cumulative Distribution
DD_l	½	$P(DD \mid \text{unaffected}) = 1.00$	1.0
dD_l	½	$P(Dd \mid \text{unaffected}) = 0.10$	1.0

So the decision rules are established as follows.

If phenotype of offspring is affected:

If $0.0 \leq$ random number < 0.00, then genotype => DD_l.
If $0.00 \leq$ random number ≤ 1.00, then genotype =>dD_l.

(note that if the offspring is affected, the probability of genotype *DD* is zero).
If phenotype if offspring is unaffected:

If $0.0 \leq$ random number < 1.00, then genotype => DD_l.

(note that an unaffected individual is much more likely to have genotype *DD* than *Dd*). In both cases, the unaffected, noncarrier mother could have transmitted either the "left" or "right" copy of her *D* allele. To simplify the example, we will presume that we have established—also by selection of random numbers—that the mother has transmitted her "left" allele, which is designated as D_l.
To assign genotypes for individuals 3 and 4, select random numbers. For individual 3, random number 0.7604 is selected so assigned genotype is dD_l. For individual 4, random number 0.9773 is selected so assigned genotype is
DD_l (Fig. A7.4*C*).

Step 3: Assign marker genotypes (and therefore linkage phase) to founders. For founders, the genotypes are assigned according to Hardy–Weinberg expectations based on allele frequencies provided by the user. The linkage phase, however, is set by determining which of the alleles is associated with the disease allele, which is again performed by selecting random numbers.

Decision rules:

If $0 \leq$ random number < 0.40, then allele => 1.
If $0.40 \leq$ random number < 0.70, then allele => 2.
If $0.70 \leq$ random number ≤ 1.00, then allele => 3.

For individual 1, for the "left" allele, the random number is 0.33; for the "right" allele, the random number is 0.87. So, the genotype for this individual is 13. Since his trait genotype is *dD*, the phase for the trait and marker loci is established as *d*1/*D*3, so that the disease allele is established to be "segregating" with the '1' allele.

For individual 2, for the "left" allele, the random number is 0.62; for the "right" allele, the random number is 0.19. So, the genotype for this individual is 21. Since her trait genotype is *DD*, the phase for the trait and marker loci is established as *D*2/*D*1 where the '2' allele is segregating with the "left" copy of *D* (Fig. A7.4*D*). Note, however, that this individual does not carry the disease allele. Since she is homozygous for the normal allele at the trait locus, she is uninformative for linkage analysis.

Step 4 Assign marker genotypes to offspring conditional on trait phenotypes and established frequency of recombination between the trait and marker locus. Because the model has been established such that the trait and marker loci are linked and our simulation has determined the disease locus genotype for each individual, we know which marker allele the affected father has transmitted to each of his offspring *unless* there has been a recombination event between the trait and marker locus. (We further earlier assigned the transmission of the D_1 disease–marker allele combination to both children from the mother). The Monte Carlo simulation process will allow us to mimic the effects of recombination by establishing decision rules for transmission of marker allele from affected parent to child who inherited the 'd' allele from him. We recall that the user has specified that the true recombination fraction between the trait and marker loci is 0.05.

If $0 \leq$ random number < 0.95, then transmit *d*1 (a nonrecombinant gamete).
If $0.95 \leq$ random number ≤ 1.0, then transmit *d*3 (a recombinant gamete).

Similarly, if the affected father transmitted the *D* allele to an offspring, the decision rules would be:

If $0.0 \leq$ random number < 0.95, then transmit *d*3 (a nonrecombinant gamete).
If $0.95 \leq$ random number ≤ 1.00, then transmit *d*1 (a recombinant gamete).

For individual 3, random number 0.7 is selected, so he receives allele '1' from his affected father; for individual 4, random number 0.98 is selected, so she receives allele "3" from her affected father (Fig. A7.4*E*)

The identical process is utilized to determine which of the maternal alleles is transmitted. For individual 3, the selected random number is 0.93 and for individual 4, the selected random number is 0.09, so each has inherited a nonrecombinant gamete from the mother; in other words, each is assigned the '2' allele at the marker locus.

B. Reduced Penetrance

Step 1: Calculate genotypic probabilities for founders, and calculate the probabilities of genotypes given known phenotype of a founder. The first step proceeds identically in assigning probabilities for trait genotypes for founders; however, since the penetrance of the trait allele is 0.90, the conditional probabilities for the genotype given the known phenotype are:

$p(\text{affected} \mid DD) = 0.00$	$p(\text{unaffected} \mid DD) = 1.00$
$p(\text{affected} \mid Dd) = 1.00$	$p(\text{unaffected} \mid Dd) = 0.00$
$p(\text{affected} \mid dD) = 1.00$	$p(\text{unaffected} \mid dD) = 0.00$
$p(\text{affected} \mid dd) = 1.00$	$p(\text{unaffected} \mid dd) = 0.00$

***Step 2:* Simulate trait genotypes of founders.** Again, the conditional probabilities for genotype given phenotype must be calculated.
In summary (see also Fig. 7.5*B*):

	Conditional Probability	Cumulative Distribution
$p(DD \mid \text{unaffected})$	1.000000	1.000000
$p(Dd \mid \text{unaffected})$	0.000000	1.000000
$p(dD \mid \text{unaffected})$	0.000000	1.000000
$p(dd \mid \text{unaffected})$	0.000000	1.000000
$p(DD \mid \text{affected})$	0.000000	0.000000
$p(Dd \mid \text{affected})$	0.499750	0.499750
$p(dD \mid \text{affected}\}$	0.499750	0.999500
$p(dd \mid \text{affected})$	0.000500	1.000000

And the two sets of decision rules for assigning genotypes to given phenotypes are as follows.

If the founder's phenotype is affected, then:

If $0.0 \leq$ random number < 0.000000, then genotype => *DD*.
If $0.000000 \leq$ random number < 0.499750, then genotype => *Dd*.
If $0.499750 \leq$ random number < 0.999500, then genotype => *dD*.
If $0.999500 \leq$ random number ≤ 1.000000, then genotype => *dd*.

Note here that if an individual's phenotype is affected, then the genotype *DD* is impossible.

If the founder's phenotype is unaffected, then:

If 0.0 ≤ random number ≤ 0.999798, then genotype => *DD*.
If 0.999799 ≤ random number < 0.999899, then genotype => *Dd*.
If 0.999899 ≤ random number < 0.999999, then genotype => *dD*.
If 0.999999 ≤ random number ≤ 1.000000, then genotype => *dd*.

Next, select random numbers for use in assigning trait genotypes to founders. For individual 1, random number 0.9007 is selected so trait genotype is *dD*; for individual 2, random number 0. 6581 is selected so trait genotype is D_lD_r.

Step 3: Assign trait locus genotypes to offspring given their phenotypes and the parental trait locus genotypes. Since the established parental mating type is unchanged from the preceding example, the two possible offspring genotypes remain the same. We will also make the same simplifying assumption that the mother transmits the "left" allele.
If the offspring phenotype is affected, then:

Possible Offspring Genotypes (*g*)	*P*(g | parents)	Conditional Probability	Cumulative Distribution
DD_l	½	*p*(*DD* | affected) = 0.0	0.0
dD_l	½	*p*(*Dd* | affected) = 1.00	1.0

If the offspring phenotype is unaffected, then:

Possible Offspring Genotypes (*g*)	*P*(g | parents)	Conditional Probability	Cumulative Distribution
DD_l	½	*p*(*DD* | unaffected) = 1.00	0.91
dD_l	½	*p*(*Dd* | unaffected) = 0.10	1.00

So the decision rules are established as follows:

If phenotype of offspring is affected:

If 0.0 ≤ random number < 0.00, then genotype => *DD*.
If 0.00 ≤ random number ≤ 1.00, then genotype => *Dd*.

(note that if the offspring is affected, the probability of genotype *DD* is zero).
If phenotype if offspring is unaffected:

If 0.0 ≤ random number < 0.91, then genotype => *DD*.
If 0.91 ≤ random number ≤ 1.00, then genotype => *Dd*.

(note that an unaffected individual is much more likely to have genotype *DD* than *Dd*). Again, for simplicity we assume that the mother has transmitted her "left" allele, which is designated as D_l.

To assign genotypes for individuals 3 and 4, select random numbers. For individual 3, random number 0.3371 is selected so assigned genotype is dD_l. For individual 4, random number 0.9802 is selected so assigned genotype is dD_l (Fig. A7.4*F*).

From here, the assignment of marker alleles to founders and transmission to offspring proceeds as described in the full penetrance example.

8

The Collection of Biological Samples for DNA Analysis

Jeffery M. Vance
Division of Neurology
Department of Medicine
Duke University Medical Center and Health System
Durham, North Carolina

This chapter provides practical knowledge on sample collection in genetic studies, so that an individual can make informed decisions concerning the options available for sample collection at various levels of cost and expertise.

ESTABLISHING THE GOALS OF THE COLLECTION

It is very important to the success of any family study for the investigator to clearly define the goal of the investigation. If the intent of the study is to perform genomic screening and gene localization, then a relatively large and consistent amount of DNA is required; usually such a supply is obtained through the collection of whole blood. If only a few specific tests are planned, the smaller amount of DNA obtain-

Approaches to Gene Mapping in Complex Human Diseases, Edited by Jonathan L. Haines and Margaret A. Pericak-Vance. ISBN 0-471-17195-6

able from a buccal sample may be appropriate. Mutational analysis in unknown disorders is greatly enhanced by the availability of mRNA (Noguchi et al., 1995); thus lymphoblasts or tissue samples from selected affecteds may be desired in these instances. The geographic location of families and patients, and the technical background of those collecting the samples are also important items to consider. For example, an individual in a remote location, where venipuncture would not be practical, may easily be included in a study if a mailed buccal sample is used instead.

TYPES OF DNA SAMPLE COLLECTION

Venipuncture (Blood)

Venipuncture should preferably be done using a 21-gauge needle or larger. While a 23-gauge needle can be used, in our experience the risk of hemolysis is greater. Because field studies often occur in less than optimum conditions, we use butterfly needles, which are easier to handle than Vacutainer® needles, to collect our samples.

Gustafson et al. (1987) demonstrated that DNA is most stable at room temperature when acid citrate dextrose (ACD) or EDTA is used as the anticoagulant. DNA is not stable in heparin, and residual heparin has been reported to interfere with *Taq* polymerase as well (Kirby, 1990). Therefore, use of this anticoagulant should be avoided.

The volume of blood required for sampling depends on the purpose of the study. There are many different methods for DNA extraction, differing primarily in the volume of blood they can process easily, the toxicity of the components, and the number of steps or pieces of equipment required. DNA is obtained only from the lymphocytes (red cells are without a cell nucleus); therefore the amount of DNA in a blood sample can vary from individual to individual. Perhaps this contributes partially to the wide variation in the literature of the estimates of the amount of DNA in a milliliter of whole blood, ranging in the majority of cases from 25 to 40 μg DNA per milliliter of whole blood (Kirby, 1990).

Immediately after collection, the collection tube should be gently inverted 5–10 times to ensure proper mixing of the anticoagulant. Omission of this step is a common mistake that leads to clotted samples. If slated for DNA extraction, the blood should be placed at 4°C as soon as possible. This slows the metabolism of the living cells. However, samples to be used for tissue culture need to be held at room temperature, since chilling these samples will reduce the transformation success.

Receiving samples by mail also requires planning. Blood samples are best sent by an overnight carrier. This is absolutely true for samples destined for transformation, which need to be received and processed as soon as possible, certainly within 2–3 days of collection. However, it may be cheaper for samples marked for DNA extraction to be sent by postal service, where time is important but not as critical. There are specific regulations for sending blood products via mail both within the

United States and internationally. The investigator should check with the service selected to find out the specific rules that apply.

Buccal Brushes

In the last few years buccal brushes have become increasingly popular for collecting DNA specimens (Richards et al., 1994). These noninvasive devices do not require the technical skills needed for venipuncture and eliminate the chance of an accidental needle stick occurring during the collection of the sample. Prior to extraction, the brushes are very stable. They are particularly good for young children, for obese patients, and those situations calling for a smaller amount of DNA. Disadvantages are also present. The amount of DNA can vary significantly from one individual to another: typically, even under the best collection conditions, it is significantly smaller than the quantity collected by venipuncture. However, this may not be a problem unless a large genomic screening project is planned. The DNA from buccal samples usually is isolated only crudely using NaOH. Thus, the presence of proteases and nucleases makes the sample much less stable than DNA purified from blood, so it must be kept at 4°C or frozen at all times. Since only DNA consisting of small fragment sizes will be available using this method, it is not suitable for techniques requiring large fragments, such as pulsed field gel electrophoresis (PFGE).

Before a sample is collected, the individual washes his or her mouth out *gently* with water to remove large particles. The brush is then twirled as it is firmly moved over the inner surface of one cheek for at least 30 seconds. A second brush is then used against the other cheek for another 30 seconds. These steps are critical, since many failures in buccal sampling appear to occur at this stage. It has been our experience, as well as the experience of others, that one is more likely to obtain sufficient DNA from buccal collection when the samples are collected under the supervision of an experienced worker than when samples are collected by the subjects themselves, without supervision. It may be that many unsupervised subjects perform just a cursory brushing which will not collect enough cells for a successful PCR analysis. To alleviate this problem, we use the increased brushing times above.

We have kept buccal brushes after collection of a sample in their container tube for over a month with no detrimental effect on PCR results. Thus, they are good for field studies (where the samples cannot be sent to the laboratory within several days), for extensive field trips, and for collecting samples from patients in isolated locales.

Dried Blood

Dried blood samples can also be used as a source of DNA when a relatively small number of PCR reactions are required (Guthrie and Susi, 1963; McCabe, 1991). Matsubara et al. (1992) have suggested that dried blood can be a source of mRNA as well. We use these samples as a method to check for sample mix-up, so that if problems arise later in the genotyping analysis we will have a second sample to test. From some individuals—children, for example—one can also supply samples via fingerpricks. At the time of venipuncture, a small amount of blood is placed on relatively inexpensive "Guthrie cards" (Guthrie and Susi, 1963). These are then dried and stored in photo albums at room temperature. Since these samples are obtained at the time of collection, we can be assured that they do indeed represent the DNA of the intended participant. They also provide backup DNA on an individual.

Tissue

Both fresh (frozen) and fixed tissues can provide samples for study. If fresh tissue from biopsies or autopsies is quickly frozen in liquid nitrogen, it can be used in mutational analysis during molecular studies. It is useful to observe and plan for those opportunities where these samples may be obtained. Fixed tissue may provide samples on affected or deceased individuals for genotyping or association studies. In most cases, paraffin-embedded tissue (PET) will be the source of DNA for genetic study (Kosel and Graeber, 1994). Such samples, used routinely for analysis of surgical biopsies and subsequently archived, represent a resource for retrospective study. Recently, however, recently questions of informed consent have been raised concerning the use of some of these samples (Marshall, 1996) (see also Chapter 6).

The central problem of fixed tissue is that the fixatives themselves degrade DNA. Thus, the size of the products of polymerase chain reaction (PCR) obtainable from fixed tissue is usually limited. Therefore, obtaining results from fixed tissue can be difficult and should be reserved for individuals where no other DNA source is available. Greer et al. (1991) and Smith et al. (1987) studied the variation in the size of PCR products that can be obtained using different fixatives and found that 95% ethanol routinely allowed the largest PCR fragments to be obtained; OmniFix and acetone are next in order of preference. Buffered neutral formalin produced the smallest PCR fragments.

DNA EXTRACTION AND PROCESSING

Blood

The methods to extract DNA are numerous. These methods usually differ with respect to (1) the volume of blood they can easily handle, (2) the number of steps (and therefore the number of samples that can be effectively processed), (3) toxicity, (4) the equipment needed, and (5) the purity (a crude prep for PCR or a more traditional extraction). Since each author believes his or her method to be "simpler" than earlier approaches, the reader should chose the method that best fits present needs and laboratory facilities. Usually a pellet is isolated which consists mainly of lymphocytes, the only component of blood that contains DNA. Centrifugation in the presence of a medium that lyses the red blood cells but leaves the lymphocytes intact can accomplish this isolation. An alternative to this approach is to spin the blood in a proprietary medium whose specific gravity causes the blood to separate into visible fractions from which lymphocytes can be harvested (buffy coat).

In either case, the enriched pellet is then treated by one of several methods to lyse the remaining lymphocytes and isolate the DNA. A primary method for many years has been a mixture of sodium dodecyl sulfate (SDS) and proteinase K to rupture cell walls and degrade protein (Kirby, 1990). To remove other proteins and peptides from the DNA, the sample is treated to a standard phenol–chloroform extraction. Finally it is ethanol-precipitated in the presence of a simple salt (Kirby, 1990). Several other techniques, seeking to avoid the toxicity of phenol and chloroform, either substituted compounds for these chemicals (Johns and Paulus-Thomas, 1989; Planelles et al., 1996) or used a different approach, such as phase separation using guanidine thiocyanate (Chomczynski, 1993) or isolation employing a high salt concentration (Miller et al., 1988). Methods for PCR for small amounts also include proteinase K digestion (Kawasaki, 1990). Several commercial products are available including those based on guanidine thiocyanate like DNAzol (Molecular Research Center) and the popular method sold by Puregene.

The quality of the DNA can matter, again depending on the study. For restricted fragment length polymorphism (RFLP) or enzyme restriction analysis, the quality of the DNA should be very high to ensure good digestion results. However, since the bulk of genetic analyses is now done by PCR, small amounts of DNA can be easily obtained without significant purification. Such is the case in buccal brushes. In these cases the most important properties of an extraction method are freedom of the DNA from contaminants that would interfere with Taq polymerase and the ability to produce a clean product. However, the highest quality DNA possible is the most suitable for genotyping studies and also provides the most stable sample for long-term storage.

Quantitation

Traditionally, absorbence at 260 nm wavelength is the method for quantitation of DNA, although fluorescence may be used as well. An optical density (OD) of 1.0 corresponds to 50 μg double-stranded DNA per milliliter and 40 μg/mL for single-stranded DNA or RNA. Therefore, the concentration for double-stranded DNA (μ/mL) is $OD_{260} \times 50 \times$ dilution factor. However, one should remember that significant amounts of RNA in the sample can increase the OD spectrophotometric readings, producing inflated concentrations.

The purity of the DNA sample is usually determined using the ratio of the absorbence at OD_{260}/ OD_{280}. While several factors can affect this ratio, DNA without significant contamination usually has a ratio of approximately 1.8. Ratios significantly less than 1.7 suggest contamination by protein or other contaminants like phenol. Manchester (1995) suggests using a commercially pure DNA sample as a standard for DNA measurements. This excellent suggestion will allow the correction of inaccuracies of the individual spectrophotometer, which can be significant enough to lead to errors in concentration estimates and purity ratios.

Tissue Culture

Historically the primary reason for immortalizing lymphocytes (lymphoblasts) was to provide a long-term source of DNA even if a subject was unavailable for additional samplings. When the primary method used for genetic analysis was RFLP, this step was very useful, since up to 10 μg of DNA was used for one restriction digest. However, since a single genotype obtained by PCR requires only 30 ng or less of DNA in our laboratory (Ben Othmane et al., 1992), a standard draw of 10–30 mL of blood provides a sufficient amount of DNA for most analyses. Therefore, to transform all samples is usually a significant waste of effort and resources. Finally, even in the most experienced laboratories, transformation will occasionally fail, especially if the blood samples are more than a few days old. In this case the entire sample committed to the transformation is lost to the study. Nevertheless, the prudent investigator may elect to maintain lymphocytes for the purpose of ensuring a supply of DNA for certain key individuals in a study. Such individuals would include one or two affecteds for later mutational analysis in the study, elderly family members, individuals with critical recombination events, or individuals defining a haplotype in a family.

Another use of lymphoblasts or lymphocytes is to provide large molecular weight fragments (≥ 2 megabases) of DNA for use in PFGE. Standard extraction shears DNA into pieces that are too small (< 50 kb) for this analysis. This damage can be avoided by embedding whole lymphoblast cells (10^6–10^8 cells) in agarose prior to purification (Smith et al., 1988). The protein is then extracted, leaving the DNA intact and preventing any shearing.

Finally, lymphocytes or lymphoblasts may be an easily obtainable source of RNA for studies of mutations in a gene. While point mutations can be identified in ordinary extracted DNA, the process requires knowledge of exon–intron boundaries, whose delineation can be very time–consuming. However, if the gene under study is expressed in lymphocytes, PCR can be used to directly amplify the mRNA to study the gene product (reverse transcriptase PCR) (Foley et al., 1993).

One deterrent to the routine transformation of lymphocytes into lymphoblasts is the high cost of tissue culture. However, one alternative does exist that can significantly reduce the cost of transformation, but still provide it as an option, namely, freezing lymphocytes in dimethyl sulfoxide (DMSO) in nitrogen prior to transformation (Louie and King, 1991; Pressman and Rotter, 1991). Intact cells can be kept in long-term storage until the investigator decides whether that sample will require transformation. If the sample is not needed for transformation, it can then be extracted for DNA. However, one disadvantage of this approach is that freezing before transformation will kill some cells and therefore provide less for the transformation. Thus, the failure rate for delayed transformation is higher than for direct processing. This could lead to the loss of the entire sample from individuals when they no longer are available for repeat sampling. Thus, if transformation of a specific individual is known to be needed for an analysis, it may be better to do it initially, for if failure occurs, the individual can be quickly sampled again. The techniques for tissue culture and lymphocyte transformation, which are beyond the space limitations of this chapter, can be found in many manuals (e.g., Doyle, 1990).

Buccal Brushes

Buccal brushes can be stored after collection at room temperature for weeks or at −80°C for an extended time, prior to cell lysis. However, once the brushes have been processed, they are potentially unstable owing to protease and nuclease contamination, and should be kept frozen for long-term storage or at least maintained at 4°C when not frozen.

The primary question always asked concerning buccal samples is "How much DNA will I get?" This is, unfortunately, difficult to answer. We routinely use 1–2 μL of a 660 μL total sample in a 10 μL PCR reaction. This volume limits the use of buccals for genomic screening to secondary samples. However, we have performed experiments to test the potential usefulness of buccal samples as primary samples as well. All samples for the experiment were collected by experienced physician assistants on patients aged 20–30. While some samples could be diluted 50-fold and still provide an excellent PCR product, samples from other individuals had to be used at full strength for successful results. Therefore, the variability of DNA concentration, even when collected under controlled conditions, as well as the potential for degradation, lead us at present to recommend the use of buccal samples only for secondary samples when large genomic studies are performed. However, they remain excellent samples for isolated patients, patients in whom venipuncture is difficult, candidate gene analysis, genotype confirmation, smaller studies, and as an additional source of DNA.

Dried Blood Cards

Many techniques are now available for DNA extraction from dried blood on Guthrie cards. Clearly, there is enough DNA on the cards to provide many PCR reactions (Del Rio et al., 1996). The problem is processing the sample in a way that allows Taq polymerase to efficiently access the DNA. This increases the manipulations required for the analyses and limits the practical number of PCR reactions that can be performed using a card. Many techniques have been published employing boiling, heat cycles, sonication, and autoclaving to denature the sample prior to the PCR reaction (Carducci et al., 1992; Raskin et al., 1993; Fishbein et al., 1995). In addition, re-PCR of an initial reaction can be used to bolster the amount of PCR product, although aberrant bands are also increased. We have used the method of Del Rio et al. (1996) with excellent results.

Fixed Tissue

There are many published protocols for the extraction of DNA from fixed tissue. The different approaches reflect the difficulty of isolating DNA from these tissues compared to other tissue sources. In our lab we use a modified technique in which the paraffin is removed with xylene, the residual tissue washed with ethanol, and a one-square-centimeter piece treated with proteinase K. After digestion, the resulting lysate is used without further purification in a PCR reaction (De Souza et al., 1995).

SAMPLE MANAGEMENT

Management of samples can be critical to the success of any project. In our laboratory samples are initially signed in using a genotype form. This allows data to be entered for the sample, as well as collected for quality control issues (volume, presence of hemolysis, etc.). At this time, the sample is assigned a sample number, which is used for labeling all subsequent analyses, rather than the family information. The samples for extraction can be held at 4°C before freezing at –80°C. If desired, an aliquot of plasma can be taken prior to freezing. This freezing of whole blood provides a "holding buffer" that allows the sample to be stored until extraction can occur. We have successfully extracted whole-blood samples that were frozen for over one year. A flowchart of management of samples is shown in Figure 8.1.

Several factors affect the success of DNA extraction. The volume of the sample and its effect on the chosen extraction method should be considered. After 3 days at room temperature, the amount of DNA that can be successfully extracted from an ACD sample begins to diminish, decreasing rapidly after 5 days (Gustafson et al., 1987). Therefore, it is important to place the sample at 4°C and extract the DNA as

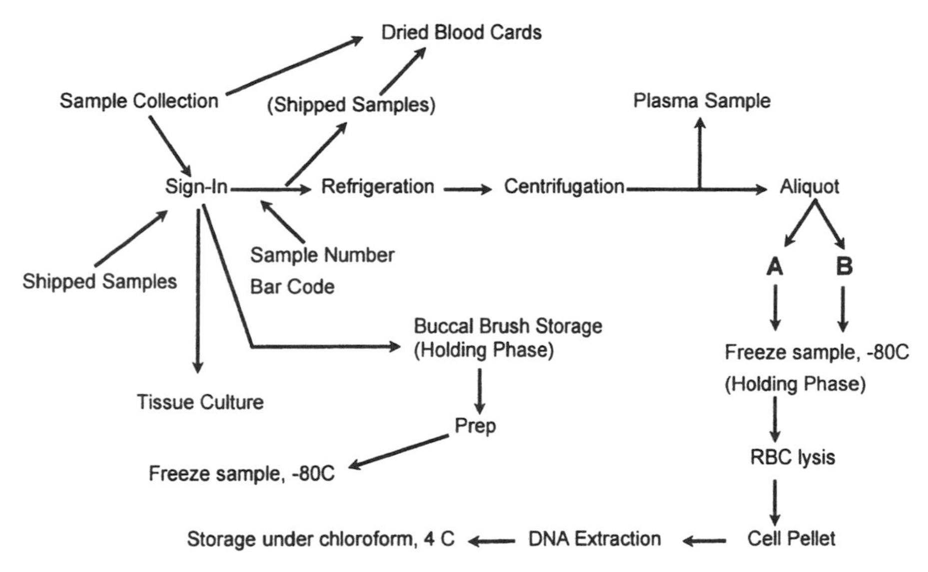

Figure 8.1 *Flowchart for processing of biological samples.*

soon as possible. When energy available for the living cells begins to be depleted, the cells will die and the DNA will become degraded. If the sample cannot be extracted quickly, it should be stored at –80°C. In addition, light will degrade the DNA, and the samples should be kept in the dark (Kirby, 1990). We have also found that hemolysis is correlated with decreasing DNA yield. It is not clear that the hemolysis itself is affecting the extraction. Rather, it seems more likely that the factors causing the red blood cell hemolysis have also affected the viability of the white blood cells. Clotting of tubes occurs occasionally, usually as a result of improper mixing. This can affect DNA yield as well, depending on the extraction method chosen.

Once the DNA has been extracted, we store the samples in a small vial in PCR TE (10 mM Tris, 0.2 mM EDTA, sterile filtered) at 4°C with an added drop (5 μL/mL) of chloroform (Kirby, 1990). We have stored DNA for over 10 years in this manner without any obvious degradation. The chloroform sterilizes the sample and then evaporates with time. If this technique is used, it is important to avoid vials with any component (e.g., a rubber gasket) that will dissolve in the presence of organic solvents, contaminating the DNA. Less than 2% of our 20,000+ extracted samples have displayed significant evaporation over time, and almost all that did evaporate could be reconstituted by adding buffer.

It is debatable whether any advantage is incurred by freezing DNA at –80°C for long-term storage. Freezing and thawing do cause degradation, especially of larger molecular weight fragments. However, this diminution of quality probably has little effect on today's PCR-based technology, which utilizes smaller fragments. In addition, depending on the number of samples, cost and freezer space can quickly become a problem in storage.

INFORMED CONSENT/SECURITY

DNA samples are not like other types of stored tissue, and their potential for abuse is a concern. A sample of DNA is intensely personal and should be treated as a confidential source of information, similar to a medical record. The investigator must assess the consent forms used for DNA storage and sampling and ensure that participants understand the purpose and limitations of that DNA bank. It is the investigator's responsibility to the participants in each study that appropriate security be maintained, both for the physical samples and for the data surrounding them. Detailed discussions of the issues in DNA banking have been the topic of several recent reviews (Yates et al., 1989; Clayton et al., 1995 see also Chapter 6).

REFERENCES

Ben Othmane K, Ben Hamida M, Pericak-Vance MA, Ben Hamida C, Blel S, Carter SC, Bowcock AM, Petruhkin K, Gilliam TC, Roses AD, Hentati F, Vance JM (1992): Linkage of Tunisian autosomal recessive Duchenne-like muscular dystrophy to the pericentromeric region of chromosome 13q. *Nat Genet* 2:315–317.

Carducci C, Ellul L, Antonozzi I, Pontecorvi A (1992) DNA elution and amplification by polymerase chain reaction from dried blood spots. *BioTechniques* 13:735–737.

Chomczynski P (1993): A reagent for the single-step simultaneous isolation of RNA, DNA and proteins from cell and tissue samples. *BioTechniques* 15:532–536.

Clayton EW, Steinberg KK, Khoury MJ, Thomson E, Andrews L, Kahn MJ, Kopelman LM, Weiss JO (1995): Informed consent for genetic research on stored tissue samples. *JAMA* 274:1786–1792.

De Souza AT, Hankins GR, Washington MK, Fine RL, Orton TC, Jirtle RL (1995): Frequent loss of heterozygosity on 6q at the mannose 6-phosphate/insulin-like growth factor II receptor locus in human hepatocellular tumors. *Oncogene* 10:1725–1729.

Del Rio SA, Marino MA, Belgrader P (1996): Reusing the same bloodstained punch for sequential DNA amplifications and typing. *BioTechniques* 20:970–974.

Doyle A (1990): Establishment of lymphoblastoid cell lines. In: Pollard JW, Walker JM, eds. *Animal Cell Culture, Methods in Molecular Biology.* Totowa, NJ: Humana Press.

Fishbein WN, Foellmer JW, Davis JI, Kirsch IR (1995): Detection of very-rare-copy DNA in 0.2 mL dried human blood blots. *Biochem Mol Med* 56:152–157.

Foley KP, Leonard MW, Engel JD (1993): Quantitation of RNA using the polymerase chain reaction. *Trends Genet* 9:380–385.

Greer CE, Lund JK, Manos MM (1991): PCR amplification from paraffin-embedded tissues: Recommendations on fixatives for long-term storage and prospective studies. *PCR Methods Appl* 1:46–50.

Gustafson S, Proper J, Bowie EJW, Sommer SS (1987): Parameters affecting the yield of DNA from human blood. *Anal Chem* 165:294–299.

Guthrie R, Susi A (1963): A simple phenylalanine method for detecting phenylketonuria in large populations of newborn infants. *Pediatrics* 32:338–343.

Johns MB, Paulus-Thomas JE (1989): Purification of human genomic DNA from whole blood using sodium perchlorate in place of phenol. *Anal Biochem* 180:276–278.

Kawasaki ES (1990): Sample preparation from blood, cells, and other fluids. In: Innis MA, Gelfand DH, Sninsky JJ, White TJ, eds. *PCR Protocols: A Guide to Methods and Applications*. San Diego, CA: Academic Press, pp. 146–152.

Kirby LT (1990): *DNA Fingerprinting: An Introduction*. New York: Stockton Press, pp. 51–74.

Kosel S, Graeber MB (1994): Use of neuropathological tissue for molecular genetic studies: Parameters affecting DNA extraction and polymerase chain reaction. *Acta Neuropathol (Berlin)* 88:19–25.

Louie LG, King M-C (1991): A novel approach to establishing permanent lymphoblastoid cell lines: Epstein–Barr virus transformation of cryopreserved lymphocytes. *Am J Hum Genet* 48:637–638.

Manchester KL (1995): Value of A_{260}/A_{280} ratios for measurement of purity of nucleic acids. *BioTechniques* 19:208–209.

Marshall E (1996): Policy on DNA research troubles tissue bankers. *Science* 271:440.

Matsubara Y, Ikeda H, Endo H, Narisawa K (1992): Dried blood spot on filter paper as a source of mRNA. *Nucleic Acids Res* 20:1998.

McCabe ERB (1991): Utility of PCR for DNA analysis from dried blood spots on filter paper blotters. *PCR Methods Appl* 1:99–106.

Miller SA, Dykes DD, Polesky HF (1988): A simple salting out procedure for extracting DNA from human nucleated cells. *Nucleic Acids Res* 16:1215.

Noguchi S, McNally EM, Ben Othmane K, Hagiwara Y, Mizuno Y, Yoshida M, Yamamoto H, Carsten G, Bönnemann CG, Gussoni E, Denton PH, Kyriakides T, Middleton L, Hentati F, Ben Hamida M, Nonaka I, Vance JM, Kunkel LM, Ozawa E (1995): Mutations in the dystrophin-associated protein (γ-sarcoglycan in chromosome 13 muscular dystrophy. *Science* 270:819–822.

Planelles D, Llopis F, Puig N, Montoro JA (1996): A new, fast and simple DNA extraction method for HLA and VNTR genotyping by PCR amplification. *J Clin Lab Anal* 10:125–128.

Pressman S, Rotter JI (1991): Epstein–Barr virus transformation of cryopreserved lymphocytes: Prolonged experience with technique. *Am J Hum Genet* 49:467.

Raskin S, Phillips JA 3rd, Krishnamani MR, Vnencak-Jones C, Parker RA, Rozov T, Cardieri JM, Marostica P, Abreu F, Giugliani R (1993): DNA analysis of cystic fibrosis in Brazil by direct PCR amplification from Guthrie cards. *Am J Med Genet* 46:665–669.

Richards B, Skoletsky J, Shuber AP, Balfour R, Stern RC, Dorkin HL, Parad RB, Witt D, Klinger KW (1994): Multiplex PCR amplification from the CFTR gene using DNA prepared from buccal brushes/swabs. *Hum Mol Genet* 2:159–163.

Smith CL, Klco SR, Cantor CR (1988): Pulsed-field gel electrophoresis and the technology of large DNA molecules. In: Davies K, ed. *Genome Analysis: A Practical Approach*. IRL Press Oxford, England

Smith LJR, Braylan RC, Nutkis JE, Edmundson KB, Downing JR, Wakeland EK (1987): Extraction of cellular DNA from human cells and tissues fixed in ethanol. *Anal Biochem* 160:135–138.

Yates JRW, Malcolm S, Read AP (1989): Guidelines for DNA banking. Report of the Clinical Genetics Society working party on DNA banking. *J Med Genet* 26:245–250.

9

Methods of Genotyping

Jeffery M. Vance and Kamel Ben Othmane

Division of Neurology
Department of Medicine
Duke University Medical Center and Health System
Durham, North Carolina

One of the critical components of any gene mapping study is obtaining the genotypes for the genetic markers used to test for linkage. This chapter provides the reader with practical knowledge of different methods and techniques used in genotyping, including options, advantages, and disadvantages.

A BRIEF REVIEW OF MARKERS USED FOR GENOTYPING

Several different genotyping markers have been utilized in recent years in performing linkage analyses. These are listed in Table 9.1 and described in the subsections that follow.

Restriction Fragment Length Polymorphisms (RFLPs)

Introduced in 1978 (Kan and Dozy, 1978; Botstein et al., 1980), RFLPs in 1982 became the first modern genotyping markers to be used in a successful linkage (Huntington disease) (Gusella et al., 1983). They are based on a single base pair change that creates or obliterates a cleavage site for a specific restriction enzyme. The resulting variation between individuals can be detected by digestion of the DNA by that restriction enzyme. The DNA fragments produced by the digestion are subse-

Approaches to Gene Mapping in Complex Human Diseases, Edited by Jonathan L. Haines and Margaret A. Pericak-Vance. ISBN 0-471-17195-6

TABLE 9.1 Comparison of Genotyping Markers

Markers	Sample Amount	Ease of Comparison Between Families	Speed of Detection	Labor Intensity	PCR?	Typical Heterozygosity
RFLP	μg	Easy	Several days to a week	++++	No	< 0.4
VNTR	μg	Difficult	Several days to a week	++++	Some	> 0.6
Microsatellites	ng	Easy	Hours	+	Yes	> 0.7

quently electrophoresed in an agarose gel and transferred to a membrane (Southern blot) (Southern, 1975). A labeled (usually ^{32}P) DNA probe that overlaps the restriction fragments of interest is hybridized to the membrane. The resulting variability in fragment size (alleles) is then detected using X-ray film (Vance et al., 1989). Major drawbacks of this technique include the large amount of DNA that is needed (2–10 μg, requiring the establishment of cell lines for the studied individuals), the low heterozygosity (Botstein et al., 1980) of the markers (usually < 0.4), and the large amount of laborious effort required and therefore relatively sluggish throughput obtained (often requiring 1–2 weeks before results are obtained).

Variable Number of Tandem Repeat (VNTR) Markers

The low heterozygosity characteristic of genotyping markers was greatly increased by the identification of VNTRs in 1985 (Nakamura et al., 1987). This new class of probes, also known as *minisatellites*, is made of specific tandem sets of consensus sequence that vary between 14 and 100 base pairs in length. Although some can be converted for use with the polymerase chain reaction (PCR) (Mullis et al., 1986), most require probe hybridization like RFLPs. They are remarkably polymorphic, with a high heterozygosity rate in the population, and are commonly used for paternity testing. However, this high heterozygosity rate, coupled with their relatively large size, often makes comparisons difficult when researchers are attempting to identify allele sizes between individuals.

Short Tandem Repeats (STRs) or Microsatellites

More recently, the use of PCR for the analysis of microsatellite polymorphisms has dramatically improved the rapidity and efficiency of genotyping. Initially described by Weber and May (1989) and Litt and Luty (1989), STRs are widely and evenly distributed in the genome, and are relatively easy to score. They can produce genotypic data not in weeks, but in hours. To detect the variability that exists between individuals at these STRs, two unique sequences are determined on each side of the STR. These short sequences (approximately 20 bp) are then used to synthesize

DNA primers, which are subsequently employed in a PCR reaction to amplify the DNA that lies between them (including the repeat). The variable number of repeats in the STR produces PCR fragments of different sizes, which can be easily detected by means of denaturing sequencing gels. STRs require only very small (nanogram) amounts of DNA and can easily provide heterozygosities greater than 0.70. While the number of repeated motifs in microsatellites vary, the most useful microsatellites consist of a repeated sequence motif of two (dinucleotide), three (trinucleotide) or four (tetranucleotide) bases.

Finally, it was mentioned earlier that the basis of RFLPs is usually a single base pair change. While these single base pair changes are numerous, as mentioned above they are often not particularly informative. Recently, however, silicon chips or "DNA chips" have been used in mutation analysis (Hacia et al., 1996). This technology has the potential to overcome the heterozygosity problem by allowing the very rapid analysis of very large numbers of these single base pair changes or SNPs (single nucleotide polymorphism). While still in development, the DNA chips, if successful, could provide yet another major step in the evolution of genotyping.

MICROSATELLITES

Source of Markers

Information on several thousand microsatellite markers is now electronically accessible to investigators through databases such as the Genome Database, the Whitehead Institute, the Marshfield Institute, the Cooperative Human Linkage Center, the Utah Marker Development Group, Entrez (at NCBI, the National Center for Biotechnology Information), Research Genetics, and Généthon (see Chapter 16 and volume Appendix). Much effort has been expended to use these highly polymorphic microsatellites to provide the investigator with valuable maps for linkage analysis. However, it is important to remember that all genetic and physical maps are not constructed alike, or with equal confidence. This is discussed in more detail in Chapter 16.

Dinucleotide Repeats

The dinucleotide CA (GT) is the most common repeat, with a highly polymorphic form of the repeat occurring on average one every 0.4 cM (Weber, 1990). To date, dinucleotide repeats have been the workhorse of microsatellite genotyping. Their polymorphic status usually depends on two factors: the size of the repeat and whether a perfect (no interruptions) or imperfect repeat is present. The majority of markers with 15 or more perfect repeats are polymorphic (Weber, 1990). While powerful for linkage analysis, dinucleotide repeats can have several technical drawbacks. Since the alleles are only two base pairs apart, it can be difficult to distinguish one allele from another. More importantly, many dinucleotide repeats have a "stutter" that produces a background ladder of bands two base pairs apart. This feature can make scoring difficult, particularly in differentiating homozygotes from heterozygotes.

Trinucleotide Repeats

Although certain trinucleotide repeats that cause disease (LaSpada et al., 1994) have received much attention, the majority of these markers are quite stable, have heterozygosities similar to dinucleotide repeats, and have been incorporated into recent maps. They also generally have fewer stutter bands than dinucleotide repeats. Fewer trinucleotide than dinucleotide repeats, have been described, however.

Tetranucleotide Repeats

Although less numerous than dinucleotide repeats, tetranucleotide repeats are quite polymorphic (Edwards et al., 1991). These larger repeat motifs typically generate unique PCR bands, without a laddering artifact, greatly facilitating the identification of alleles. The clarity of the results makes the application of computer algorithms to determine molecular weights much more feasible. Accurate allele identification is particularly important in the study of complex disorders, where the lack of significant family structure decreases the likelihood of detecting genotyping errors by inconsistencies of segregation.

PCR AND GENOTYPING

The optimization of the PCR reaction is one of the most critical factors in efficient genotyping. The reduction of background bands and maximum production of product provides clarity in identifying alleles. Therefore, the investigator needs to optimize laboratory conditions for genotyping so that they will provide a flexible environment for as many different PCR primers as possible.

Laboratory and Methodology Optimization

In PCR the amplification of the desired target continually competes with nonspecific priming. In addition, the primer dimerization reaction, in which one of the primers is substituted for the target DNA and extended, can be a problem. To help eliminate nonspecific binding problems, the PCR cycle times should be reduced as far as possible without significantly reducing the amount of product. In addition, extended incubations at high temperatures should be avoided, since the half-life ($t_{½}$) of Taq polymerase rapidly decreases above 93°C. Therefore, shortening the denaturing time increases the amount of available Taq for subsequent cycles and reduction of the annealing and extension times will increase specificity and decrease background bands. The actual times used are dependent on type of thermocycler, tubes, and PCR conditions. Some laboratories have found two-step cycle times to be particularly useful in reducing cycle times. In this approach, after a significant amount of template has been amplified, longer initial cycle times are followed by shorter cycles. Depending on the detection method used and on the quality and specificity of the primers, some reactions yield properly visualized products after only 25 cycles. Other markers require additional cycles for satisfactory results. Beyond 35 cycles, the reaction is near the plateau phase, and additional cycles tend to increase the background by preferentially amplifying spurious products. In addition, thin-walled tubes often allow faster cycle times and thus may produce a larger quantity and more specific product than the thick, larger tubes of the past.

Standard Taq polymerase is minimally active at room temperature. It is therefore recommended to start the thermal cycling immediately after the addition of reagents to the reaction, to minimize these spurious products. Alternatively, variations of the hot start method (Chou et al., 1992) can be employed, in which at least one of the essential reagents is withheld from the reaction mixture until the temperature of the reaction exceeds the annealing temperature of the primers. The missing reagent can be added individually to each reaction, in the thermocycler block, once the desired temperature has been reached. However, this approach is not practical in dealing with a large number of samples and is prone to cross-contamination. A possible solution to this problem consists in using a solid wax barrier to separate the retained reagent from the bulk of the reaction until the first cycling step melts the wax, or in the use of wax-embedded PCR reagents (Bassam and Caetano-Anollés, 1993). Recently, a form of Taq (Taq gold, Perkins-Elmer), which does not have significant activity at room temperature, has become available as has a form of Taq coated with an antibody (Platinum® Taq, Life Technologies). Use of these products reduces nonspecific activity and thus increases specificity without any additional reagents.

Perhaps the greatest concern in optimizing a laboratory is controlling PCR contamination, which can significantly reduce the consistency and reliability of genotyping reactions. While false positives can be contained by the use of UV irradiation, uracil DNA glycosylase, and physical barriers, false negatives and inhibition of the PCR reaction by previously amplified material are more difficult problems. Neiderhauser et al. (1994) demonstrated that intact and partially degraded amplified products (amplicons) and primer artifacts from previous reactions are able to inhibit subsequent reactions, probably by primer competition and DNA blocking. In fact, microsatellite reactions may generate more false negative problems than other types of PCR contamination. Although the markers used in genotyping may be different, the repeat motifs are usually highly similar, allowing the core repeat to block the template and inhibit the reaction.

Several basic precautions may help minimize the contamination problem. These include pipeting reagents into small aliquots for single use, frequently changing gloves, and avoiding splashes (Kwok and Higuchi, 1989). However, the single most important guideline in preventing contamination is the complete physical separation between a first location (where the reagents are stored and set up) and a second location (where the thermocycling is done and the post-PCR product is handled and stored). Hence, either separate rooms should be employed for pre- and post-thermocycler reactions or reagent storage and setup performed under a biosafety hood equipped with a UV germicidal lamp. If UV decontamination is used, it is important not to expose the mineral oil to the UV irradiation, since this will cause the PCR reaction to be inhibited by induced radicals interfering with Taq polymerase (Gilgen et al., 1995).

Optimization of Reagents

In general, one strives to obtain the most economical and flexible set of conditions in which most primers will perform adequately. We perform genotyping PCR reactions on 30 ng of DNA in a final volume of 10 μL (Ben Othmane et al., 1992). A standard PCR buffer is composed of 20 mM Tris-HCl, and 50 mM KCl. A combined deoxynucleotide triphosphates (dNTP) final concentration of 0.8 mM and a Mg^{2+} concentration of 1.5 mM are usually optimal for a wide range of primers. A surplus of Mg^{2+} reduces the PCR specificity and increases the laddering artifacts, while a shortage of magnesium reduces the reaction efficiency. Since there is a mutual titration between Mg^{2+} and dNTP, it is important to always remember that a significant change in the Mg^{2+} concentration usually requires a similar modification in the amount of dNTP. A final concentration of 0.5 μM of each primer is optimal for most markers. An excess of primers will result in the generation of artifacts.

If the amount of PCR product is a problem, an increase may be obtained by increasing the dNTP concentration and adjusting the Mg^{2+} concentration accordingly. Optimization of the PCR conditions for that primer pair should also increase specificity of the reaction and therefore the amount of desired product as well. Additionally, Taq extender PCR additive (Stratagene) may be added to the general reaction. We have found that this practice significantly enhances the yield, and it has been reported to increase the specificity of PCR products. Tetramethylammonium is another reagent that can be used to increase the amount of specific PCR product (Chevet et al., 1995). Since, however, its effect is to increase the thermal stability of AT base pairs, its utility varies from primer to primer and it should be tested using a battery of concentrations, similar to dimethyl sulfoxide (DMSO) (see Other Tools, below).

"I Can't Read a Marker, What Should I Do?"

Despite the best general conditions, some primers provide poor results. What to do to improve them and how hard to try are common questions. Optimizing the PCR conditions for a microsatellite can be a time-consuming task and therefore should be attempted only when genotypes from the marker in question are necessary and cannot be replaced. Suggestions are offered in the subsections that follow.

Get Another Marker. If one is just screening, obtaining a better marker in the same chromosomal region is usually the better part of valor. However, if it is necessary to use *that* marker, several options are open.

Optimize Temperature, Mg^{2+}, and pH. Temperature and pH generally are the most useful factors to initially optimize for an individual set of primers, although other factors can be occasionally important as well (Innis and Gelfand, 1990). In our experience, Mg^{2+} concentration, although potentially important, is less useful in optimizing a specific primer set. To optimize temperature, the touchdown or step-down PCR techniques are particularly useful (Hecker and Roux, 1996). Optimizing pH and Mg^{2+} is probably best done by means of one of several kits on the market. A typical optimization scheme may include a sequential testing of different annealing temperatures, usually between 50 and 65°C, Mg^{2+} concentrations between 1 and 3.5 mM, and pH conditions between 8.5 and 10. Optimally, one would wish to test these factors simultaneously. In practice, however, one condition usually stands out as the key factor affecting a primer set's performance.

Redesign the Primer. It is widely recognized that a crucial factor in determining the success of a PCR reaction is the design of primer sequences. Poorly designed primers lead to nonspecific PCR with background, ghost, and stuttering bands. In our laboratory redesigning a primer is probably one of the most useful optimization steps, and as time goes along, we tend to redesign a primer sooner rather than later, especially as the cost of oligomer synthesis continues to decrease.

Usually the original clone sequence encompassing the repeat is deposited in GenBank (see NCBI entry in volume appendix) and therefore can be easily accessed. Several commercial programs are available to aid in primer selection. However, efficient primers can often be redesigned if simple criteria are followed. The primers should be as close to the repeat as possible: the smaller the product, the easier it is to read in standard denaturing gels. An optimal primer length is about 20 bp, and while longer primers may improve the PCR stability, they are rarely necessary. The G/C content should be kept near 50%, and the base distribution should be random to minimize secondary structure and self-complementation. The last 3–7 base pairs in the 3′ end of the primer are crucial to the success of the annealing and extension steps of the PCR. Complementarity of the primers to each other, particularly in the 3′ end, needs to be avoided, since it may induce primer dimerization (Saiki, 1990).

Other Tools. Dimethyl sulfoxide (DMSO) has been used to enhance PCR for some primer pairs, increasing the amount of product and primer specificity. The actual mechanism of this improvement is in debate (Filichkin and Gelvin, 1992; Baskaran et al., 1996). The primer set can be run against a battery of 1, 2.5, 5.0, 7.5, and 10% DMSO to see whether an improvement of product specificity can be obtained. DMSO reduces the activity of Taq, so concentrations over 10% are not generally useful.

MARKER SEPARATION

While some authors have used nondenaturing conditions to visualize polymorphisms (Buzas and Varga, 1995), on the whole most researchers select a denaturing (sequencing) setup for its greater sensitivity (Ben Othmane et al., 1992). Various points to be considered in marker separation are presented next.

Apparatus

The selection of the type of apparatus is important because uniform heat distribution is as critical to genotyping as it is to sequencing. Units also can differ in terms of width, which affects both the number of samples and detection schemes. We have found units from C.B.S. Corp., San Diego, CA to work well.

Acrylamide

It is very important to have a high-quality matrix for clean, sharp bands. We use freshly made acrylamide solutions because, in our experience, premade solutions do not provide the desired consistency. Acrylamide substitutes and variations are also available but are usually more costly. If wrapped to prevent evaporation, gels can be made up several days ahead of time and stored at 4°C.

One of the most common problems in pouring gels is keeping the plates clean. The routine use of cerium oxide (Millard and de Couet, 1995) is highly recommended.

Gel Formation

A key factor in this regard is maintaining a standard thickness of the gel during polymerization. Our laboratory uses clamps (Hoefer, SE 6003) on three sides instead of tape, since a millimeter of difference in gel width is enough to prevent straight lanes.

Fixing

The fixing often used in sequencing is not necessary for genotyping gels.

Loading

We prefer to use slot gel combs, utilizing 64, 96, or 108 well slots on 40 cM gels (Marshfield Machine Corp., Marshfield, WI or C.B.S. Corp., San Diego, CA). We find shark-tooth combs too difficult to read when large volumes of data are at hand. Smaller wells (144 slots) can also be utilized, but these can be confusing to read for any markers that have aberrant bands. The choice of comb size is also dependent on the volume of sample that must be loaded for detection, and obviously the number of samples to be screened.

There are many variations in loading PCR samples on a gel. While loading can be accomplished with a single pipet, it is much faster with a multichannel pipet such as the one available from the Hamiliton Company. A 12-syringe apparatus can load every third well of a 108-well comb, allowing the loading of a 96-well plate, greatly speeding the process.

Finally, light is a key component in accurately loading small wells. We have found a simple headband light, similar to the ones used in spelunking, both useful and relatively inexpensive for loading.

The subsections that follow describe some of the variations to standard loading of samples.

Multiloading. This is the simple act of loading different aliquots of the same marker on the same gel in multiple loads, separated by 15–30 minutes, providing many layers of the same marker for analysis on one gel. Multiloading works well for markers with few aberrant bands and can greatly increase the number of genotypes generated by a technician.

Multiplexing. This approach has itself several variants. In one, many different primer sets (2–50) are amplified individually by PCR, then mixed and run in a single lane of a gel. These overlapping fragments are then Southern-blotted to a nylon membrane. This is subsequently hybridized with a single labeled primer of each set (Litt et al., 1991), which provides the specificity for the detection.

The second technique is similar to the first, except the primers are fluorescently labeled. One to five primers are then placed in a single lane, and detection depends on the different emissions of the fluorescent moiety (Ziegle et al., 1992). This approach requires a laser for detection.

The third technique is employed regularly in our laboratory. Instead of separate PCR reactions and subsequent pooling of the PCR products, a single PCR amplification is performed with three to eight primer sets amplified together in one tube (Fig. 9.1). Primers are chosen prior to amplification based on the molecular weight of their products and their PCR compatibility. They are then loaded into a single lane on the gel and separated according to molecular weight. This approach can be combined with fluorescence to provide multiple multiplex sets that can be run in the same lane, a combination of the second and third techniques given here.

This single tube multiplex technique can use the identical conditions employed for single primer PCR, especially when radioactivity is used for detection. The high sensitivity of this isotope allows satisfactory results for less than optimum PCR conditions for many of the primers. However, other detection schemes (e.g., silver stain or fluorescence) may be less forgiving. In these situations, multiplexes may work more optimally when two to three times the dNTP concentration used for single primer pair reactions and a 2–3 mM concentration of Mg^{2+} are employed.

In setting up a single tube multiplex set, the primers chosen should be run on a small number of test samples as a single reaction, as well as in combinations. This allows one to identify interferences due to spurious bands that may fall within the same molecular weight as the alleles from other primers in the set. It also demonstrates which primer sets produce weak or exceptionally strong products when multiplexed. Typically one marker in the multiplex set will have a significantly higher PCR proficiency, relative to the rest of the markers. This will cause it to dominate the reaction early in the amplification. Thus, the amount of this primer set added to the multiplex mixture may need to be adjusted accordingly.

The advantage of single tube multiplexing is the increase in genotyping output and reduction cost that can be accomplished without high-priced equipment, as three to eight markers can be analyzed at once. It does, however, require advance work and therefore is really most appropriate for large laboratories with extensive

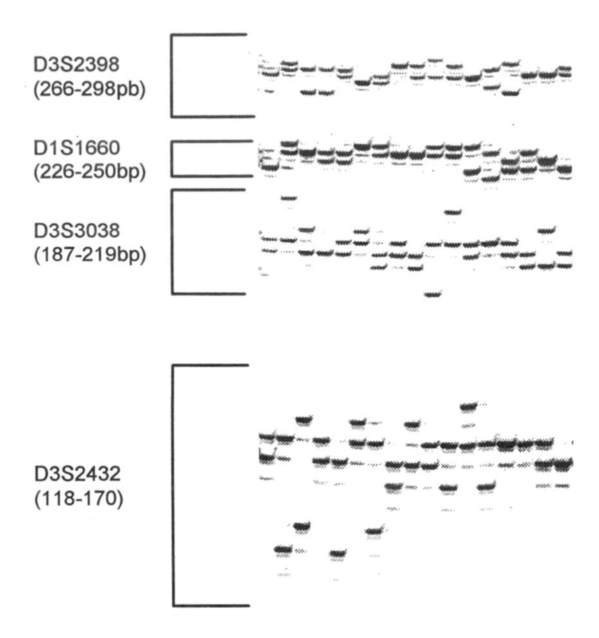

Figure 9.1. *Genotyping multiplex: this portion of a multiplex genotyping gel demonstrates the use of running multiple markers in a single PCR reaction and gel. Primers were grouped according to product size, with one primer prelabeled with Hex, seperated after PCR on a Baserunner for 1.5 hours and visualized using a Molecular Dynamics FluorImager SI. Each column represents a single PCR of genomic DNA.*

genotyping. Recently, Jim Weber has made available his version 9 genomic multiplex screening sets (see Marshfield Web site, Volume appendix), which use this single tube multiplex technique.

DNA Pooling

DNA samples from all affected may be pooled together and processed as a unique template in a single PCR reaction, a practice that greatly reduces the number of working samples and significantly accelerates the genotyping process (Sheffield et al., 1994). In a similar manner, an equal number of unaffected members are pooled and used as a control. A shift in allele frequencies toward a single homozygous allele in the affected DNA pool, relative to controls, may indicate linkage (the intensity of each allele band will be directly proportional to its relative frequency in the pooled population). This efficient strategy is particularly recommended as an initial screening tool for autosomal recessive conditions. It has been estimated to reduce by as much as 75% the number of markers that will require subsequent pedigree linkage analysis (personal communication, J. Haines).

DETECTION METHODS

The methods to detect microsatellites are numerous. Factors to consider in choosing a method include the number of samples that will be analyzed, cost, and whether radioactivity is an option.

Radioactive Methods (^{32}P or ^{33}P)

Two primary methods exist. In one, radioactively labeled dNTP is *incorporated* into the PCR reaction (Ben Othmane et al., 1992); in the second, *hybridization* with a labeled primer occurs after the cold microsatellite product has been separated by electrophoresis and Southern-blotted (Southern, 1975). To improve the sharpness of the bands, Zagursky et al. (1992) used ^{33}P instead of ^{32}P; the reduced emission energy that results has safety merits as well. However, the high cost of the larger isotope reduces its practicality for most investigators, particularly those in the USA.

The use of radioactivity provides the highest sensitivity of any of the detection methods. However, the regular availability of ^{32}P can be a problem in some locales. The safety requirements and waste management protocols that necessarily accompany the use of this method can be additional drawbacks. The incorporation method is fast, requires no special equipment, and when coupled with a multiplex method, has a high output capability. It also requires minimal manipulation. It has been one of the most common and popular methods. The hybridization method is less straightforward, since its sensitivity requires the additional step of efficient hybridization of the labeled primers. But many primers can be incorporated into a single lane, meaning that fewer gels are required; and since only the primer is labeled, this method in general has less background than the incorporation method.

Silver Stain

Silver staining is a method of detection well known to protein chemists. The application of silver staining of nucleic acids in polyacrylamide gels was introduced in 1981 (Merril et al., 1981; Bassam et al., 1991; Von Deimling et al., 1993). It is now available commercially for DNA, and the process is relatively straightforward. After marker separation, the gel is subjected to several staining steps, with the eventual DNA–silver product developed chemically.

Silver staining is a sensitive method that is a low-cost alternative to radioactivity. It is relatively fast (gels can be developed in approximately an hour). However, its reactions are time dependent and it requires more hands-on technician time than radioactivity. Gel plates may have to be committed to the silver stain method, since a gel adhesive is applied to the plates to stabilize the gel through the washing steps. In addition, the gels are "developed" at a specific point and fixed. Therefore, light or dark bands can be a problem if wide variation appears in the gel. Gels can be stored, but permanent images are readily made using X-ray duplicating film or photography. Silver staining is an excellent alternative to radioactivity-based methods.

Fluorescence

Fluorescence is the fastest genotyping detection technique, and the most amenable to automation. Two basic formats are available: static and real-time scanning. Real-time scanning has two additional formats, plate or capillary electrophoresis. All call for intensive capital investment because lasers are required.

Static Scanning. The application of this technology to genotyping was developed recently in our laboratory (Vance et al., 1996). Multiple gels can be run at once, in different locations, and then scanned separately on a fluorescent laser scanner, with a scanning time of only a few minutes per gel. Analysis of the data can take place at one of many networked stations, concurrent with scanning. Primers can be prelabeled with a fluorescent moiety (Fig. 9.1), or the PCR product can be stained following electrophoresis using Sybrgold® (Molecular Probes) for 10–20 minutes, and then scanned. The image is downloaded to a networked computer system for image analysis and allele determination (BioImage).

Real-Time Scanning. In this approach the electrophoresis is done directly on the detector, which is a DNA sequencer, and the data are then analyzed by incorporated software. The equipment is usually provided as a single system. All primers must be prelabeled. Several commercial "packages" are available.

Advantages and Disadvantages. Fluorescence is the fastest method available. In addition, because it can be automated, it has the highest potential throughput. The main disadvantage is the large capital outlay required, which makes it prohibitively expensive for smaller laboratories.

There are distinct differences between the formats available. Real-time scanning has the the fastest output, but there are distinct disadvantages in some cases. Because electrophoresis must be performed on the machine, and only one project can be electrophoresed at once, the machine itself becomes a bottleneck for laboratories that run many projects. Thus one of the advantages of static scanning is that because detection is uncoupled from electrophoresis, thus an essentially unlimited number of projects can be detected by a single machine. Also, much genotyping involves "redos" and inconsistency correcting, which can be done easily on static equipment without the interruptions in flow of traffic that occur in real-time scanning. In addition, real-time scanners require prelabeled primers, so previously synthesized primers are not useful. This is not a limitation of static scanners, since a relatively inexpensive fluorescent dye such as SybrGreen® or Sybrgold® can be used following electrophoresis. Currently some real-time scanners (Applied Biosystems) allow more isofluors per lane than static scanners, hence can analyze more markers per lane. This greater capacity allows a standard to be placed in each lane as well. In our experience, however, in-lane standards are not needed for accurate molecular weight determination, especially with tetranucleotides.

Newer Scanning Technologies. Currently, capillary scanners are much slower than either static or real-time scanners but have the advantage of higher automation potential because it is not necessary to hand-load the gels. This format is being refined, with only single capillary units currently on the market (Applied Biosystems). However, significant progress in genotyping efficiency and speed in the near future should come from these systems when the recently announced units with multiple capillaries become available (Perkin-Elmer, Molecular Dynamics). "DNA chip" technology (Chee et al., 1996; Hacia et al., 1996) is still in development. Proponents predict that an entire genome could be screened in days using several chips containing single base pair polymorphisms. Whether this technology will reach so great a potential for genotyping is not yet known, but it would seem most useful for initial screening where the polymorphic oligo sets would be fixed.

DATA MANAGEMENT

Data management is one of the true keys to success in genotyping, especially for complex disorders where the number of genotypes required for the analyses will be in the hundreds of thousands. There are at least four key factors, which are discussed next.

Objectivity

The importance of objectivity in reading genotypes cannot be overemphasized. No matter how sophisticated a system, there is always data that is not as clear as one would like, and prior knowledge of how the genotype should be (or how one wishes it to be) has swayed many an "objective" decision. Objectivity is enhanced by identifying samples by sample number, not an individual's name or family ID number. This convention also is important for maintaining the confidentiality of results for family members. Genotypes should be read independently and without the use of pedigrees.

Genotype Integrity

Sample integrity in genotyping can be enhanced by using a gel key, which is a grouping of samples that always places them in the same order. This is very useful in repeated screening of the same family data, and eliminates individual technician placements that can lead to sample switching. We have employed bar code systems to enter genotyping numbers to reduce coding errors as well. Data management systems such as PEDIGENE (Haynes et al., 1988, 1995) are invaluable in maintaining both clinical and genotypic data.

Scoring

The keys to accurately reading genotypes are many, although "garbage in, garbage out" is probably applicable here in terms of the time spent in optimizing PCR products. Pattern recognition is the key, especially in dinucleotide primers. In reading gels, the best rule of thumb is "If in doubt, don't read it, redo it." This can create a conflict if the individual reading a gel is also responsible for running it.

Standards

Since population allele frequencies are needed for most analyses, alleles should be read by their molecular weight. Two types of standards are needed for genotyping. One standard is an "in-gel" standard that allows the molecular weight of the product to be determined, providing a genotype. The second standard is a "between-gel" standard, to assure that the same individual always produces the same size genotype for a specific marker. This precaution is necessary because there is always variability between gels in actual running conditions and between individuals in interpretation of results. In addition, PCR itself can often vary in a "run" by the addition of one base (Smith et al., 1996), actually changing the molecular weight for that group.

For the first standard we place a sequencing ladder such as that generated using M13, a commercially available ladder, or any individualized set of known PCR products, in at least three lanes distributed across the gel. If the molecular weights of alleles are determined manually, then a sequencing ladder is most useful. More automated allele determination, with the use of software algorithms, will allow ladders with greater spacing between bands to be utilized.

For the gel-to-gel standard, two to three individuals (CEPH 1331-01, 1331-02, or 1347-02) are amplified with each run and checked to ensure that the same results are obtained each time. Cell lines or DNA on these individuals are available from the Coriell Institute for Medical Research (volume appendix). This has become a general standard approach for most genotyping laboratories. Internal allele sizing methods may not be consistent from laboratory to laboratory with different standards and genotyping techniques, and this amplification not only allows the same family or group of data to be run on different gels within a single laboratory, but also allows data from several laboratories to be efficiently combined for analysis.

REFERENCES

Baskaran N, Kandpal RP, Bhargava AK, Glynn MW, Bale AE, Weissman SM (1996): Uniform amplification of a mixture of deoxyribonucleic acids with varying GC content. *Genome Res* 6:633–638.

Bassam BJ, Caetano-Anollés G (1993): Automated "hot start" PCR using mineral oil and paraffin wax. *BioTechniques* 14:30–34.

Bassam BJ, Caetano-Anollés G, Gresshoff PM (1991): Fast and sensitive silver staining of DNA in polyacrylamide gels. *Anal Biochem* 196:80–83.

Ben Othmane K, Ben Hamida M, Pericak-Vance MA, Ben Hamida C, Blel S, Carter SC, Bowcock AM, et al. (1992): Linkage of Tunisian autosomal recessive Duchenne-like muscular dystrophy to the pericentromeric region of chromosome 13q. *Nat Genet* 2:315–317.

Botstein D, White RL, Skolnick M, Davis RW (1980): Construction of a genetic linkage map in man using restriction fragment length polymorphisms. *Am J Hum Genet* 32:314–331.

Buzas Z, Varga L (1995): Rapid method for separation of microsatellite alleles by the Phast-system. *PCR Methods Appl* 4:380–381.

Chee M, Yang R, Hubbell E, Berno A, Huang XC, Stern D, Winkler J, Lockhart DJ, Morris MS, Fodor SP (1996): Accessing genetic information with high-density DNA arrays. *Science* 274:610–614.

Chevet E, Lemaitre G, Katinka MD (1995): Low concentrations of tetramethylammonium chloride increase yield and specificity of PCR. *Nucleic Acids Res* 23:3343–3344.

Chou Q, Russell M, Birch DE, Raymond J, Bloch W (1992): Prevention of pre-PCR mis-priming and primer dimerization improves low-copy-number amplifications. *Nucleic Acids Res* 20:1717–1723.

Edwards A, Civitello A, Hammond HA, Caskey CT (1991): DNA typing and genetic mapping with trimeric and tetrameric tandem repeats. *Am J Hum Genet* 49:746–756.

Filichkin SA, Gelvin SB (1992): Effect of dimethyl sulfoxide concentration on specificity of primer matching in PCR. *BioTechniques* 12:828–830.

Gilgen M, Hofelein C, Luthy J, Hubner P (1995): Hydroxyquinoline overcomes PCR inhibition by UV-damaged mineral oil. *Nucleic Acids Res* 23:4001–4002.

Gusella JF, Wexler NS, Conneally PM, Naylor SL, Anderson MA, Tanzi RE, Watkins PC, et al. (1983): A polymorphic DNA marker genetically linked to Huntington's disease. *Nature* 306:234–238.

Hacia JG, Brody LC, Chee MS, Fodor SPA, Collins FS (1996): Detection of heterozygous mutations in *BRCA1* using high density oligonucleotide arrays and two-colour fluorescence analysis. *Nat Genet* 14:441–447.

Haynes CS, Pericak-Vance MA, Hung W-Y, Deutsch DB, Roses AD (1988): PEDIGENE—A computerized data collection and analysis system for genetic laboratories. [Abstract] *Am J Hum Genet* 43:A146.

Haynes CS, Speer MC, Peedin M, Roses AD, Haines JL, Vance JM, Pericak-Vance MA (1995): PEDIGENE: A comprehensive data management system to facilitate efficient and rapid disease gene mapping. *Am J Hum Genet* 57:A193.

Hecker KH, Roux KH (1996): High and low annealing temperatures increase both specificity and yield in touchdown and stepdown PCR. *BioTechniques* 20:478–485.

Innis MA, Gelfand DH (1990): Optimization of PCRs. In: Innis MA, Gelfand DH, Sninsky JJ, White TJ, eds. *PCR Protocols: A Guide to Methods and Applications*. San Diego, CA: Academic Press, pp. 3–12.

Kan YW, Dozy AM (1978): Polymorphism of DNA sequence adjacent to human beta-globin structural gene: Relationship to sickle mutation. *Proc Natl Acad Sci USA* 75(11):5631–5635.

Kwok S, Higuchi R (1989): Avoiding false positives with PCR. *Nature* 339:237–238.

La Spada AR, Paulson HL, Fischbeck KH (1994): Trinucleotide repeat expansion in neurological disease. [Review]. *Ann Neurol* 36:814–822.

Lander ES, Botstein D (1987): Homozygosity mapping: A way to map human recessive traits with the DNA of inbred children. *Science* 236:1567–1570.

Litt M, Luty JA (1989): A hypervariable microsatellite revealed by in vitro amplification of a dinucleotide repeat within the cardiac muscle actin gene. *Am J Hum Genet* 44:397–401.

Litt M, McPherson MJ, Quirke P, Taylor GR, eds. (1991): *PCR: A Practical Approach*. New York: IRL Press (Chapter 6, PCR of TG Microsatellites. pp. 85–99).

Merril CR, Goldman D, Sedman SA, Ebert MH (1981): Ultrasensitive stain for proteins in polyacrylamide gels shows regional variation in cerebrospinal fluid proteins. *Science* 211:1437–1438.

Millard D, de Couet HG (1995): Preparation of glass plates with cerium oxide for DNA sequencing. *BioTechniques* 19:576–576.

Mullis K, Faloona F, Scharf S, Saiki R, Horn G, Erlich H (1986): Specific enzymatic amplification of DNA in vitro: The polymerase chain reaction. *Cold Spring Harbor Symp Quant Biol* L1:263–273.

Nakamura Y, Leppert M, O'Connell P, Wolff R, Holm T, Culver M, Martin C, et al. (1987): Variable number of tandem repeat (VNTR) markers for human gene mapping. *Science* 235:1616–1622.

Neiderhauser C, Hofelein C, Wegmuller B, Luthy J, Candrian U (1994): Reliability of PCR decontamination systems. *PCR Methods Appl* 4:117–123.

Saiki RK (1990): Amplification of genomic DNA. In: Innis MA, Gelfand DH, Sninsky JJ, White TJ, eds. *PCR Protocols: A Guide to Methods and Applications*. San Diego, CA: Academic Press, pp. 13–20.

Sheffield VC, Carmi R, Kwitek-Black AE, Rokhlina T, Nishimura D, Duyk GM, Elbedour K, et al (1994): Identification of a Bardet–Biedl syndrome locus on chromosome 3 and evaluation of an efficient approach to homozygosity mapping. *Hum Mol Genet* 3:1331–1335.

Smith JR, Carpten JD, Brownstein MJ, Ghosh S, Magnuson VL, Gilbert DA, Trent JM,Collins FS (1996): Approach to genotyping errors caused by nontemplated nucleotide addition by Taq DNA polymerase. *Genome Res* 5:312–317.

Southern EM (1975): Detection of specific sequences among DNA fragments separated by gel electrophoresis. *J Mol Biol* 98:503–517.

Vance JM, Nicholson GA, Yamaoka LH, Stajich J, Stewart CS, Speer MC, Hung W-Y, et al. (1989): Linkage of Charcot–Marie–Tooth neuropathy type 1a to chromosome 17. *Exp Neurol* 104:186–189.

Vance JM Slotterbeck B, Yamaoka L, Haynes C, Roses AD, Pericak-Vance MA (1996): A fluorescent genotyping system for multiple users, flexibility and high output. American Society of Human Genetics 46th Annual Meeting, San Francisco, November, p. A239.

Von Deimling A, Bender B, Louis DN, Wiestler OD (1993): A rapid and non-radioactive PCR based assay for the detection of allelic loss in human gliomas. *Neuropathol Appl Neurobiol* 19:524–529.

Weber JL (1990): Informativeness of human (dC-dA)n–(dG-dT)n polymorphisms. *Genomics* 7:524–530.

Weber JL, May PE (1989): Abundant class of human DNA polymorphisms which can be typed using the polymerase chain reaction. *Am J Hum Genet* 44:388–396.

Zagursky RJ, Conway PS, Kashdan MA (1992): Use of ^{33}P for Sanger DNA sequencing. *BioTechniques* 11:36–38.

Ziegle JS, Su Y, Corcoran KP, Nie L, Mayrand E, Hof LB, McBride LJ, et al. (1992): Application of automated DNA sizing technology for genotyping microsatellite loci. *Genomics* 14:1026–1031.

10

Database Design for Gene Mapping Studies

Carol Haynes and Colette Blach
Section of Medical Genetics
Center for Human Genetics
Duke University Medical Center
Durham, North Carolina

INTRODUCTION

Success of a genetics project or any data-intensive project is enhanced by the quality and amount and accessibility of information that is available. The likelihood of finding data to answer a pressing research question or to draw a conclusion concerning a hypothesis depends on the ability to store and retrieve the appropriate data in a timely fashion. Clearly this requires selecting in advance what data to store and choosing the software and hardware that will handle the data storage and retrieval requirements. Functionality and usability must be designed into the database system. The projected size of the project and the expected number of users also directly affect the design. This chapter specifies the steps required to produce a functional database for a genetics project.

ELEMENTS OF A DATABASE

A database is a collection of similar information. With this definition, a small database can be stored on a personal computer with, for example, Microsoft Excel

Approaches to Gene Mapping in Complex Human Diseases, Edited by Jonathan L. Haines and Margaret A. Pericak-Vance. ISBN 0-471-17195-6

(a spreadsheet application), or a large database can reside on a multiprocessor Sun workstation running, for example, Sybase (a database management system). The development process is independent of size. All database development requires five components, each reviewed here. The components are the users and contributors, the computer hardware, the database programs, the database information, and communications.

Development

Personnel. The human requirements consist of database users, contributors, and database support personnel. For a small database, the same person may be the sole user, contributor, and support technician. The database must be designed from the requirements of the users; a successful database will be able to provide answers to the anticipated questions of the users. Additionally, the database should make data entry intuitive for the contributors. Ideally, the database should be flexible enough for the database support personnel to add functions as requirements change.

Hardware. The hardware specified consists of the computer or computers, printers, backup tape drives, scanners, and network devices. Computers range in size and cost from single PC units to large multiprocessor systems. The computer must be sized to handle the database requirements while taking into consideration cost limitations. Because all equipment fails at one time or another, each laboratory must decide how much—or how little—computer downtime will be tolerated. Hardware service contracts can be purchased to provide immediate service and replacement equipment, if necessary. Large laboratories with time-critical data may require hardware redundancy.

Database. The database is composed of the data and a database engine. The data to be stored must be specified, and ways to interrelate data must be provided. The database engine is a software product purchased to store and manage data on the existing hardware. Examples of database engines are Lotus and Excel spreadsheets, and Access, Oracle, and Sybase database management systems (DBMS). The database engines provide functions to add, remove, and update data, and query functions to produce reports and summaries according to user-specified conditions.

Programs. The developers, working with the database users, design the forms to be used and reports to be generated. A program that interacts with the database can be used to simplify data entry and database query writing. Programs include contact forms for the entry of new data into the database and report formats for the data output. These programs may consist of spreadsheet macros, or of scripts in a database language. The code is written and extensively tested to eliminate errors ("bugs"), and then the program is deployed.

Because the field of genetics is advancing rapidly, the database is not likely to remain static. Changes will be needed to correct errors found by users after deployment, and additional functions will be needed to accommodate new laboratory and clinical procedures. Both large and small databases require that this additional development be done on a test version of the database. Note that sufficient computer disk space must be allotted to accommodate both the test and production databases.

Communications

Finally, to allow many users access to the data, network access may be desired. This requires additional computer hardware (such as an internal Ethernet card) and some type of networking software. Newer operating systems (e.g., Microsoft Windows NT) now include networking software. As a general rule, networking software adds a level of complexity that requires the services of a professional (in-house or consultant) to set up and maintain the system. The advantages for a multicomputer laboratory, however, are enormous and well worth the added expense.

DESIGNING AND BUILDING A DATABASE

All database development follows a similar process from the initial stage of realization that a database is required to the database's final deployment. The process begins with the design of the database, which must be done with specific users, contributors, and specific functions in mind. This step is often called requirements collection and analysis. Once the requirements have been established, they must be converted into a conceptual schema, or database design. With the requirements and scale of the project specified, the hardware and database engine are selected. Next, the database fields are established for the known database engine. Indexes may be specified here to speed up queries.

The programs or macros are written to provide the functions required. For example, input screen or report generating programs are written. If the data will be used over more than one computer, a method of communication is devised between computers. Finally, after the components have been tested together, the functioning database is deployed.

This chapter is a primer for database development for a genetics database and will discuss each of these stages of the genetics database development.

Stages

Goal Setting. Correctly specifying the database requirements is the key element in designing a successful schema. For a one-person project, the user would simply write down a list of the questions he or she would like the database to answer; for a large project, the database designer will want to interview the prospective users to determine their needs. The designer must determine how much and what types of information will be stored, and what types of questions the database users will be asking.

Questions commonly asked about individuals and their samples include the following:

1. Should the database have the ability to calculate allele frequencies from the marker genotypes?
2. Is there a need to sort genotype results by locus chromosome number?
3. Is the data used for monitoring technical productivity (e.g., how many markers are run in a given time period?)?
4. Is a family associated with a single diagnosis, or multiple diagnoses?
5. Will the database be expected to list affected sibling pairs or avuncular relationships?
6. Will an individual's diagnosis be stored in addition to the family diagnosis?
7. Will the database be used as a patient mailing list for newsletters or follow-up letters?

Typically, genetic data can be divided into clinical data and laboratory data. Clinical data may include clinical visit information, with the patient's status monitored over time. Laboratory data (e.g., the marker genotypes for each individual sample) must be coded by number, not name, to avoid bias and to maintain confidentiality.

The obvious advantage of collecting both laboratory and clinical information is the ability to search for disease subtypes compared to marker information. Is there an early-onset form of the disease, or is there more that one pattern of manifestation of a disorder, possibly attributed to a different genetic marker?

Platform Selection. Before the database can be designed, the platforms must be selected. The hardware and software together are known as the program's platform. Careful selection is required to identify a platform adequate for current needs and for anticipated future needs, but not extravagantly expensive and difficult to use. Selection of hardware, operating system, and software is like a jigsaw puzzle; the pieces fit together in only a limited number of ways (Table 10.1). Only a few operating systems will run on each hardware platform, and only a few databases will

TABLE 10.1 Operating System Characteristics

	Single User	Networked Computers	File Server or Distributed Database
Users	1	2+	10+
Hardware	Personal computer	Personal computer, workstation	Workstation, file server, or larger
Operating system	Windows 95, Windows 98, Windows 3.1, Mac OS, OS/2	Windows NT, Unix, Linux	Windows NT, Unix, Linux
Multiuser facilities	Limited	Extensive	Extensive
System administration	Minimal	Required	Required

run on each operating system. The obvious challenge, therefore, is to select compatible hardware, operating systems, and programs that will suit the user's needs.

Hardware Selection. The computer that will store the database must be fast enough to respond to the user's database questions (called queries) in a reasonable amount of time. Each user must decide what "reasonable" means in his or her setting. The hard drive must be large enough to store the database and to allow for growth. The computer must have enough memory for the program to respond quickly. Since all computer hardware will eventually fail, the computer must have a method to store the data in a protected place. This backup may be on diskettes for small amounts of data, or on tapes or remotely mounted hard drives for larger amounts of data.

If there is more than one user, networking computers may be required to share data. A local area network (LAN) can be set up for systems in close physical proximity using Ethernet. Ethernet communications cards cost about the same as a modem card and are easy to configure. Ethernet communications are 100 times faster than dial-in communications, but network administration software (e.g., MS Windows NT, Novell Lan Manager) is required to manage the network. Therefore a knowledgeable person must be recruited to set up and maintain the network.

Cost is a prime consideration for selection of a computer. Most computers vary in cost depending on size of storage, speed of processor, amount of memory, size of monitor, and additional devices such as modems and tape backups.

Figure 10.1 depicts the three main characteristics desired in a hardware and software platform. The basic rule of thumb is that only two of these points can be optimized at any one time. In other words, an inexpensive computer with a flexible database system that will allow maximum customization is achievable, but performance will be sacrificed. A system can be fast and flexible, but it will not be inexpensive. A fast, cheap system will not have all the elaborate database options available.

Although the newest computer hardware and software will be out of the question for many users, it is not advisable to begin a new database project using software based on an outdated operating system. A current operating system, such as Windows 95 (rather than Windows 3.1) creates a far better platform for the project; database software that is designed specifically for that operating system will run faster. The minimum guidelines for hardware configuration (memory and CPU) then become by default the minimum guidelines for the operating system chosen.

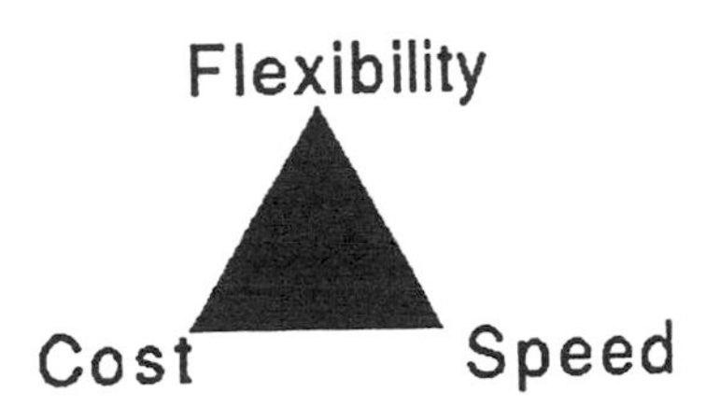

Figure 10.1. *The three main characteristics of hardware and software.*

Disk space is currently inexpensive, so it is best to purchase the most disk space possible. Software programs never shrink in size, and the new database programs use 25–50 megabytes (MB) of disk space before the first piece of data is entered. A small database (50 variables) with approximately 5000 records will be in the 1–2 MB range. Memory prices continue to decrease, and more memory improves performance. The minimum memory required for Windows 95 (8 MB) will suffice for only the most elementary database queries; even spreadsheet calculations will be slow. Therefore, 32 MB is recommended, although additional memory makes multitasking more efficient.

Software Selection

Operating Systems. Software includes the operating system, the database engine, and the programs. The operating system controls the file system and may allow the computer to run more than one program at a time. Some operating systems allow two or more users to share the computer's hardware. At present, common operating systems are Unix, Windows 3.1, Windows NT, Windows 95, Mac OS, OS/2, and Warp.

Since each operating system runs only on specific hardware platforms, the selection of an operating system for any particular hardware platform is greatly limited. Operating systems differ in degree of security, multiple-user and multiple-tasking abilities, database programs and engines available, performance, and requirements for system administration.

When more than one user can be logged on to a computer simultaneously, the computer's operating system is defined as "multiuser." Multitasking is the ability for more than one program to execute at the same time on a single computer (e.g., a spreadsheet and a word processor may be displayed simultaneously in two windows on the same computer screen).

Most multiuser operating systems require a system administrator to control computer operations. The system administrator establishes who can log on to the computer, assigns disk storage space to users, and assigns security access to certain files and file systems.

The systems administrator may handle data backup and the restoring of lost or corrupted data. Multiuser operating systems, such as Unix and Windows NT, or a LAN, require a system or network administrator.

Database Selections. As Table 10.2 shows, spreadsheets and databases can both be used for information storage. Selection is based on the complexity and amount of the data to be processed. In any case, data is stored in tables. Each column, containing different types of information, is called a field. Each row (or record) contains information on different individuals.

Excel and Lotus are two of many spreadsheet products. In addition to storing rows and columns of data, they have facilities for sorting data, viewing data that meets certain conditions, listing a limited number of fields, and counting and averaging results. With instructions for the spreadsheet called a macro, two tables can be

TABLE 10.2 Database Characteristics

Characteristic	Database Size		
	Small	Medium	Large
Simultaneous users	1–4	3–10	10^+
Number of records	< 1000	1000–100,000	$100,000^+$
Number of tables	< 4	4–15	10^+
Programs (not a definitive list—examples only)	Excel and Lotus spreadsheets	Access, dBase, FoxPro databases	Sybase, Informix, Oracle databases
Visual interface	Visual Basic or equivalent	Visual Basic, Visual C++	Powerbuilder, Oracle, or other 4GL languages
Security	Limited	Extensive	Extensive
Database administration	Minimal	Required	Required

joined together. For example, a table with the patient's diagnosis, and a table with the patient's marker information can be joined to display both on the same line for a patient. Spreadsheets are thus quite useful for small amounts of data (only several to a few thousand rows of data being kept) and when only a limited number of reports or queries are expected. However, if more data, more diverse types of data, or more and more complex data queries are expected, a formal database structure is probably required.

There are several common database structures used today. The flat file, such as a spreadsheet, stores the data in a single table. But many complex situations cannot be modeled successfully by a flat file. Next, hierarchical databases represent situations such as an airplane flight reservation database characterized by multiple-tier parent–child relationships ("parents" are represented by flight numbers and "children" are represented by passengers). This hierarchical arrangement allows for high-speed searches of lists of passengers on a single flight. But this tiered parent–child approach does not represent many situations. Also, in this example, the hierarchical database produces a very slow search from child to parent—for example, when the airline tries to find out the plane on which a particular passenger is flying.

Networked databases consist of records and sets. Records store the data, and sets consist of a $1{:}n$ relationship between two records. Data does not need to strictly follow a hierarchical model, making this complex model more flexible than hierarchical.

The relational database model has a simple data structure of tables. Each table has one or more fields which uniquely identify the record. Tables can be joined together through common fields on the tables, forming relations. This simple design allows for modeling complex relationships, including recursion. Relational databases, the most frequently used general-purpose databases, are the basis of dBase,

Paradox, Access, Sybase, Oracle, and many other commonly available database packages.

Object-oriented databases are a newer architecture. The entire object-oriented concept is a radically different design philosophy. Unlike traditional database architectures, the data and the functions are bound together into objects. The objects communicate with each other through message passing. Objects can inherit information from other objects. For example, an application that stores shapes and manipulates them could have an object type of Rectangle. But a more specific shape, a Square, has all the properties of a Rectangle but has additional limitations and properties. As a new architecture, object-oriented databases are principally used in database theory research and are just beginning to have mainstream applications.

Database Security. Databases and spreadsheets provide for security by encrypting data and using password accesses. In larger projects, however, these measures are often insufficient. Certain groups of database users may need access to patient sample numbers and the genotype results based on those samples; but to preserve patient confidentiality, those users should not have access to the patient names. In this case, the schema would be designed to place the patient name in a separate table from other identifying patient information, and access to this name table would be limited to a few authorized individuals. Access also might need to be limited in situations involving clinical data and genotype information. The physicians and physicians' assistants who examine the patients to gather clinical data might be denied access to genotype information, to blind clinicians to the lab data and avoid bias. This is particularly important when a clinician is trying to confirm a crossover detected during genetic analysis.

Database programs provide protection to keep the database synchronized; this means that when two users attempt to read or modify a single data record simultaneously, the database will ensure that only one user is allowed access to that record. The more sophisticated database systems will allow one user to view a record while it is being modified by another user.

A spreadsheet lacks the foregoing abilities to handle larger amounts of data, provide extensive security, or keep the data synchronized. But for storage of a small amount of data, the more extensive facilities of a database may not be needed.

SCHEMA DESIGN

Hardware and software are not all that is required to create a database. The schema is the description of the database that must be developed before the actual database tables can be created.

Tables 10.3 and 10.4 show an example schema. The schema does the following:

1. Lists what tables are stored. A table has rows (records) of data, and columns (fields) that specify variable information. For example, a table of patient information might have the individual's name, sex, race, date of birth, blood

TABLE 10.3 Scheme for the INDIVIDUAL Table[a]

Field Name	Data Type	Field Size	Rule	Description
ID Number	Text	11		Identification number = primary key variable
FamNumber	Int	6	>0	Family number
IndNumber	Int	6	>0	Individual number
LastName	Text	25		Last name
FirstName	Text	20		First name
SampNumber	Text	8		Blood sample number
SampDate	Date	(default)		Date sample collected
Sex	Text	1	M/F	Sex (M, Male; F, Female)
Race	Text	1	W/B/H/O	Race (W, white; B, black; H, Hispanic; O, other)
DOB	Date	(default)		Date of birth
DOD	Date	(default)		Date of death
AffStatus	Text	3		Disease affection status
OnsetAge	Int	3	> 0, < 110	Age of disease onset

[a]Indexes: ID#_Index, indexed by ID number; Sample_Index, indexed by sample.

sample number, and disease affection status. Because several individuals may have the same name, a unique key is needed. Here we have assigned an individual ID number to uniquely identify a single patient.

2. Identifies the data field size. Choosing 10 characters for the last name will force a name like "Porterfield" to be truncated, causing loss of data. The designer must choose to store a numeric field as a real (floating-point) number or as an integer. Fields can have associated "rules," which force data to be entered according to those rules. In Table 10.3, for example, the variable "sex" will allow only one character to be entered, and that character must be either M or F; any other character will be rejected.
3. Specifies which fields have indexes. Like the index in a book, an index on a database field allows for rapid retrieval of information.

TABLE 10.4 Schema for the GENOTYPE Table[a]

Field Name	Data Type	Field Size	Rule	Description
ID Number	Text	11		Identification number: primary key variable
MarkName	Text	20		Marker locus name
DateRun	Date	(default)		Date gel run
TechRun	Text	20		Name of technician running gel
AlleleBand1	Text	6		Allele 1 reading
AlleleBand2	Text	6		Allele 2 reading

[a]Indexes:
ID#_Index, indexed by ID Number; Marker_Index indexed by MarkName.

4. Describes the relationships between tables. In this example, an individual may be genotyped several times for various genetic markers. In Table 10.3 "INDIVIDUAL" represents the identifying information for individuals. In Table 10.4 "GENOTYPE" represents the results of each genetic marker for the individual by ID number and marker name. A researcher who wants to learn the age of individuals in this study who were typed for a particular genetic marker would relate the two tables through individual ID numbers.

Figure 10.2 shows how sample data might look in a spreadsheet format.

IMPLEMENTATION

Once a platform has been chosen and the schema set, it is time to implement the design. Generally, users fill out preconfigured data contact forms, which then are keyed into the database. In a small group, the same person might accomplish both tasks. In a larger group, however, several users will generate contact forms, all of which will be keyed into the database by one or more data entry technicians (DET). A major task of implementation is assuring the integrity of each data entry.

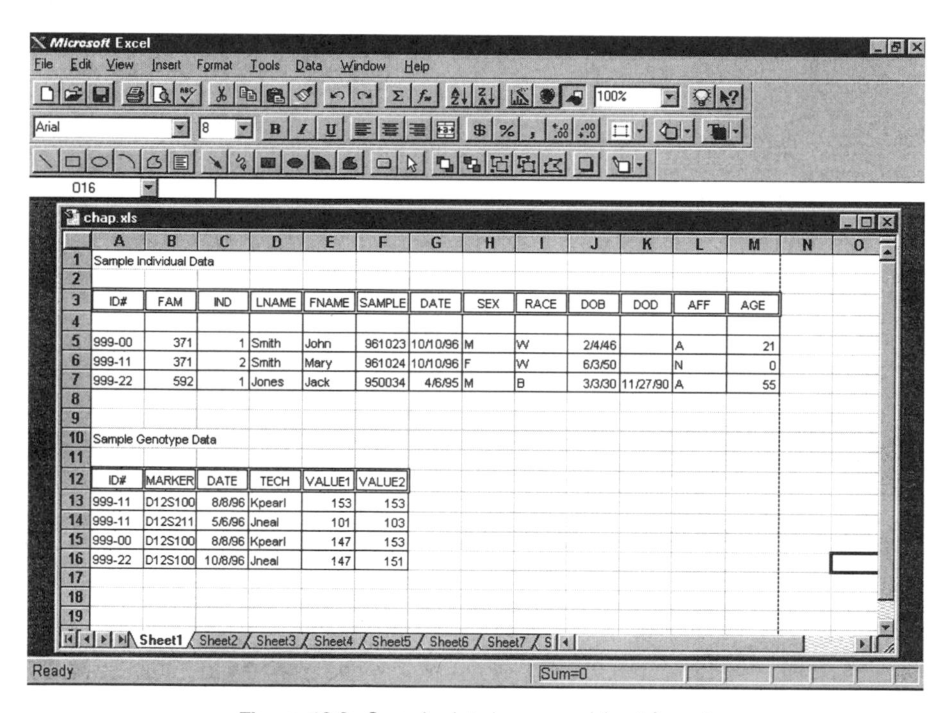

Sample Individual Data

ID#	FAM	IND	LNAME	FNAME	SAMPLE	DATE	SEX	RACE	DOB	DOD	AFF	AGE
999-00	371	1	Smith	John	961023	10/10/96	M	W	2/4/46		A	21
999-11	371	2	Smith	Mary	961024	10/10/96	F	W	6/3/50		N	0
999-22	592	1	Jones	Jack	950034	4/6/95	M	B	3/3/30	11/27/90	A	55

Sample Genotype Data

ID#	MARKER	DATE	TECH	VALUE1	VALUE2
999-11	D12S100	8/8/96	Kpearl	153	153
999-11	D12S211	5/6/96	Jneal	101	103
999-00	D12S100	8/8/96	Kpearl	147	153
999-22	D12S100	10/8/96	Jneal	147	151

Figure 10.2. *Sample data in a spreadsheet format.*

Data Integrity

Unambiguous contact form design is the key to successful data entry, and therefore to the integrity of the data. If users have trouble filling out the form, there is little hope that the data will transfer successfully to the database. Also, if comment space on the form is insufficient, users will start writing in the margins; if there is no "Other" box to check, users who want to give responses not offered on the form may skip questions. Such misuses of the contact form present problems for the data entry technician; he or she should not have to guess what the user had in mind. Careful review of the first group of contact forms before data entry will avoid problems later.

Double Entry

To avoid data entry errors, many sites utilize the concept of double entry. This originated with keypunch data, where one DET would do the actual keypunching, and a second DET would enter the same data at a keypunch machine that punched no holes in cards but, rather, checked to make sure that the new entry matched the original entry. This idea can be replicated with database entry software. In another method of checking data, the first DET prints out the entered data and another technician checks the printout against the original contact form. Note that it is not useful for the same person to do the checking; a misread word or phrase is likely to be perceived similarly the second time by the same person. The advantage of double entry over printout checking is that the second DET is not biased by the entry of the first DET, since the second worker is entering the data from the original contact form. In the printout checking scenario, the second DET already sees how the first person interpreted a word or phrase, and this can affect how the supposed checker views the answer.

Other Data Entry Options

Current technology allows data entry forms to be designed which the clinician or researcher fills out by hand and the form is then fed through an optical scanner and the data transmitted electronically directly to the database. This avoids transcription errors and is an extremely efficient mode of data entry. Personal Digital Assistants (PDAs) work in a similar fashion to eliminate the DET: the information can be entered in the field on the PDA and the data transmitted directly to the host computer.

Data Locking

Data locking is not an issue for a single user database, since only one person is accessing the database at a time. On a multiuser system, however, it is important that only one person be allowed to change a particular record at the same time. A multiuser system is one in which the database resides on a common disk, available to several computers simultaneously. A local area network is a good example; the computers are networked together, and many users have access to the same files. Most commercial databases (e.g., Access, Sybase) have built-in locking, which assures data integrity by restricting access to records being altered to a single user. Spreadsheets do not allow this type of on-the-fly record locking; however, the newest versions of spreadsheet programs do give the owner of a worksheet some flexibility in locking and password-protecting cells.

Validation and Incongruities

The old saying "garbage in, garbage out" is still true; unless the database is routinely checked for general validity, errors will accrue. There are many different types of validation, ranging from internal checks to assure that all tables and primary keys are synchronized, available on large, commercial systems, to the queries that a user should set up to ensure that the data makes sense. For data items that have specific responses, the user should configure the database to allow only certain codes to be accepted as responses to these items (e.g., M or F for the "sex" data item). Some data items have a range of possible responses; age is an example of this type of variable. The data item might be "age at disease onset," and the valid age range for this disease could be 10–90 years; the user would configure the database to accept only responses in the range 10–90.

More elaborate types of validation are at the core of a good database. A data item may "depend" on another data item (e.g., affection status and age of disease onset). If the "affection status" data item is entered as "affected," then the database should expect the "age of disease onset" data item to be answered (even if the answer is coded as "unknown age"); conversely, if "affection status" is entered as "not affected," there should not be an "age of disease onset."

Dates should always be validated; the date a blood sample was received should not be earlier than the starting date of the study, the date of birth of a parent should always be earlier than the date of birth of his or her children, and so on to avoid date compatibility problems in the year 2000 (also called the Y2K problem) always enter the full four digits of the year and verify that your database software stores all four digits. Every data item should be scrutinized for possible dependencies and specific responses, and these restrictions should be coded into the schema design.

Data validation can be built into the schema design; many types of validation can be built into data entry. For instance, the data entry program can require that only M or F be entered for sex; any other response will be rejected and the user not allowed to proceed until an acceptable response has been entered. Value ranges can be added to a table field to eliminate impossible values. Some types of validation, however, cannot be coded in this way. In many cases, careful consideration of monthly data summaries can be useful validation tools. If the database tracks genotypes, and the monthly summaries show too many (or not enough) genotypes produced for a month (based on laboratory projections), the first step would be to make sure that all the data had been entered properly—and promptly. Likewise, if the summary shows more (or fewer) affected individuals than the user expects, perhaps the DET was confused about how to enter the affections status.

The key here is to know "your" data; unusual summary reports are a sign of potential problems with database validity.

Backup Management

A primary part of the database implementation is setting up a suitable backup strategy. Large database systems (e.g., Sybase, Oracle) provide "rollback" capability, which allows the user to return the database to its status at a particular point in time. This feature is particularly useful when a database becomes corrupted for some reason, perhaps during a hardware crash. Spreadsheets and small-scale databases do not have this type of capability; the user will need to reenter all the information entered since the last backup of the database. Databases of all types must periodically be stored on offline media; the following guidelines should be followed.

1. Backups should be done whenever so much data has been added (or so many changes made) that reentry in the event of a disk failure or system crash would be unpleasant. Generally, daily backups are the rule.
2. Keep multiple backups. Have a separate set of disks or tapes for each day of the week, so that if a problem with the database that occurs on Tuesday is not noticed until Wednesday, Monday's copy will still be available. Large systems will want to keep several weeks' worth of backups at all times. Then, keep one full backup of the database each month as an archive file. It can be quite useful to look back 6 months to find a record that was deleted or a the premodification version of table.
3. Maintain off-site backups. Keep a backup copy of the database in an off-site location in case of a major disaster, such as fire or flood.
4. Periodically test the backups. Tapes and diskettes can become worn out, so test a different tape or diskette from your daily rotation periodically to be sure that the database can be restored from the media being used.
5. Keep the tapes in a secure location. The backup files have the same patient information as the online database, so be sure that unauthorized users do not have access to the backup information.

INTERNETWORKING

Internet access has become an important part of any genetic study. The wide range of information available on the Internet can save individual researchers time and energy. Detailed chromosome maps, abstracts of research papers, and many large, public genetic databases are accessible in real time through the Internet. The volume appendix lists many of the most useful sites, while Chapter 16 discusses their use.

Several steps must be taken to access this wealth of information. First, the computer must have either an Ethernet communications card (for use with installed Ethernet) or a high-speed modem (for Internet access over phone lines), both of which are available at most computer stores. In a university or large-business setting, Internet access is generally available via installed Ethernet, and thus only an Ethernet card is needed to connect into existing Ethernet connections. The Ethernet card is then configured with an Internet Protocol (IP) address, assigned by the person who oversees the installed Ethernet. Communications software (e.g., Reflections, LAN

Workplace) is then needed to start the Ethernet connection. The transmission speed of standard installed Ethernet is 10 megabits per second (Mbps); however, 100 Mbps is now available and, as the price drops, will become standard.

Other users will have to contract with one of the many commercial Internet service providers currently accepting new accounts. The provider assigns an IP address, generally provides the communications software, and gives the client a telephone number to dial (using a high-speed modem), to afford access to the Internet via a SLIP or PPP connection. Serial line Internet protocol (SLIP) and point-to-point protocol (PPP) provide a connection to the Internet using analog telephone lines. By encapsulating IP packets for serial transmission, these two protocols mimic an Ethernet connection, albeit at a much slower transmission rate. This rate is determined by the speed of the mode, typically 28.8 Kbps, although 56 Kbps is becoming widely available at a reasonable price. Another type of protocol uses a digital rather than analog telephone line. Integrated Services Digital Network (ISDN) allows for high-speed (64 Kbps for one-channel transmission, 128 Kbps for two-channel transmission), high-bandwidth IP packet transmission. Currently, this type of connection is more popular in Europe than in the United States for reasons of cost and because a single transmission standard is lacking.

After the Ethernet connection has been obtained, networking software must be installed and customized. Many commercial Internet browsers include other desirable features such as e-mail, Telnet, and FTP capabilities. A browser displays on the screen documents accessed from the World Wide Web and allows the user to connect from computer link to link. Electronic mail (e-mail) enables users to send messages around the world. These messages can have attached spreadsheet or word processor documents, and e-mail can even be used to send messages directly to a fax machine. Telnet capability allows the user to remotely connect to another networked computer (provided the user has an account on that machine) as if locally connected. FTP (file transfer protocol), a method of transferring files between computers on the Internet, provides a means of sharing data between computers. In "anonymous" mode, the user does not have to have an account on the remote computer. All these tools are invaluable for staying in touch with colleagues, joining newsgroup discussions, and keeping current with the latest maps and research abstracts.

CONCLUSION

The elements of database management described in this chapter can be applied to almost any gene mapping project. Data management is a critical component of any such undertaking, particularly projects using the large numbers of families and genetic markers so typical of efforts to map genetically complex traits. The advances in both hardware and software technology have made it possible to accurately enter, store, query, and output massive amounts of data using well-defined database structures. A well-designed database system will greatly ease the analysis and interpretation of the data, and make projects run more smoothly and efficiently.

11

Genomic Screening

Jonathan L. Haines
Department of Molecular Physiology and Biophysics
Program in Human Genetics
Vanderbilt University School of Medicine
Nashville, Tennessee

INTRODUCTION

For any common and genetically complex trait, there are two possible approaches to mapping the underlying gene(s). The first is the *random genomic screening* approach, which tests for linkage of the trait to (mostly) anonymous polymorphic markers chosen primarily for their ease of genotyping and their location within a genetic map. The second is the *directed genomic screening* approach, which tests for linkage and/or association to specific polymorphic markers chosen primarily for their location within or very near functional genes or previously identified regions from a random genomic screen. While the former is the primary practice currently, the great increase in the library of known genes is making the latter approach appropriate under certain circumstances, and it will certainly become more widely applicable in the future (Risch and Merikangas, 1996; Scott et al., 1997).

RANDOM GENOMIC SCREENING

Random genomic screening can be defined as testing the chosen trait for linkage to polymorphic markers spread throughout the genome. A good random genomic screen will attempt to cover the entire human genome using markers evenly spaced

Approaches to Gene Mapping in Complex Human Diseases, Edited by Jonathan L. Haines and Margaret A. Pericak-Vance. ISBN 0-471-17195-6

across a given genetic map. It does not require any knowledge of the function of any genes, or of the biology of the trait in question. In fact, practitioners tend to revel in the fact that they (and most everyone else) do not know anything about the functional relationship between genes and the trait. Traditionally in humans and many model organisms, the known biology of a trait was used to find the gene(s) in question. Because finding a gene was now being used as a tool to dissect the biology of the trait, this approach was erroneously termed "reverse genetics" for a number of years. This nomenclature fortunately has fallen into disuse.

Genetic Maps

Several qualities of the markers and maps must be considered when one is trying to choose the best combination for a screen. The best genetic markers (Chapter 9) are easy to genotype, have a high heterozygosity, and are confidently mapped into an overall genetic map. The tetranucleotide microsatellite markers tend to have the best of these qualities. Information about most markers is readily available from public databases, as discussed in Chapter 16. The best genetic maps have a dense set of markers, have the markers placed within small intervals and with high confidence, and have been carefully error-checked.

The strategy for generating genetic maps is straightforward (Beutow, 1994). parametric lod score analysis can be used, since virtually all markers are inherited in a simple codominant fashion. To ease genotyping and analysis, a simplified pedigree structure consisting of nuclear families with many children, both parents, and usually all four grandparents, is selected. The Centre d'Étude du Polymorphisme Humain (CEPH) has amassed a large collection of such families that is used by virtually all research groups (Dausset et al., 1990). This arrangement has the great advantage that genotypic data generated by each group can be combined for joint and therefore more comprehensive analysis.

While several sources of maps are available, three global genomic mapping projects (Généthon, the Cooperative Human Linkage Center, and the Utah Marker Development Group) are the most generally useful. The maps generated by Généthon consist of over 5000 dinucleotide markers mapped across the entire genome, with an average spacing of about 2 cM (Dib et al., 1996). This is the densest set of markers, and perhaps the most widely used set of maps. The information, including marker, map, and genotypic data, are available through the facility's Web site (see volume appendix for URLs of this and other Web sites mentioned here). The maps generated by the Cooperative Human Linkage Project (CHLC) consist of over 3600 mostly tri- and tetranucleotide markers, with an overall spacing of about 1 cM (Murray et al., 1994). The information, including marker, map, and LINKAGE format datafiles, are available on the World Wide Web. The maps generated by the Eccles Institute of Human Genetics in Utah consist of tetranucleotide markers, with an overall spacing of 10–15 cM. The information, including some marker primer information, maps, and LINKAGE format datafiles, can be found from the institute's Web site.

Additional maps of individual chromosomes have been published under the banner of the CEPH consortium. These maps consist of an amalgam of both older RFLP and VNTR markers along with a minority of microsatellite markers. The CHLC has made a modest effort to integrate mapping data from these different groups. In addition, the Marshfield Medical Research Foundation (MMRF) has produced both screening and comprehensive maps of all the chromosomes using data from all these sources. The MMRF maps are being updated regularly, and are now probably the most useful maps available.

Despite the refocusing of the efforts of the Human Genome Initiative toward high-resolution physical maps and large-scale sequencing, the efforts in genetic mapping are not complete, and many problems remain in using the currently available maps. These maps were generated using either a core of four CEPH families (Utah), or eight families (Généthon, CHLC), or in some cases an expanded set of 15 families (CHLC). Thus there is limited segregation information, and the possible resolution of marker locations is limited. Comprehensive maps such as those from Marshfield place all markers in the best possible position regardless of the strength of the statistical support for that position. Thus any order derived from such maps should be used cautiously. Despite claims of coverage at the 1 or 2 cM level, the confidently placed markers are spaced at intervals ranging from 2 to 25 cM. It must be remembered that any claims of a particular spacing are based on an *average* spacing, with a large standard deviation. Thus many of the "mapped" markers do not have a unique position in the map but may occupy any of several positions, often spanning a 20–30 cM interval, making their use difficult, particularly in later stages of fine mapping.

Most global genomic screens rely on one of several standard sets of markers (generally termed a *mapping set*). Such sets have been developed by Dr. Jim Weber at the Marshfield Medical Research Foundation, by Généthon, by Applied Biosystems Inc., and within other individual large laboratories (J. E. Davies et al., 1994). Again, these sets generally were developed for an average 10 cM coverage, with many intervals of 20–30 cM, and thus (depending on the power of the data set: Chapter 7), leading to no more than 80–90% coverage of the genome.

Homozygosity Mapping

Under certain situations the special design of homozygosity mapping can be used. In this approach, single affected individuals can be used for mapping, under the assumption that each person has inherited the same ancestral chromosomal segment surrounding the gene in question. Random genomic screening can be performed in essentially the same manner as for other types of families.

Random Association Studies

It has recently been asserted (Risch and Merikangas, 1996) that the current random screening approach using anonymous microsatellite markers will be supplanted by a random screening approach using only polymorphisms within genes. While this prediction may ultimately prove true, it bears a bit of scrutiny. First, while even at a

1 cM screening level, only 3000–3500 microsatellite markers would need to be screened, to screen each gene will require screening between 50,000 and 100,000 markers (the range of estimates of the number of genes in the human genome), a much more daunting task. In addition, while the 3000–3500 microsatellite markers currently exist (although only about 1500 in a well-mapped format), only a few hundred polymorphisms within genes have been described. In addition, most of these polymorphisms are only diallelic, leading to a general loss of potential information. It has been calculated (K. Davies, 1996) that approximately 2.2 times as many diallelic polymorphisms as microsatellite polymorphisms need to be genotyped, leading to a total genotyping of 110,000–220,000 polymorphisms.

Techniques to overcome the sheer volume of genotyping are being developed, such as microchip technology (Chapter 9, Chee et al., 1996). However, the polymorphisms do not yet exist and are not likely to be identified in sufficient volume until the human genome is substantially sequenced, in 5–10 years. In addition, the real bottleneck may not be genotype generation, but data analysis, where potential errors must be identified and tested, the statistical dependency between genotypes within the same or nearby genes must be considered, and the problem of what P value constitutes a true rare event must be solved.

Significance and the *P*-Value Problem

Much has been written about what constitutes an appropriate P value to permit any interesting result to be declared as significant (e.g., Thomson, 1994; Lander and Kruglyak, 1995). In the more classical forum of single gene disorders, a lod score of 3.00 has generally been accepted as "significant," meaning, of course, that such a score would be a rare occurrence by chance alone (Chapter 12). The score of 3.00 was originally chosen for considering linkage to a single marker, not for a general screening of a multitude of markers. In a practical sense, a score of 3.00 has stood up quite well, even when many markers have been tested. However, some correction for multiple testing should be employed (Chapter 12).

One approach that has been suggested in the context of single gene disorders (Ott, 1991) is to correct the lod score by adding the $\log_{10}$ (number of tests performed) to 3.00 to provide a corrected score. Thus if two markers were tested, the new criterion would be 3.30 $[3.00 + \text{Log}_{10}(2)]$; if 10 markers were tested, the new criterion would be 4.00; and if 350 markers (approximately a 10 cM screen) were tested, a new criterion of 5.54 would be needed. This approach is highly conservative because it ignores completely the dependency between markers due to linkage. A conservative score is not inherently bad, since the chance of a false positive result is greatly reduced. The problem is that the power to detect a true linkage suffers in the process, so that the chance of obtaining a true positive result may also be small.

The dependency (or correlation) between markers depends on the strength of their linkage to each other. Thus any markers within 50 cM are correlated to some extent, with that correlation becoming stronger as the markers move closer together. How to address this variable dependence theoretically is not clear, and so general rules of thumb and simulation efforts have been exploited. If no correction for multiple tests is made, a nominal score of 3 is used; but this is clearly a liberal choice, increasing power but also false positive results. If correction for every test is made, ignoring the correlations between markers, a nominal score near 5.5 is needed. This is clearly conservative. Thus an appropriate score should fall somewhere between 3.0 and 5.5.

Because the most commonly used family sampling strategy in genetically complex diseases is the collection of sib pairs, most discussions have surrounded *P* values rather than lod scores, since *P* values are the usual output from sib pair analysis programs. These are generally based on the chi-square distribution, and so the exact conversion depends on the number of degrees of freedom in the data. However, in general:

$$\text{lod} = \frac{\chi^2}{4.605}$$

Thomson (1994) proposed three categories of "interesting" results: weak, moderate, and strong (Table 11.1). She argues that weak results (e.g., nominal $P < 0.05$) must occur in three independent data sets (an overall $P \approx 0.0001$), moderate results ($P < 0.01$) must occur in two data sets (overall $P \approx 0.0001$), and strong results ($P < 0.001$) must occur in a single data set. Lander and Kruglyak (1995) proposed that the three categories be termed suggestive (e.g., a nominal $P = 0.0007$), significant ($P = 0.00002$), and highly significant ($P = 0.0000003$). At first these schemes may seem disparate, but the goals of these classifications are somewhat different. Thomson is concerned with determining which regions to examine further, and so suggests less stringent *P* values and accepts that some false positive results are likely. This is appropriate if a two-stage screening strategy is being used. Lander and Kruglyak are concerned with reducing false positive rates so that only true positives are likely to be identified, an approach more appropriate if a single stage screening strategy is being used. It is also interesting to note that Lander and Kruglyak's range of lod scores, from 2.2 to 5.4, is quite similar to that

Table 11.1 Proposed *P* Values for a Genomic Screen

Thomson (1994)	Nominal *P* Value (adjusted *P* Value) [lod score]	Lander and Kruglyak (1995)	*P* Value [lod score]	This Book	*P* Value [lod score]
Weak	0.05 (0.0001) [2.9] (3 × 0.05)	Suggestive	0.0007 [2.2]	Interesting	0.03 [1.0]
Moderate	0.01 (0.0001) [3.0] (2 × 0.01)	Significant	0.00002 [3.6]	Very interesting	0.0009 [2.0]
Strong	0.001 (0.001) [2.1] (1 × 0.001)	Highly significant	0.0000003 [5.4]	Provisional linkage	0.00003 [3.0]
				Confirmed linkage	0.0000008 [4.0] (from at least two independent data sets)

obtained when the problem of multiple tests is ignored or complete independence of the multiple tests is assumed.

It is unfortunate that there has been so much concentration on the false positive rate (e.g., type I error, or *P* values), since this is only half of the story. It is just as important to know what the rate of false negatives is (e.g., type II error, power). It does little good to use a very stringent *P* value to declare significance if there is little hope that even a true linkage would reach that level. Power studies are discussed in more detail in Chapter 7, but some general points need to be repeated. Any calculation of power requires the assumption of an underlying genetic model. While an adequate genetic model can be defined for most single gene disorders, this is simply not possible for most complex traits. Thus any power estimates must be interpreted carefully, and used as only general guidelines. In our experience, it is unlikely that any genetically complex trait will yield a *P* value of 0.0000003 in an initial screen unless there is an extremely strong effect of a single gene.

We would advocate an approach modestly different from either of these, based on empiric data and the advantages of a multistaged approach to screening (see below). Table 11.1 presents these criteria. A nominal *P* value of 0.03 from a single data set would be considered *interesting,* while a nominal *P* value of 0.0009 from a single data set (0.03 from two independent data sets) would be considered *very interesting.* With a *P* value of 0.000027, a provisional linkage could be declared but not considered confirmed unless an independent data set generated a nominal *P* value of 0.03 (e.g., an overall *P* value of 0.0000008). These values are roughly equivalent to lod scores of 1, 2, 3, and 4.

DIRECTED GENOMIC SCREENING

In contrast to random genomic screening approach, a directed genomic screening approach may be used. By *directed genomic screening* we simply mean that the polymorphisms chosen for analysis are not spread either randomly or evenly across the genome, but instead are concentrated in certain areas based on additional information. The additional information that define these candidate regions can come in two different forms.

Functional candidate regions are defined by the genes that reside within them. If something is known of the biology of the trait, then genes affecting that biology become candidates. For example, multiple sclerosis is an autoimmune disease in which the myelin sheathes around nerves are attacked and often destroyed. This information suggests that certain genes, such as the HLA genes, T-cell receptor genes, and the myelin basic protein gene are prime candidates for analysis. The strength and weakness of this approach arises from the confidence placed on the role of these genes. If the evidence is strong that a direct role is played, only one or a few such genes may need to be tested. If the evidence is more circumstantial, then many genes may have an equal chance of being involved, and not much has been gained over a random genomic screen.

At the phenotypic level, what is known of the physiology may suggest a scheme for stratification (subsetting) of the data such that a single gene (or at least a reduced set of genes) may be responsible. For example, for many years neurofibromatosis was considered a single disorder, manifesting both peripheral and auditory nerve tumors. Later it became clear that the peripheral and auditory tumors rarely occurred together and likely represent two different forms. This eased the linkage analysis, with each form being linked to different chromosomes (Barker et al., 1987; Rouleau et al., 1987).

Locational candidate regions are defined simply by where the polymorphism that generates an interesting result resides, not by any potential function it might have. Thus if an initial genomic screen has been performed, and several "interesting" regions have been defined, any polymorphism within such a region becomes a locational candidate.

DATA SETS

The type of sample available for collection and analysis affects not only the power of the analysis, but also the types of statistical analysis and the density of markers to be screened. If large, extended families with multiple affected individuals are available, then lod score analysis and more traditional positional cloning can be anticipated. If a substantial portion of families have one or a few affected relatives other than sibs, an affected relative member method of analysis may be appropriate. If only sib pairs are available, sib pair analysis will be done by default. An ancillary question arises: If more than just sib pairs are available, is anything lost by concentrating only on sib pairs? In other words: Can sib pairs be used for everything? This is a difficult question, for the answer lies with the true genetic model, which is unknown. It may be more powerful, for example, to identify loci that are necessary (although not necessarily sufficient) for a trait by using cousin pairs rather than sib pairs. If there is substantial genetic heterogeneity, it may be more difficult to identify any one of the loci if all different types of families are analyzed together. For example, if some of the underlying genes interact in a multiplicative manner, then sib pairs are much more likely than extended relative pairs to carry these genes. Thus using sib pairs would generate a more etiologically homogeneous data set. In the absence of better guidelines, the best approach may be to collect whatever is easiest and most cost effective. Better answers to this question await large-scale simulation studies using complex genetic models as a substrate.

Determining the power of a data set is a critical step in designing the screening approach (Chapter 7). Although the underlying genetic model is unknown, and thus exact power calculations are impossible, some rules of thumb can be employed. Any sample of fewer than 40 sib pairs is unlikely to detect even moderately strong genetic effects, inasmuch as simulation studies with single gene disorders indicate this to be a lower bound. Sample sizes of several hundred sib pairs are likely to detect moderate effects. If the sample is small, then a more dense map of genetic markers needs to be analyzed because each marker will provide useful information over a smaller genomic region. While most studies aim for a genomic screen with a 10 cM average spacing of markers, a large data set may use a 20 cM screen, and a small one a 5 cM screen.

The strength of the genetic effect being mapped relates to the unknown underlying genetic model and has an obvious effect on power (Chapter 7). For example, if the relative risk to siblings (λ_s; Chapters 4, 13) is large (e.g., 40), it will be easier to detect the genetic effect than if the λ_s is small (e.g., 2). It should be remembered that the λ_s is an overall measure of the entire genetic effect. Thus if multiple genes are involved, each gene will carry some portion of the overall risk. If the expected genetic effects are small, then large sample sizes and stratification toward homogeneity should be pursued.

One alternative to using a large sample of families is to concentrate on a genetically homogeneous population, such as Amish or Finnish populations, or Ashkenazi Jews. There are several advantages to these populations, including a potential reduction in the number of segregating susceptibility genes, the subsequent strengthening of the relative effect of the remaining susceptibility genes, and the likelihood that linkage disequilibrium exists and may ease the mapping process (Chapter 15). Thus within these populations it may be easier to find at least one susceptibility gene. Not all susceptibility genes are likely to be segregating, however, and generalization of the results to a wider, outbred population may not be possible. Thus the decision to concentrate on, or even to use, such a population rests with the relative ease of collection and the probability that any result will be meaningful outside that population.

SINGLE STAGE VERSUS MULTISTAGE SCREENING

Although there has been some debate on this issue (Lander and Schork, 1994; Elston et al., 1996), there is a growing consensus that a multistage strategy is best. In a single stage approach, the balance between false positive and false negative results must fall heavily toward reducing the false positive result. This requires a large initial data set genotyped for every marker in the screen, and most likely requires that a more dense map be used. In contrast, a multistage approach is usually more cost efficient, since power becomes the major consideration in the first stage. Thus a smaller number of individuals and genotypes can be used in stage 1, with the explicit goal not of confirming a linkage but of reducing the number of candidate regions to a fraction of the whole genome (usually $< 5\%$). The penalty is that a second stage of genotyping and analysis of a second data set will be required to examine all the candidate regions, most of which will be false positive results.

REFERENCES

Barker D, Wright E, Nguyen K, Cannon L, Fain P, Goldgar D, Bishop DT, Carey J, Baty B, Kivlin J (1987): Gene for von Recklinghausen neurofibromatosis is in the pericentromeric region of chromosome 17. *Science* 236:1100–1102.

Beutow KH (1994): Construction of reference genetic maps. In: Dracopoli NC, Haines JL, Korf BR, Norton CC, Seidman CE, Seidman JG, Moir DT, Smith DR, eds. *Current Protocols in Human Genetics.* New York: Wiley, Chapter 1, Unit 1.5.

Chee M, Yang R, Hubbell E, Berno A, Huang XC, Stern D, Winkler J, Lockhart DJ, Morris MS, Fodar SPA (1996): Accessing genetic information with high-density DNA arrays. *Science* 274:610–613.

Dausset J, Cann H, Cohen D, Lathrop M, Lalouel JM, White R (1990): Centre d'Étude du Polymorphisme Humain (CEPH): Collaborative genetic mapping of the human genome. *Genomics* 6:575–577.

Davies JE, Kawaguchi Y, Bennett ST, Copeman JB, Cordell HT, Pritchard LE, Reed PW, Gough SC, Jenkins SC, Palmer SM (1994): A genome-wide search for human type 1 diabetes susceptibility genes. *Nature* 371:130–136.

Davies K (1996): To affinity . . . and beyond! Editorial. *Nat Genet* 14:367–370.

Dib C, Faure C, Fizames D, Samson N, Drouot A, Vignal P, Millasseau P, Marc S, Hazan J, Seboun E, Lathrop M, Gyapay G, Morissette J, Weissenbach J (1996): A comprehensive genetic map of the human genome based on 5,264 microsatellites. *Nature* 380:152–154.

Elston RC, Guo X, Williams LV (1996): Two-stage global search designs for linkage analysis using pairs of affected relatives. *Genet Epidemiol* 13:535–558.

Lander ES, Kruglyak L (1995): Genetic dissection of complex traits: Guidelines for interpreting and reporting linkage results. *Nat Genet* 11:241.

Lander ES, Schork NJ (1994): Genetic dissection of complex traits. *Science* 265:2037–2048.

Murray JC, et al. (1994): A comprehensive human linkage map with centimorgan density. Cooperative Human Linkage Center (CHLC). *Science* 265:2049–2054.

Ott, J (1991): *Analysis of Human Genetic Linkage.* Baltimore: Johns Hopkins University Press.

Risch N, Merikangas K (1996): The future of genetic studies of complex human disorders. *Science* 273:1516–1617.

Rouleau GA, et al. (1987): Genetic linkage of bilateral acoustic neurofibromatosis to a DNA marker on chromosome 22. *Nature* 329:246–248.

Scott WK, Pericak-Vance MA, Haines JL (1997): Genetic analysis of complex diseases. *Science* 275:1327.

Thomson G (1994): Identifying complex disease genes: Progress and paradigms. *Nat Genet* 8(2):189–194.

12

Lod Score Analysis

Jianfeng Xu and Deborah A. Meyers
Center for the Genetics of Asthma and Complex Diseases
University of Maryland School of Medicine
Baltimore, Maryland

Margaret A. Pericak-Vance
Section of Medical Genetics
Center for Human Genetics
Duke University Medical Center
Durham, North Carolina

INTRODUCTION

Genetic linkage refers to the tendency of alleles from two loci to segregate together in a family if they are located physically close to each other on a chromosome. The extent of linkage is a function of the distance between the two loci, which can be measured by the number of crossovers between the two loci among the observed meioses (i.e., recombination fraction, θ). There is "complete linkage" if there is no recombination ($\theta = 0$), some degree of linkage if the number of recombinants is less than half of the number of observed meioses ($\theta < 0.5$), and "no linkage" if the observed recombination fraction is 50% ($\theta = 0.5$) (Chapters 2, 4). Although any estimate of a recombination fraction less than 50% ($\theta < 0.5$) is indicative of linkage, genetic studies require a measure of whether there is *significant* evidence for linkage.

Lod score analysis is a likelihood-based parametric linkage approach to estimate the recombination fraction and the significance of the evidence for linkage. The lod

Approaches to Gene Mapping in Complex Human Diseases, Edited by Jonathan L. Haines and Margaret A. Pericak-Vance. ISBN 0-471-17195-6

or lod score represents $\log_{10}$ of the ratio for two likelihoods, the likelihood of observing a particular configuration of a trait and a marker locus in a family (or set of families) assuming linkage (i.e., $\theta < 0.5$), and the likelihood of observing the same configuration of the two loci within the same family (or set of families) assuming no linkage (i.e., $\theta = 0.5$). The traits at each locus can be qualitative (disease status, microsatellite marker allele size) or quantitative (disease or marker measurement), and the method can be extended to include multiple loci. This method was originally developed for traits with known mode of inheritance and allele frequencies (the genetic model; ϕ), where an unbiased (correct) θ can be estimated (Morton, 1955).

There are four major advantages of lod score analysis over other methods described in this book:

Statistically, it is a more powerful approach than any nonparametric method.

It utilizes every family member's phenotypic and genotypic information.

It provides an estimate of the recombination fraction.

It provides a statistical test for linkage and for genetic (locus) heterogeneity.

When the mode of inheritance is unknown, as in genetically complex traits, the lod score may still be calculated but depends on several assumed parameters: the recombination fraction, trait allele frequencies, the penetrance values for each possible disease phenotype, and the marker allele frequencies. It should be remembered that any test of linkage using a parametric approach is a test of *all* the assumptions, with linkage being only one. Failure to find linkage thus could be due to a misspecification in any of these parameters; it does not prove lack of linkage. If a parametric lod score analysis is undertaken when the genetic model is not known, it is critical that the results be interpreted cautiously. Despite these potential difficulties, lod score analysis has been successfully applied to several complex traits, including breast cancer (Hall et al., 1990) and Alzheimer disease (St. George-Hyslop et al., 1992; Schellenberg et al., 1992; Pericak-Vance et al., 1991; Pericak-Vance et al., 1997).

The purpose of this chapter is to acquaint the reader with the effects of the many parameters necessary for lod score analysis and with approaches for applying this method to the analysis of genetically complex traits.

TWO-POINT ANALYSIS

The definition and calculating formulas for two-point linkage analysis were given in Chapter 4 and are not repeated here. While formulas (4.3) and (4.4) cover both the phase-known and phase-unknown situations, each one assumes that all genotypic and phenotypic data are available on all individuals. As a result of the complicating factors of uninformative markers and unavailable genotypic data (usually because individuals are deceased or otherwise unavailable for DNA sampling, or from incomplete laboratory genotyping), however, in most human pedigree data it is not possible to simply count recombinants and nonrecombinants. Yet as long as the genetic model can be specified, there is computer software such as LINKAGE (Lathrop et al., 1984), FASTLINK (Schaffer et al., 1994), VITESSE (O'Connell and Weeks, 1995), and GENEHUNTER (Kruglyak et al., 1996), which can be used to calculate the lod scores. These programs use a variety of computing algorithms for their calculations, as discussed in the primary papers.

Lod scores (Z) can be calculated at any values of θ between 0 and 0.5, but are conventionally reported at $\theta = 0, 0.01, 0.05, 0.1, 0.2, 0.3$, and 0.4. Since the likelihoods of independent families are multiplied to accumulate a total likelihood for a sample, log-likelihoods or lod scores are simply summed over all independent families at any specified θ. The maximum lod score Z_{max} can be acquired by maximizing Z over a range of recombination fractions, and the recombination fraction that gives Z_{max} is the maximum likelihood estimate (MLE) of θ. Note that Z_{max} values cannot be summed across pedigrees, since they will likely occur at different θ.

A lod score of 3.0 or more has been traditionally considered strong evidence for linkage when two-point analysis is used with a known mode of inheritance. This critical value corresponds to 1000:1 odds for linkage, which means that the observed data is 1000-fold more likely to arise under a specific hypothesis of linkage than under the null hypothesis of no linkage (e.g., independent assortment of the chromosomes). The 1000:1 odds are not, however, equivalent to a P value of 0.001. Because of its statistical properties, the lod score can be converted to a chi-square statistic by multiplying by $2(\log_e 10) \approx 4.6$, which asymptotically follows a chi-square distribution with 1 degree of freedom (df) under the null hypothesis of absence of linkage (Ott, 1991). Thus the asymptotic significance level, α, of lod score of 3.0 is equivalent to a P value of 0.0001 ($\chi^2 = 4.6 \times 3 = 13.8$). The reason for applying this much smaller significance level, compared to the conventional statistically significance level of 0.05 or 0.01, is the low prior probability of two loci being linked. This can be illustrated using the following argument. Assume that the human genome is 30 Morgans long (3000 cM) and thus can be divided into sixty 50 cM segments. Since the locus must exist in one segment, there is a 1/60 = 0.02 chance that the locus resides in any one segment. Thus testing for linkage to any single locus has only a 2% prior chance of success. Applying Bayes' theorem, with a lod score of 3.0, the posterior probability of linkage is 95%.

When a genome-wide approach is used in the linkage study, the critical value of the lod score should be raised to account for the testing of multiple markers. A lod score of 3.3, corresponding to $P = 5 \times 10^{-5}$, is equivalent to the recommended genome-wide significance level of 5% (i.e., statistical evidence expected to occur 0.05 times in a genome scan when there is absence of linkage) (Lander and Kruglyak, 1995). Except for the candidate gene approach, in most cases the genome-wide criteria should be used in a linkage study (but see Chapter 11 for more discussion of this issue).

EXAMPLE OF LOD SCORE ANALYSIS AND INTERPRETATION

A linkage study was carried out for bipolar disorder, and markers on chromosome 18q using 28 multiplex bipolar (bipolar I, II, and recurrent unipolar) families were selected for apparent unilineal transmission of bipolar disorder (Stine et al., 1995). Among the 28 families, 11 are paternal pedigrees in which the father or at least one of the father's sibs is affected, and 16 are maternal pedigrees in which the mother or at least one of her sibs is affected. Although the mode of inheritance for bipolar disorder is unknown, a dominant model allowing for phenocopies and incomplete penetrance was used in the analysis, which is consistent with the results of most family studies and segregation analyses (Chapter 5). The disease allele frequency was assumed to be $q_a = 0.025$, which reflects the population prevalence of approximately 5% [i.e., $2(1 - q)q + q^2 = (2 \times 0.975 \times 0.025) + (0.025 \times 0.025) = .0494$ assuming Hardy–Weinberg equilibrium]. The penetrance for gene carriers (f_{aa}, f_{Aa}) was assumed to be 0.85, which is a very subjective number but represents the results from many family studies of bipolar disorder. The penetrance for non–gene carriers (f_{AA}) was assumed to be 0.004, which represents a phenocopy rate of approximately 10%, the proportion of phenocopies among all affected individuals (i.e., $0.004 \times (.975)^2 = .0038/[.85(.0494) = .0905$). The allele frequencies for the marker, D18S41, were calculated based on the independent individuals in the data set (i.e., grandparents, married-in individuals).

Table 12.1 presents the two-point linkage analysis for bipolar disorder and marker D18S41 under a dominant model. The highest lod score is only 0.28 at $\theta = 0.30$. Using the classical criterion for linkage (lod = 3.0), one can easily conclude that there is no significant evidence for linkage. This is also consistent with the nonparametric linkage result of affected sib pair analysis (Chapter 13) using the mean test (96 affected sib pairs, mean proportion of IBD sharing of 0.55 ± 0.29, $P = 0.036$). However, unlike nonparametric analysis (where the summary statistics provide limited information for further investigation), the lod score approach provides a rich source of information and thus requires more consideration in interpretation. It is important to understand the assumptions made in such an analysis and the possible impact on the observed results.

EFFECTS OF MISSPECIFIED GENETIC PARAMETERS IN LOD SCORE ANALYSIS

These lod scores were obtained using one prespecified genetic model. For a genetically complex disorder where factors such as incomplete penetrance, phenocopies, genetic heterogeneity, and oligogenic inheritance are involved, the assumed genetic model is most likely to be incorrect in at least some aspects. If the assumed genetic model is wrong, the true picture of linkage can be disguised, leading to either false positive or false negative evidence for linkage, since the lod score is the function of both the recombination fraction and the genetic model. This is especially important when negative lod scores are seen. The impact of misspecified genetic parameters on the lod score is complicated and depends on several factors, such as the true underlying disease model, the parameters that are misspecified and the extent of misspecification, and the pedigree structures. Results from theoretical calculations and simulation studies generally agree that the power to test linkage is highly sensitive

Table 12.1 Two-Point Linkage Analysis for Bipolar Disorder and D18S41

Genetic Model	Pedigree	Recombination Fraction, θ						
		0.00	0.01	0.05	0.10	0.20	0.30	0.40
Dominant	Total	−6.92	−5.29	−2.29	−0.74	0.27	0.28	0.02
	9	0.77	0.76	0.67	0.67	0.37	0.10	0.05
	11	0.52	0.51	0.46	0.40	0.26	0.14	0.04
	15	−0.32	−0.33	−0.36	−0.39	−0.38	−0.26	−0.11
	22	−0.20	−0.16	−0.04	0.05	0.11	0.09	0.03
	30	0.00	0.00	0.00	0.00	0.00	0.00	0.00
	39	−0.89	−0.74	−0.43	−0.25	−0.09	−0.03	0.00
	45	−0.98	−0.85	−0.55	−0.36	−0.16	−0.06	−0.02
	58	0.10	0.10	0.08	0.07	0.04	0.02	0.00
	61	−0.68	−0.57	−0.31	−0.16	−0.03	0.01	0.01
	64	1.61	1.58	1.44	1.26	0.90	0.52	0.17
	66	−1.20	−1.00	−0.62	−0.40	−0.18	−0.07	−0.02
	74	−1.63	−1.28	−0.77	−0.51	−0.24	−0.10	−0.02
	75	0.02	0.02	0.02	0.01	0.01	0.00	0.00
	79	1.30	1.27	1.15	1.00	0.68	0.35	0.09
	84	0.00	0.00	0.00	0.00	0.00	0.00	0.00
	106	−1.20	−1.01	−0.61	−0.37	−0.14	−0.04	−0.01
	111	0.01	0.01	0.00	0.00	0.00	0.00	0.00
	112	0.00	0.00	0.00	0.00	0.00	0.00	0.00
	113	−0.08	−0.08	−0.06	−0.05	−0.03	−0.01	−0.00
	115	−1.17	−0.91	−0.50	−0.30	−0.13	−0.06	−0.01
	116	−0.78	−0.71	−0.51	−0.36	−0.18	−0.07	−0.02
	117	−0.34	−0.33	−0.32	−0.32	−0.30	−0.23	−0.12
	121	0.57	0.56	0.51	0.44	0.30	0.16	0.05
	122	−0.70	−0.63	−0.43	−0.27	−0.11	−0.03	0.00
	123	0.03	0.03	0.04	0.04	0.03	0.01	0.00
	203	0.22	0.21	0.19	0.15	0.09	0.04	0.01
	204	−1.01	−0.88	−0.57	−0.37	−0.17	−0.07	−0.02
	217	0.00	0.00	0.00	0.00	0.00	0.00	−0.01

to the degree of dominance (dominant vs. recessive), somewhat sensitive to marker allele frequencies, slightly sensitive to the penetrance, and not very sensitive to the disease allele frequency. However, the estimation of the recombination fraction may be strongly affected by an error in any genetic parameter (Clerget-Darpoux et al., 1986).

To help illustrate the impact of misspecifying each genetic parameter on the power to detect a linkage or the type I error (false positive), computer simulations using the 28 bipolar families were performed. One thousand replicates were generated under either complete linkage ($\theta = 0.00$) or no linkage ($\theta = 0.50$) between a disease locus and a marker with six alleles. The disease locus was simulated using a dominant model with a disease allele frequency (q_a) of 0.02, penetrances for gene carriers (f_{aa}, f_{Aa}) of 0.85 and a penetrance for non–gene carriers (f_{AA}) of 0.004. The marker was simulated using allele frequencies of 0.02, 0.02, 0.18, 0.18, 0.10, and

0.50. Each replicate was analyzed using the simulated model as well as other models representing model misspecification: four models with different disease allele frequencies, four models with different penetrances, two models with different dominance, and one model in which all marker alleles were assumed to be equally frequent. When one parameter was evaluated, the other parameters were fixed at their simulated values, except when the parameter of dominance was evaluated, where the allele frequency of the disease gene was changed accordingly to have the same population prevalence. The impact of misspecifying the genetic model on the lod score analysis was evaluated using the mean maximum lod scores (Z_{max}) and mean estimated MLE (θ). The results at $\theta = 0.00$ reflect the power to detect linkage, and at $\theta = 0.50$ reflect the type I error.

Impact of Misspecified Disease Allele Frequency

Although the true disease allele frequency was 0.02, the data were analyzed using the disease allele frequencies of 0.001, 0.1, 0.2, and 0.3, a sequence that generates population prevalences of 0.2, 19, 36, and 51%, respectively (Table 12.2). This represents a wide variation of disease allele frequencies. In the case of no linkage, there is little difference in the mean maximum lod score (from 0.11 to 0.12). In the case of linkage, however, the mean maximum lod score decreases when the disease allele frequency is either underestimated or overestimated. When the frequency is underestimated (0.001), the mean maximum lod score slightly decreases (from 9.14 to 8.88) and MLE (θ) slightly increases (from 0.01 to 0.03). When the frequency is overestimated (0.1), the mean maximum lod score decreases to 8.15, and MLE (θ) decreases its value to 0.00. An increased disease allele frequency may have an impact on the lod score in two ways: either increasing the probability of affected parents being homozygous or increasing the probability that the disease allele is introduced into the pedigree through married-in individuals instead of through a single founder. As seen in the Table 12.2, when the frequency is varied at a reasonable range (0.001–0.1), the impact of misspecification of the disease allele frequency on the lod score and on the estimate of recombination fraction is generally small.

Table 12.2 Impact of Misspecifying Disease Gene Frequency on Lod Score Analysis

Model Used for Analysis				Model Used for Simulation: $q_a = 0.02$, $f_{AA} = 0.004$, $f_{Aa} = f_{aa} = 0.85$			
				$\theta = 0.00$ (power)		$\theta = 0.50$ (type I error)	
Q_0	f_{AA}	f_{Aa}	f_{aa}	Mean Z_{max}	Mean MLE (θ)	Mean Z_{max}	Mean MLE (θ)
0.02	0.004	0.85	0.85	9.14	0.01	0.11	0.43
0.001	0.004	0.85	0.85	8.88	0.03	0.11	0.43
0.1	0.004	0.85	0.85	8.15	0.00	0.12	0.41
0.2	0.004	0.85	0.85	6.88	0.00	0.12	0.40
0.3	0.004	0.85	0.85	5.79	0.00	0.13	0.38

Table 12.3 Impact of Misspecifying Disease Dominance on Lod Score Analysis

Model Used for Analysis				Model Used for Simulation: $q_a = 0.02$, $f_{AA} = 0.004$, $f_{Aa} = f_{aa} = 0.85$			
				$\theta = 0.00$ (power)		$\theta = 0.50$ (type I error)	
Q_0	f_{AA}	f_{Aa}	f_{aa}	Mean Z_{max}	Mean MLE (θ)	Mean Z_{max}	Mean MLE (θ)
0.02	0.004	0.85	0.85	9.14	0.01	0.11	0.43
0.02	0.004	0.425	0.85	7.51	0.00	0.11	0.41
0.02	0.004	0.004	0.85	1.17	0.21	0.08	0.43

Impact of Misspecifying Disease Dominance

In general, misspecified disease dominance has a large impact on the lod score (Clerget-Darpoux et al., 1986). This effect is particularly serious when a dominant disease is misspecified as a recessive disease (Table 12.3). In the bipolar example, the mean maximum lod score decreases from 9.14 to 1.17, and the MLE (θ) increases from 0.01 to 0.21. For example, when a disease that is inherited in a dominant fashion is analyzed under a recessive model, the random segregation of alleles from the non-disease-allele-carrying parent will be scored (half the time) as a recombination between the disease and marker locus. In addition, the affected parent will be considered homozygous for the disease locus and thus uninformative for linkage. A misspecified disease dominance has little impact on lod score when there is no linkage.

Impact of Misspecified Disease Penetrances

As long as incomplete penetrance is included in the genetic model, misspecified disease penetrance has a small impact on the lod score when there is either linkage or no linkage (Clerget-Darpoux et al., 1986) (Table 12.4). As the ratio of penetrances between gene carriers and non–gene carriers decreases, the mean maximum lod score decreases, since a low ratio decreases the certainty of whether an affected

Table 12.4 Impact of Misspecifying Disease Penetrance on Lod Score Analysis

Model Used for Analysis				Model Used for Simulation: $q_a = 0.02$, $f_{AA} = 0.004$, $f_{Aa} = f_{aa} = 0.85$			
				$\theta = 0.00$ (power)		$\theta = 0.50$ (type I error)	
Q_0	f_{AA}	f_{Aa}	f_{aa}	Mean Z_{max}	Mean MLE (θ)	Mean Z_{max}	Mean MLE (θ)
0.02	0.004	0.85	0.85	9.14	0.01	0.11	0.43
0.02	0.001	0.85	0.85	9.16	0.01	0.11	0.43
0.02	0.01	0.85	0.85	9.08	0.01	0.11	0.43
0.02	0.05	0.85	0.85	8.40	0.00	0.11	0.42
0.02	0.05	0.85	0.85	6.78	0.00	0.11	0.40

individual is a gene carrier and an unaffected individual a non–gene carrier. In the presence of linkage, most individuals are nonrecombinants, so the low ratio results in a decrease in power. However, when there is some degree of phenocopy and incomplete penetrance, selection of a low ratio is a conservative strategy.

Often, however, the clinical phenotype is complex and is confounded by such factors as phenocopies (phenotypes that are due to some other etiology), variable expressivity, and penetrance. Hence, complete elimination of phenotypic misclassification is not feasible. When this is the case, a specific probability of misclassification can be incorporated into the genetic model along with the penetrance parameters. Analogous to a penetrance parameter, which defines the probability of expressing the trait given the disease allele, a misclassification parameter defines the probability of expressing the trait given a normal allele. The need to decide whether to include or exclude a misdiagnosis parameter, for example, arises in studies of late-onset (mean age of onset > 60 years) familial Alzheimer disease. The diagnosis in Alzheimer disease can be confirmed only upon autopsy that is obtained only in a minority of cases. Thus the clinical diagnosis of AD in a living person does not always occur with the pathological findings, with error rates in the range of 15% having been reported.

Inclusion of a misdiagnosis parameter protects against distortion of the estimate of θ, but it results in loss of power in the overall study. Thus, it is best not to rely solely on using a misdiagnosis parameter, but to carefully consider the diagnostic criteria at the pedigree-sampling stage. Standard criteria for diagnosis, outlined before sampling begins and subsequently applied to all participants, contributes significantly to the long-range success of the study.

Impact of Misspecifying Marker Allele Frequency

Misspecified marker allele frequencies do not always have a large impact on the mean maximum lod score (see Table 12.5). This can be explained by the small degree of misspecification (the worst cases are the frequencies for allele 1 and 2, which were misspecified as 0.16 instead of 0.02, and allele 6 which was misspecified as 0.17 instead of 0.5) and the small number of family members missing genotype data. When there are many family members without genotype data, incorrect estimates of marker allele frequencies can have a large impact on the lod score (Ott,

Table 12.5 Impact of Misspecified Marker Allele Frequency on Lod Score Analysis

	Model Used for Simulation: $q_a = 0.02$, $f_{AA} = 0.004$, $f_{Aa} = f_{aa} = 0.85$			
	$\theta = 0.00$ (power)		$\theta = 0.50$ (type I error)	
Marker Allele Frequency	Mean Z_{max}	Mean MLE (θ)	Mean Z_{max}	Mean MLE (θ)
Same as simulated	9.14	0.01	0.11	0.43
Equally frequent	9.12	0.01	0.12	0.43

1991). The sensitivity is directly related to the allele frequency distribution and to the number of ungenotyped founders (those without parents) in the pedigree. The genotype for each ungenotyped founder must be estimated and the population allele frequencies are used. For example, if two affected cousins share a marker allele in common, the probability of linkage increases with the rarity of the allele. When the parents or grandparents are not available for genotyping, the probability that the allele was present only in the line of descent is calculated from the allele frequencies. If the allele is quite common, there is an increased probability that parents are homozygous for that allele or that married-in family members may have transmitted the allele, and thus there is little evidence for linkage.

GENETIC HETEROGENEITY

Examination of the lod scores for each pedigree provides an important additional piece of information. In this example, the lod scores ranged from –1.63 to 1.61 from pedigree to pedigree, which suggests that only some of the pedigrees (e.g., genetic locus heterogeneity) may be linked to this locus. In fact, for a complex disorder that is inherently heterogeneous, the ability to find linkage depends highly on how much homogeneity can be forced into the study sample by using other criteria such as examining only a specific clinical subtype or only a specific isolated population. This may be one reason for the failure of many attempts to replicate linkage findings. However, if heterogeneity is considered in an analysis, the chance of finding evidence for linkage will be increased (Faraway, 1993).

The hypothesis of homogeneity can be formally tested using two different approaches: the M test (Morton, 1956) and the admixture test (Ott, 1991). When families can be preassigned to several different groups based on some disease characteristics, such as age of onset, severity of disease, clinical features, or transmission pattern, an M test can be used to test for different recombination fractions between several groups of families, $\theta_1 \neq \theta_2 \neq \cdots \theta_n$. The hypothesis ($H_1$) of linkage and homogeneity specifies $\theta_1 = \theta_2 = \cdots \theta_n < 0.5$. Under the hypothesis of heterogeneity (H_2): $\theta_1 \neq \theta_2 \neq \cdots \theta_n$, the recombination fractions are potentially different in the different groups of families. To test H_1 against H_2, one computes $\chi^2 = 2 \times \ln(10)[\Sigma Z_i(\theta_i) - Z(\theta)]$, where $Z_i(\theta_i)$ is the maximum lod in each of the family groups, and $Z(\theta)$ is the maximum lod score over all the pedigrees. Asymptotically, under the assumption of homogeneity (H_1), there is a chi-square distribution with $(n - 1)$ df.

The admixture test (Ott, 1991) can be used when families cannot be assigned unequivocally to one or more groups based on other criteria. This is a maximum likelihood approach; in its most usual form, two family types are assumed, with α denoting the proportion of type I families and $(1 - \alpha)$ the proportion of type II families. The recombination fraction in type I families is equal to θ_1 and that in type II families is $\theta_2 = 0.5$. The hypothesis of heterogeneity is H_2: $\alpha < 1$, $\theta_1 < 0.5$, and the hypothesis of homogeneity is obtained from H_2 by a single restriction, H_1: $\alpha = 1$. Under H_2, the likelihood is maximized over α and θ_1. Under H_1, the restricted hypothesis, θ_r is obtained assuming homogeneity ($\alpha = 1$). The test of the hypothesis of homogeneity is carried out by calculating $\chi^2 = 2\,[\ln L(\alpha, \theta_1) - \ln L(\alpha = 1, \theta_r)]$, which asymptotically has a chi-square distribution with a mixture of 1 and 0 df.

An M test can be used in this bipolar data set to test the alternative hypothesis that the disease susceptibility allele in the paternal families is different from that of maternal pedigrees. This hypothesis, defined before the study was initiated, was based on earlier reports (and possibly due to mitochondrial inheritance, genomic imprinting, or X linkage). The maximum lod score $Z_i(\theta_i)$ for paternal pedigrees is 2.78 at $\theta = 0.01$, and for maternal pedigrees is 0.00 at $\theta = 0.50$. The total maximum lod score is 0.27 at $\theta = 0.30$. Under the null hypothesis of homogeneity, the $\chi^2 = 11.56$ [i.e., $2 \times \ln(10) \times (2.78 - 0.00 - 0.27)$], with df = 1. The *P* value of the chi-square is 0.00067, suggesting that the null hypothesis can be rejected. This result suggested that D18S41 may be linked to the bipolar disease locus in the paternal pedigrees, but not in the maternal pedigrees. This homogeneity test can be a very useful tool for handling the heterogeneity seen in complex disorders. However, it is vital to point out that the hypothesis should be based on solid biological and clinical information and should be stated a priori to the linkage results. This approach is not valid if the hypothetical groups are generated *after* looking at the lod scores.

An admixture test can also be applied to the bipolar data set to test the more general alternative hypothesis that two types of family exist, one linked to the marker ($\theta_1 < 0.5$) with proportion α, and another unlinked to the marker ($\theta = 0.5$) with proportion of $1 - \alpha$. The null hypothesis in this case is that all the families are linked to this marker ($\alpha = 1, \theta_r < 0.5$). The maximum likelihood for the alternative hypothesis $L(\alpha, \theta_1)$ is 19.82, with $\alpha = 0.25$ and $\theta_1 = 0.00$, and the maximum likelihood for the null hypothesis $L\,(\alpha = 1, \theta_r)$ is 1.86 with $\theta_r = 0.30$. Under the null hypothesis, the $\chi^2 = 4.73$ [i.e., $2(\ln 19.82 - \ln 1.86)$], with df of the mixture 0 and 1. The nominal *P* value of this is 0.015. Initially this result might suggest that the hypothesis of homogeneity can be rejected at the 5%, but not at 1% type I error level. However, the lod score under heterogeneity is only 1.30 (i.e., $\log_{10}19.82$) and does not reach the classically needed lod score of 3.0.

This conflict demonstrates that in practice the significance of such a finding is difficult to assess. While theory assumes that the likelihood functions follow a chi-square distribution, in reality this is not known with certainty. Thus it is very difficult to assess the evidence for heterogeneity if the overall lod score is less than 3.0. There are two schools of thought concerning the interpretation of admixture results. One argument treats the test of linkage and the test of heterogeneity as separate tests in a two-stage process. Thus an overall lod score of 3 is necessary to reject the null hypothesis of no linkage. If that hypothesis is rejected, then the additional hypothesis of heterogeneity can be tested. The alternative argument tests linkage and homogeneity simultaneously. In this situation, an overall lod 3 difference between homogeneity with no linkage and heterogeneity with linkage is sufficient. The former argument clearly produces a more conservative test.

The power of these statistical tests of homogeneity is influenced by a number of factors, including the distance between the linked marker and the disease allele locus and the magnitude of the lod score generated. In general, where the recombination fraction between a disease gene and the marker is larger, there is less power to detect heterogeneity. Also, the smaller the lod score generated by the linkage analysis (either because of smaller pedigree size or lower informativeness in a family), the more difficult it becomes to have sufficient power to detect significant heterogeneity. Larger, more informative families will yield more power than a series of smaller families in discriminating multiple disease loci. One advantage of the admixture test is that it does not require predividing families and thus does not require any prior knowledge about possible heterogeneity. If there is a valid reason to predivide the samples, however, the admixture test is less powerful than the M test.

In summary, heterogeneity can be a major confounding factor in the gene mapping process. Once heterogeneity has been confirmed, one must ask what criteria should be used to classify a family as "linked" for molecular follow-up purposes. The admixture test can also calculate the posterior probability (i.e., probability after considering additional information—in this case, lod score results) that a family is of the linked type, but this is still just a statistical estimate and is based on only the families included in that particular set of data. These results may be useful when crossover information is used for physical mapping of a disease gene region. It is necessary to make sure that the family is of the linked type, since a single mistake in determining crossovers could result in considerable unnecessary expense of laboratory time and effort.

STRATIFICATION BASED ON CLINICAL CRITERIA

As discussed with respect to the M test for homogeneity, an important consideration in analyzing a disease is the possibility of sample stratification based on clinical subtypes. Such strategies are often used to generate a more homogeneous sample under the assumption (which may or may not be true) that by narrowing the definition of the trait, one will be including only individuals more likely to have similar genetic etiologies leading to the disease phenotype. This is one way *a priori* to address the inherent heterogeneity in genetically complex traits. This approach has been highly successful in mapping genes in Alzheimer disease (AD), breast cancer, and cardiovascular disease (Jeunemaitre et al., 1992; Corder et al., 1993; Futreal et al., 1994; Miki et al., 1994; Pericak-Vance et al., 1991, 1994). Two examples of stratification variables are given.

1. *Age of Onset* Examining the age of onset of the family data may give clues to underlying differences in genetic etiology. For example, early-onset familial AD is caused by three different autosomal dominant disease genes, while late-onset AD has a major risk factor in the *APOE-4* allele. Initial linkage studies in this complex trait involved stratifying the families by age at onset. This led to the successful finding of a linkage to chromosome 19 in late-onset AD (Pericak-Vance et al., 1991). A similar strategy was used in locating and cloning *BRCA1* (Hall et al., 1990; Futreal et al., 1994).

2. *Clinical Phenotype* Careful examination of the clinical phenotype may reveal a subset of patients that show a particular clinical manifestation. Families can be grouped by this characteristics, provided sufficient sample size and power for analysis remain available. Another important consideration for quantitative traits entails the sampling of individuals with extreme phenotypes. Families can be selected in which family members present in the upper 5–10% of the distribution of the trait. The underlying hypothesis here is that these individuals represent extreme phenotypes that are more likely to be the result of genetic influences.

UNINFORMATIVE MARKERS AND MISSING GENOTYPE DATA

Uninformative markers and missing genotypes scattered across the families in a data set can lead to either a false conclusion of linkage or exclusion of a true linkage. The more uninformative a marker (e.g., the lower its heterozygosity), the more likely it is to be uninformative in the critical individuals who may contribute substantially to the linkage information in a pedigree. This is particularly true for small and medium-sized pedigrees, where one or two individuals may carry most of the linkage information. By chance then, such individuals may be uninformative (e.g., homozygous), thus masking the true linkage information. Thus in some cases many unlinked pedigrees may be uninformative, leading to an inflated lod score. In other cases, many linked pedigrees may be uninformative, leading to a deflated lod score.

This problem, which is confounded when two-point analysis is used on many markers, is partly responsible for the phenomenon of having evidence for linkage to a single marker without evidence for linkage in the flanking markers. In this bipolar data set for D18S41, there are 10 uninformative families (about 36% of all the families) with lod score less than ±0.3, mostly due to homozygosity at the marker locus for an affected parent (Table 12.1). Thus caution should be used in the interpretation of the linkage results for this marker. The conclusions drawn from the linkage analysis might be very different if the complete and true linkage information from these uninformative pedigrees could be obtained.

There are several ways to guard against this problem. The first is to increase marker heterozygosity, a much easier task now that microsatellite markers are so pervasive. The second is to select relatively large pedigrees, where the impact of any one uninformative individual is diminished. Success on this score, of course, depends on the unlikely situation of having many pedigrees to choose from. The third is to carefully consider the results from flanking markers. Perhaps the best approach, however, is to perform multipoint analysis.

MULTIPOINT ANALYSIS

Multipoint lod score analysis is an extension of two-point analysis in which linkage of a disease trait is tested not to just a single marker, but to an entire map of markers. There are several types of multipoint lod score analysis, each defined differently (Ott, 1991). Among them, the map-specific multipoint lod score is the most commonly used. It is defined as

$$Z(x, \phi_0) = \log_{10}\left[\frac{L(x, \phi_0)}{L(\infty, \phi_0)}\right] \tag{12.1}$$

where $L(x, \phi_0)$ is the likelihood that a disease locus is located at a distance x on a fixed map consisting of several markers and $L(\infty, \phi_0)$ indicates the likelihood that the disease locus is not on the map (corresponding to $\theta = 0.50$; i.e., no linkage), under a correctly assumed model.

Usually $Z(x)$ is evaluated at several positions on the map from one end to the other end. Sometimes, a location score is used, which is twice the natural logarithm in equation (12.1) instead of the base_{10} logarithm. However, this can cause confusion when multipoint values are compared to two-point lod scores. To further the confusion, the term "location score" is often used in the literature to mean the base_{10} conversion.

There are two major advantages of multipoint lod score analysis. First, it provides an opportunity to impute the genotype information at an original uninformative locus via haplotype information. Thus the linkage results are less sensitive to the uninformative or missing genotype at any single marker. Using the same example of the bipolar data set, 6 of 10 previously uninformative pedigrees for two-point lod scores became informative (multipoint lod $> \pm 0.3$) in that region when one flanking marker on each side was used. In essence, multipoint analysis can extract more of the total inheritance information from the family. In one simulation study, when a rare dominant disease gene was simulated in the middle of an 18 cM map using a pedigree with two affected fourth cousins, a lod score of 2.2 (91% of its theoretical maximum) was obtained when 20 markers were analyzed simultaneously. In contrast, the highest two-point lod score was only 0.83 (34% of its theoretical maximum), and even simultaneous six-marker analysis yielded, at most, a lod score of 1.74 (72% of theoretical maximum) (Kruglyak et al., 1996).

Second, the multipoint lod score approach can be very useful to pinpoint a disease gene location in the fine-mapping of a Mendelian disorder. This is achieved by evaluating many locations $Z(x)$ on a fixed map. Any true recombinant at some location x will contribute a strongly negative lod score and decrease the possibility that this position contains a candidate locus. This significantly narrows the region in which the disease locus can exist, thus helping to better define the minimum candidate region.

Although it is less useful to use a multipoint lod score approach to pinpoint a disease locus for a complex disorder, multipoint analysis is still very useful for complex disorders if a map covers only a small region and $Z(x)$ can be evaluated at locations outside the fixed map, which is equivalent to allowing for overestimation of recombination fraction in the two-point analysis. This is essentially the same as testing for linkage to a haplotype of markers. This approach takes advantage of using haplotype information and makes the multipoint analysis as robust to a misspecified genetic model as two-point analysis. Multipoint lod scores are more informative (higher absolute values of lod score) than the results of two-point analysis, which is useful in the subsequent heterogeneity test.

THE EFFECT OF HETEROGENEITY ON MULTIPOINT ANALYSIS

As with two-point analysis, multipoint analysis can be subjected to tests of homogeneity using the M test or the admixture test. However, the interpretation of the significance level in the multipoint situation is not straightforward because it is not clear that the asymptotic properties of the chi-square distribution hold in multipoint analyses. Thus, comparison of $\log_{10}$ of the likelihoods of the various hypotheses, such as linkage versus nonlinkage and linkage and homogeneity versus linkage and heterogeneity, is used to evaluate these results (Ott, 1992). A $\log_{10}$ likelihood difference of 3 between the likelihoods under homogeneity and heterogeneity is generally used to indicate a significant result.

As in the two-point analysis, heterogeneity has a major impact on the power to correctly detect linkage using multipoint analysis. In two-point analysis, the problem of heterogeneity can be partially absorbed by an increased (overestimated) recombination fraction. However, in multipoint analysis, the flanking markers prevent the recombination fraction from floating very far; thus the whole region between markers is more frequently and incorrectly excluded. In addition to locus heterogeneity, phenocopies and incomplete penetrance can lead to the same problem. Thus extreme caution is needed in interpreting multipoint lod scores within a fixed map, especially when the map covers a very large chromosomal region. This is demonstrated in Figure 12.1, where the region holding the gene of interest is excluded (all families) when heterogeneity is ignored. When heterogeneity is taken into account (linked families), the region is clearly identified. The classic shape of the curve for multipoint analysis when heterogeneity exists is seen in Figure 12.1 for all families, where there are two humps on the outside of the map and a well on the inside. One partial solution to this general problem is to use inflated disease allele frequencies, since this allows probabilistically for nonsegregation (through parental homozygosity or dual matings) (Risch and Giuffra, 1992).

This problem is magnified for genetically complex disorders because a misspecified model (one part of which may be heterogeneity) can result in evidence for a false recombinant, which will subsequently generate a strongly false negative lod score and potentially eliminate the true region. For example, a true nonrecombinant phenocopy (non–gene carrier but with the same affected phenotype, possibly due to an environmental factor) will be a recombinant if he or she is classified as affected.

One critical factor for multipoint analysis is correctly specifying the order of the markers and the distances between the markers. The results of multipoint lod score analysis depend on the quality of the map. If it is not possible to obtain a published map for a selection of markers, and there are a large number of meioses, the order and distances can be estimated from the study sample.

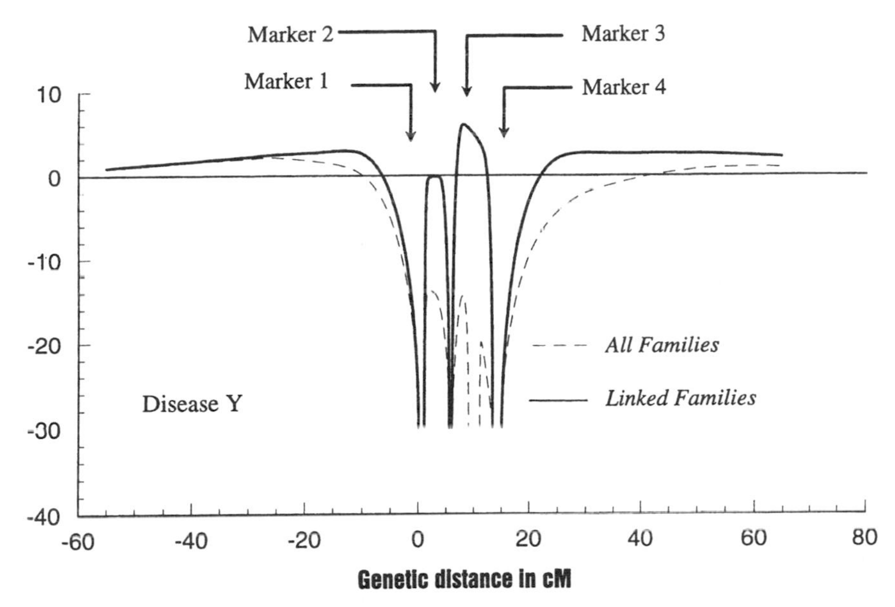

Figure 12.1 Examples of multipoint linkage analysis in the absence and presence of genetic heterogeneity. Multipoint linkage analysis between disease Y and four marker loci in the presence of heterogeneity. When all families are examined without testing for heterogeneity, evidence for significant linkage is lost. Examination of "linked" families after formal heterogeneity testing shows evidence for linkage.

PRACTICAL APPROACH FOR ANALYSIS IN GENETICALLY COMPLEX TRAITS

Many factors are important in designing, performing, and interpreting lod score analysis. These factors include the selection of genetic models, the characteristics of diseases, and the pedigree structures. An appropriate lod score analysis requires comprehensive considerations of every aspect of these factors.

Genetic Models and Parameters for Complex Disorders

The classic strategy for selecting genetic models for lod score analysis is to rely on the results of complex segregation analysis, which provides the disease allele frequency, degree of dominance, and penetrances for disease allele carriers and non–disease allele carriers (Chapter 5). Such a strategy has been successful for diseases that follow a clear Mendelian single locus segregation pattern. The results from segregation analysis may have some utility for complex disorders, and successful experiences have been reported. In the linkage study of familial breast cancer, for example, the genetic model (and parameters including disease allele frequency, age-dependent penetrances for disease allele carrier, and non–disease allele carriers) from a segregation analysis served in lod score analysis. Linkage to the region of *BRCA1* was identified (Easton et al., 1993).

However, there are some caveats associated with the application of the results from segregation analysis to a lod score analysis in genetically complex traits. First, since genetic heterogeneity is common in many complex disorders, the results from a segregation analysis are difficult to generalize to other populations and races. Even in the same population and race, different ascertainment schemes used for segregation analysis and for linkage analysis might represent two different disease subgroups. The multiplex family ascertainment scheme (families heavy loaded with affected members) used for many linkage analyses to increase linkage information must be accounted for, and the usual result is little remaining information that is useful for segregation analysis. Second, segregation analysis using a one-locus disease model does not correctly model a disease controlled by two or more disease genes (heterogeneity or epistasis). It has been shown that segregation analysis assuming a one-locus disease model can lead to evidence for a major gene effect for many two-locus models, with the estimates of the major gene parameters being very different from those of the two genes involved in the disease (Dizier et al., 1993). Although several simulation studies suggested that ignoring the effect of second gene has no significant effect on the power to detect linkage, these simulations used the correct one-locus disease model or at least the correct transmission mode in the analysis linkage (Vieland et al., 1992). In contrast, for some two-locus disease models, use of major gene parameters derived from segregation analysis under one-locus disease model may affect the detection of linkage with bias in the estimation of recombination fraction and the resulting lod score (Dizier et al., 1996). This is particularly true when the two genes have different characteristics (allelic frequency and mode of transmission). Fortunately, segregation analyses assuming a two-locus disease model can be used to investigate different two-locus modes of transmission and to yield better estimates of genetic parameters such as gene frequencies for two disease genes and penetrances for the nine distinct genotypes. It has been reported several times that in the case of oligogenic traits, segregation analysis assuming a two-locus disease model and consequent linkage analysis using the best model from the segregation analysis increased the chance of detecting linkage for one and two genes. Unfortunately, there is a substantial price to be paid. Such detailed and complex analysis requires intensive computation, and generally very large data sets. Collecting such data sets is usually a very long and expensive procedure (Chapter 5; Xu et al., 1994).

Liability Classes and Affecteds-Only Analysis

Liability classes provide an important tool for handling the complexity of common disorders in lod score analysis (Terwilliger and Ott, 1994). For example, for many genetically complex disorders the phenocopy rate increases as the disease age of onset increases, or as the disease severity decreases. In these cases, if the same phenocopy rate is given to all age groups and disease severity is not taken into consideration (as in the nonparametric analysis), either some linkage information will be lost because a high phenocopy rate is assigned (affected individuals are thus less certain to be gene carriers) or more false recombinants will be generated because a low phenocopy rate is assigned (old age-of-onset affected individuals and mild cases are more likely to be assigned as gene carriers, although they are more likely to be phenocopies). However, if different phenocopy rates are assigned to individuals dependent on their age of onset and severity of disease (i.e., different liability classes), one can maximize the linkage information from younger and severely affected individuals and minimize the impact of potential phenocopies among old-age-of-onset and mildly affected individuals.

Another method for using lod scores in genetically complex diseases is to use an affecteds-only (low-penetrance) model with a relatively frequent disease allele frequency (e.g., > 0.01). The choice of an affecteds-only model circumvents some of the problems in lod score analysis when the genetic model is unknown because assumptions concerning unaffected individuals can be eliminated. In the complex disease case where several genes may interact to cause the clinical phenotype, unaffected individuals may have one of the necessary loci but not the others, thereby presenting as recombinants in a single gene analysis. Several studies have shown that the problem with applying a model-dependent method to a model-unknown situation is not the false conclusion of linkage but rather the increased probability of missing a linkage when it is present (Clerget-Darpoux et al., 1986). Although this approach is not the optimal solution, it does allow some estimate of power in the data at hand.

Mod Score Approach

The lod score $Z(x)$, by definition, is the lod score calculated from a correctly specified model ϕ_0. A correct genetic model can be obtained for Mendelian disorders; however, it is difficult to correctly specify in genetically complex disorders. Usually a single arbitrary, and presumably wrong, ϕ is used in the lod score calculation, which is sometimes called a wrod score (wrong lod score), defined as $W(x,\phi) = \log_{10} [L(\theta = x, \phi)/L(\theta = 0.5, \phi)]$. Since the impact of the wrong model on two-point lod score is small as long as the dominance is correctly specified, practically, a wrod can be treated as a lod score $Z(x)$, with some caveats.

When the lod score is maximized over both recombination fractions θ and many genetic models ϕ, it is called a mod score, defined as $M(x) = \log_{10}\max_{\phi}[L(\theta = x, \phi)/L(\theta = 0.5, \phi)]$. Explicit mathematical proof and extensive computer simulations support the validity of the mod score approach under certain conditions (Hodge and Elston, 1994). One important condition is that the likelihood calculation be based on the correct genetic mechanism. For example, when the true underlying mechanism is a single locus with three penetrances f_i and one allele frequency q, maximizing lod scores with respect to q and f_i will asymptotically yield the true values of the parameters. However, if the disease mechanism is actually two locus, but assumed to be a single locus, this result does not hold. On the other hand, assumption of a two-locus model when the true mechanism is a one-locus setup will not invalidate the asymptotic results as long as the true one-locus model is a special case of the assumed two-locus model. The mod score can be used not only to detect linkage, but also to determine the appropriate genetic model. It has been shown that using the mod score approach to infer a true genetic model is ascertainment assumption free (Hodge and Elston, 1994).

In practice, some arbitrary designed models (one dominant and one recessive, based on reasonable segregation results or epidemiological data) can be used in the initial linkage study of complex disorders, that is, the wrod approach. When there is evidence of linkage, the mod score approach can be used to estimate the genetic model.

SUMMARY

Lod score analysis is a powerful approach to linkage analysis and has been the workhorse of disease gene mapping for many years. However, its power comes at the price of prior knowledge; the genetic model must be specified with several parameters, and incorrect specification can dramatically hurt the power of this approach. Several modifications of lod score analysis, including the affecteds-only and mod score approaches, can partially compensate for these problems and permit the adaptation of the lod score method to genetically complex diseases.

REFERENCES

Clerget-Darpoux F, Bonaiti-Pellie C, Hochez J (1986): Effects of misspecifying genetic parameters in lod score analysis. *Biometrics* 42:393–399.

Corder EH, Saunders AM, Strittmatter WJ, Schmechel DE, Gaskell PC, Small GW, Roses AD, Haines JL, Pericak-Vance MA (1993): Gene dose of apolipoprotein E type 4 allele and the risk of Alzheimer's disease in late onset families. *Science* 261:921–923.

Dizier MH, Bonaiti-Pellie C, Clerget-Darpoux F (1993): Conclusions of segregation analysis for family data generated under two-locus models. *Am J Hum Genet* 53:1338–1346.

Dizier MH, Babron MC, Clerget-Darpoux F (1996): Conclusion of lod-score analysis for family data generated under two-locus models. *Am. J. Hum Genet* 58:1338–1346.

Easton DF, Bishop DT, Ford D, Crockford GP (1993): Genetic linkage analysis in familial breast and ovarian cancer: Results from 214 families. The breast cancer linkage consortium. *Am J Hum Genet* 52:678–701.

Elston RC, Stewart J (1971): A general model for the genetic analysis of pedigree data. *Hum Hered* 21:523–542.

Faraway JJ (1993): Distribution of the admixture test for the detection of linkage under heterogeneity. *Genet Epidemiol* 10:75–83.

Futreal PA, Liu Q, Shattuck-Eidens D, Cochran C, Harshman K, Tavtigian S, Bennett LM, Haugen-Strano A, Swensen J, Miki Y, et al. (1994): *BRCA1* mutations in primary breast and ovarian carcinomas. *Science* 266:120–122.

Hall JM, Lee MK, Newman B, Morrow JE, Anderson LA, Huey B, King MC (1990): Linkage of early-onset familial breast cancer to chromosome 17q21. *Science* 264(1562):1141–1145.

Hodge SE, Elston RC (1994): Lods, Wrods and Mods: The interpretation of lod scores calculated under different models. *Genet Epidemiol* 11:329–342.

Jeunemaitre X, Soubrier F, Kotelevtsev YV, Lifton RP, Williams CS, Charru A, Hunt SC, Hopkns PN, Williams RR, Lalouel JM (1992): Molecular basis of human hypertension, role of angiotensinogen. *Cell* 71:169–180.

Kruglyak L, Daly MJ, Reeve-Daly MP, Lander ES (1996): Parametric and nonparametric linkage analysis: A unified multipoint approach. *Am J Hum Genet* 58:1347–1363.

Lander ES, Kruglyak L (1995): Genetic dissection of complex traits: Guideline for interpreting and reporting linkage results. *Nat Genet* 11:241–247.

Lathrop GM, Lalouel JM, Julier C, Ott J (1984): Strategies for multilocus linkage analysis in humans. *Proc Natl Acad Sci USA* 81:3443–3446.

Miki Y, Swensen J, Shattuck-Eidens D, Futreal PA, Harshman K, Tavtigian S, Liu Q, Cochran C, Bennett LM, Dent W, et al. (1994): A strong candidate for the breast and ovarian cancer susceptibility gene *BRCA1*. *Science* 266:66–71.

Morton NE (1955): Sequential tests for the detection of linkage. *Am J Hum Genet* 7:277–318.

Morton NE (1956): The detection and estimation of linkage between the genes for elliptocytosis and the Rh blood type. *Am J Hum Genet* 8:80–96.

O'Connell JR, Weeks DE (1995): The VITESSE algorithm for rapid exact multilocus linkage analysis via genotype set-recoding and fuzzy inheritance. *Nat Genet* 11:402–408.

Ott J (1991): *Analysis of Human Genetic Linkage,* rev. ed. Baltimore: Johns Hopkins University Press.

Ott J (1992): Strategies for characterizing highly polymorphic markers in human gene mapping. *Am J Hum Genet* 51:283–290.

Pericak-Vance MA, Bebout JL, Gaskell PC, Yamaoka LH, Hung W-Y, Alberts MJ, Walker AP, Bartlett RJ, Haynes CS, Welsh KA, Earl NL, Heyman A, Clark CM, Roses AD (1991): Linkage studies in familial Alzheimer's disease: Evidence for chromosome 19 linkage. *Am J Hum Genet* 48:1034–1050.

Pericak-Vance MA, Bass MP Yamaoka LH, Gaskell PC, Scott WK, Terwedow HA, Menold MM, Conneally PM, Small GW, Vance JM, Saunders AM, Roses AD, Haines JL (1997): Complete genomic screen in late-onset familial Alzheimer disease: Evidence for a new locus on Chromosome 12. *J Am Med Assoc* 278:1237–1241.

Risch N, Giuffra L (1992): Model misspecification and multipoint linkage analysis. *Hum Hered* 42:77–92.

Schaffer AA, Gupta SK, Shriram K, Cottingham RW (1994): Avoiding recomputation in linkage analysis. *Hum Hered* 44:225–237.

Schellenberg GD, Bird TD, Wijsman EM, Orr HT, Anderson L, Nemens E, White JA, Bonnycastle L, Weber JL, Alonso ME, Potter H, Heston LL, Martin GM (1992): Genetic linkage evidence for a familial Alzheimer's disease locus on chromosome 14. *Science* 258:668–671.

St George-Hyslop PH, Haines JL, Rogaev EI, Mortilla M, Vaula G, Pericak-Vance M, Foncin JF, Montesi MP, Bruni AC, Sorbi S, Rainero I, Pinessi L, Pollen D, Polinsky RJ, Nee L, Kennedy JL, Macciardi F, Rogaeva EA, Liang Y, Alexandrova N, Lukiw WJ, Schlumpf K, Tanzi R, Tsuda T, Farrer LA, Cantu JM, Duara R, Amaducci L, Bergamini L, Gusella JF, Roses AD, Crapper-Maclachlan DR (1992): Genetic evidence for a novel familial Alzheimer's disease locus on chromosome 14. *Nat Genet* 2:330–334.

Stine OC, Xu J, Koskela R, McMahon FJ, Gschwend M, Friddle C, Clark CD, McInnis MG, Simpson SG, Breschel TS, Vishio E, Riskin K, Feilotter H, Chen E, Shen S, Folstein S, Meyer DA, Botestin D, Marr TG, DePaulo JR (1995): Evidence for linkage of bipolar disorder to chromosome 18 with a parent-of-origin effect. *Am J Hum Genet* 57:1384–1394.

Terwilliger JD, Ott J (1994): *Handbook of Human Genetic Linkage.* Baltimore: Johns Hopkins University Press.

Vieland VJ, Hodge SE, Greenberg DA (1992): Adequacy of single-locus approximations for linkage analysis of oligogenic traits. *Genet Epidemiol* 9:45–59.

Xu J, Taylor EW, Panhuysen CIM, Prenger VL, Koskela R, Kiemeney B, LaBuda MC, Maestri NE, Meyers DA (1994): Evidence for two unlinked loci controlling for Q1. *Genet Epidemiol* 12:825–830.

13

Sib Pair Analysis

David E. Goldgar

Unit of Genetic Epidemiology
International Agency for Research on Cancer
Lyon, France

BACKGROUND AND HISTORICAL FRAMEWORK

The last 10 years have witnessed dramatic growth in the use of sib pairs in human genetic mapping. This is true from the point of view of theoretical developments and more sophisticated statistical methodology, as well as with respect to the successful application of these methods to mapping susceptibility genes for complex human traits. When we refer to complex traits, we typically divide them into two major categories: *qualitative* or dichotomous traits and *quantitative* or continuous ones. The qualitative traits most commonly studied are, of course, human diseases in which a person is either clinically affected or unaffected with the disease. The genetic complexity may occur for one or more of the following reasons:

1. The disease may be etiologically heterogeneous, with only a subset due to genes conferring high risk.
2. The disease may involve many different genetic loci that act together to cause disease.
3. A gene for the disease may predispose only in the presence of a particular environmental exposure.

Individual differences in quantitative traits such as height, blood pressure, or triglyceride levels are typically the result of several genes together with both specif-

Approaches to Gene Mapping in Complex Human Diseases, Edited by Jonathan L. Haines and Margaret A. Pericak-Vance. ISBN 0-471-17195-6

ic and random environmental components. These confounding factors in both quantitative and qualitative traits require the use of methods in which the precise mechanisms of disease causation are not required to be known; this has led to the recent interest in the methods described in this chapter.

Although the common use of these methods is relatively recent for reasons described below, it is important to remember that the concepts that underlie all sib pair methodology date back to the early part of this century. Indeed the first work describing a sib pair linkage approach in humans was a seminal paper by Penrose published in 1935 in *Annals of Eugenics,* the predecessor of *Annals of Human Genetics.* Following this paper, the major emphasis in human linkage analysis was focused on the development of likelihood-based methods in which the probability of the observed data is written as a function of the recombination fraction between two loci, and a statistical test is derived by comparing the likelihoods under the two hypotheses of linkage versus free recombination. Major contributions in this area were made by Fisher (1935), Haldane (1934), Haldane and Smith (1947), and most notably Morton (1955), who developed the concept of the lod score in the context of sequential sampling theory. These methods and their successors are described in detail in Chapter 12.

Aside from a second paper on the sib pair approach by Penrose (1953), the next major developments in sib pair analysis took place in the 1970s. The first such work was the seminal paper by Haseman and Elston (1972), which described an elegant regression-based approach to search for loci influencing quantitative traits, followed by the important work of Suarez et al. (1978), who considered the distribution of identity-by-descent status among sibs for a variety of genetic models and applied this method to the elucidation of the role of the major histocompatibility complex (MHC) in type I (juvenile onset) diabetes using a set of affected sib pairs.

Prior to the 1980s, the use of sib pairs for mapping complex disease loci was largely restricted to examination of the major histocompatibility complex (HLA) cluster of genes on chromosome 6. This limitation was motivated by three factors: the biological plausibility of involvement of immune host responses in many different diseases, the statistical association of particular antigens at this locus with a number of diseases, and historical circumstance (at the time, the HLA system was one of the few genetic markers that was sufficiently polymorphic to permit this approach).

Thus, until perhaps the mid-1980s, the use of sib pair methodology was largely confined to HLA and a few other systems (e.g., Rh blood group), with a relatively high degree of informativeness. With the advent of restriction fragment length polymorphisms (RFLPs), then variable number of tandem repeat markers (VNTRs), and most recently in the 1990s, short tandem repeat markers (STRs; microsatellites), the probability that there will be an informative genetic marker (or set of markers) near any disease gene has steadily increased (see Chapter 8). It is now the case that we have almost 100% coverage of the genome at a resolution of 1 centimorgan with genetic markers of high heterozygosity (> 0.70) (Gyapay et al., 1994; Dib et al., 1996). Because of this, the mapping of the relatively rare monogenic disorders is routine, and attention is now focused on more common disorders with more complex (or unknown): modes of inheritance. In addition, there is renewed interest in mapping quantitative traits, including those that may serve as surrogates for a disease (e.g., cholesterol levels for cardiovascular disease) or quantitative traits that have a strong genetic component and reflect normal human variation (e.g., height, finger ridge count). It is to problems of the latter two types that the application of sib pair and related methods has become standard practice in human genetic analysis.

This chapter details the basic concepts underlying these analyses, discusses methods of performing sib pair analysis for diseases or other dichotomous phenotypes, and then focuses on quantitative trait mapping using sib pairs. For methods of both types we first describe the quantification of a genetic effect, then present the method and some extensions, and examine the power of these methods for detecting linkage under a variety of conditions. Finally, we list the available software for these analyses and present several examples of the use of these methods to map complex traits.

IDENTITY BY STATE AND IDENTITY BY DESCENT

All methods that have been developed for the mapping of complex traits in human sib pairs or related relative sets depend in some way on quantification of the degree to which related individuals "share" alleles at the marker locus or loci under investigation. All such methods can be divided into two basic classes: those that depend on marker alleles shared *identical by state* (IBS) and those that rely on alleles shared *identical by descent* (IBD). The simpler of the two to define and understand is identity by state. Two alleles are said to be IBS if they are the same variant of some polymorphic system; that is, two alleles are IBS if they cannot be distinguished by means of a particular method of detection. For example, for an RFLP marker (see Chapter 9), if two individuals both exhibit a band on a gel of a given size when digested with a given restriction enzyme and hybridized to a probe, these two alleles are identical by state. Similarly, if two individuals both show the same number of repeat units for an STR marker, they are identical by state. Any two individuals, whether related or not, can share 0, 1, or 2 alleles identical by state; the two alleles possessed by any one individual can also be IBS, in which case the individual is said to be homozygous at this locus.

Identity by descent, on the other hand, depends not only on whether the alleles appear the same on a gel or by another detection method, but also on whether these alleles are derived or inherited from a common ancestor. Clearly, alleles that are shared IBD must also be identical by state; however, the converse is not true. That is, alleles that are identical by state are not necessarily IBD. Any two unrelated individuals who have the same genotype illustrate this; since they are unrelated, they by definition share no alleles IBD. Figure 13.1 shows two hypothetical examples to illustrate the differences between IBS and IBD.

In Figure 13.1*A*, in the first genotype configuration (i.e., both sibs are type '1 3'), it is clear from the parents that each affected child inherited the '1' allele

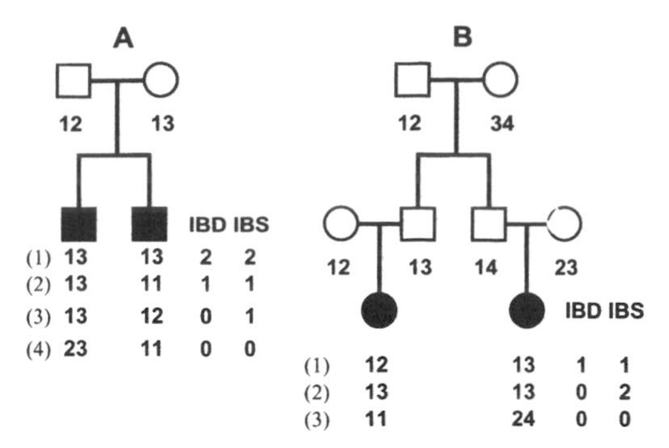

Figure 13.1 *Examples of identity by state and identity by descent. See text for discussion.*

from the father and the '3' allele from the mother. Thus both these alleles are shared identical by descent (IBD) as well as IBS. Contrast this with the situation in genotype configuration (3), in which both sibs have a '1' allele and thus share this allele identical by state. Inspection of the parental genotypes, however, indicates that in one case this allele was inherited from the father, while for the other sib it was inherited from the mother. Accordingly, this allele, while shared IBS, is not shared IBD. It is important to point out that in this situation we were only able to distinguish with certainty the origin of each allele because the parents were both genotyped and because the mating type was informative (i.e., the parents were of different heterozygous genotypes). In practice there will be cases in which one can only estimate the chance that the sibs share a particular number of alleles identical by descent. In many cases in which the parental genotypes are unknown, these estimates will depend on the frequency of the observed (and unobserved) alleles at the marker locus in question. The more polymorphic the marker, the more precisely we are able to estimate the IBD status.

For example, consider the case of two siblings who are identical homozygotes at a marker locus (i.e., they share two alleles identical by state) and the parents are unavailable. If this allele is relatively rare in the population then it is most likely that both unobserved parents are heterozygous at this locus and thus the two sibs share both alleles IBD. Conversely, if the allele is common, it is likely that at least one parent is homozygous at this locus and thus IBD status is more uncertain. A similar example for cousin pairs is shown in Figure 13.1*B*. In the second configuration, we see that although the two cousins have identical genotypes, they do not share either allele inherited from their common grandparents. Examination of the pedigree structure points out an important principle, namely, that individuals who are related through only one parent (unilineal relationships) can share at most only a single allele identical by descent. Examples of these are parent–offspring, cousins, half-siblings, and uncle–nephew. Individuals related through both parents (bilineal relationships), can share both alleles IBD. Other than siblings (and monozygotic twins), the most common bilineal relationship is double first cousins, which occur, for example, when two brothers marry two sisters; the offspring of these unions are double

Table 13.1 Expected Percentage of Affected Pairs Showing 0, 1, or 2 Alleles IBD at a Marker Locus If No Linkage Is Present

	Alleles IBD		
Pair Type	0	1	2
Monozygotic twins	0	0	100
Siblings	25	50	25
Parent–child	0	100	0
Grandparent–grandchild	50	50	0
Half-siblings	50	50	0
Uncle–nephew, etc.	50	50	0
Cousins	75	25	0
Double first cousins	56	38	6

first cousins. Other bilineal relationships occur in complex genealogies in which there are regular patterns of intermarriage between related individuals. Table 13.1 shows the expected proportion of alleles shared identical by descent for a number of common relationships.

The basic premise underlying many of the methods described later in this chapter is that if two related individuals are phenotypically similar (e.g., if both have the same disease or both have similar blood pressure values), then a genetic marker located nearby to a gene contributing to that trait, ought to be similar. The degree of similarity will depend on a number of factors. The two most significant are the overall contribution of this locus to the trait being studied and the genetic distance between the unknown (or unmeasured) gene influencing the trait and the genetic marker being tested. These issues are considered in depth in subsequent sections of this chapter.

QUALITATIVE TRAITS

Measures of Familiality

As stated earlier, the ability to detect the effects of a gene influencing a particular trait will depend on the magnitude of the effect of that gene. It may also depend on the overall contribution of genes to the trait.

For dichotomous disease traits, the most natural index of the strength of the genetic/familial component will be some measure of the extent to which relatives of affected individuals also have the disease of interest. The measure most commonly used is called the *familial relative risk* and is usually denoted by λ_R, where the subscript R denotes the degree of relationship to the index case. Most commonly, we will be concerned with the case of a relative who is a sibling or an offspring of the affected individual, although occasionally we will also be concerned with monozygotic twins or half-sibs. For example, if $\lambda_S = 3.0$ for a disease, this indicates that siblings have a threefold increased risk of the same disease compared to the risk to individuals in the general population. Expressed another way, the familial relative risk is the recurrence risk of disease in relatives divided by the population prevalence. For example, a disease in which 6% of siblings of affected individuals are themselves affected and for which the prevalence in the general population is 1/200, the value of λ_S is 0.06/0.005 = 12.0. Of course more sophisticated calculations must be performed when the disease is characterized by late and/or variable age at onset, but the concept is essentially the same. Typically, familial risks are estimated using case–control epidemiological studies in which a set of probands with the disease are either randomly selected from a population-based registry or from a consecutive series of patients, whereupon complete and detailed family histories are taken to identify the number and distribution of additional cases of disease in the family members. The same procedure is applied to relatives of age-matched controls, and the rate of disease in each group of relatives is compared, to estimate the familial relative risk, λ.

In considering other relatives, there is a predictable relationship between the λ_R values for different relationships. Following Risch (1990a), for a single locus or a model of several loci acting in an additive manner, we have:

$$\lambda_1 - 1 = 2(\lambda_2 - 1) = 3(\lambda_3 - 1) \tag{13.1}$$

where the subscripts refer to unilineal relationships of first- (parent–offspring), second- (e.g., half-sibs, uncle–nephew), and third-degree (cousins). For bilineal relationships such as sibs, the following relationship can be determined between the relative risk to monozygotic twins λ_M, and other relationships as follows:

$$\lambda_M = 4\lambda_S - 2\lambda_1 - 1$$

These are general relationships that apply to any genetic model, including both dominant and recessive. However, if the trait does not have a large recessive component, then approximately, $\lambda_S = \lambda_1$, and $\lambda_M - 1 = 2(\lambda_S - 1)$, or equivalently $\lambda_M = 2\lambda_S - 1$.

For models in which two or more loci are involved in disease susceptibility, we can distinguish two basic mechanisms of how the loci interact. The first comprise *additive models* in which the penetrances (i.e., allele-specific risks of disease) associated with the multilocus genotypes across different loci can be modeled as sums of factors for each genotype at each locus. It is important to note that this is different from the locus-specific penetrances themselves being additive; rather the pattern of the multilocus penetrances can be expressed by adding across the genotypes at each locus. For relatively rare genes, this model approximates the situation of genetic heterogeneity in which each locus is sufficient to cause disease on its own. The relationship between the overall familial risk λ_R and that attributable to each locus individually is given by (for two loci) $\lambda_R - 1 = \lambda_{R1} - 1 + \lambda_{R2} - 1$.

The second model, *the multiplicative or epistatic model,* occurs when the multilocus penetrances (or perhaps more exactly, the ratio of the high- and low-risk penetrances) can act multiplicatively across loci. In this model, individuals/families generally require the higher risk alleles at both loci to be affected. For a multiplicative model with N loci, for any degree of relationship R, the overall familial risk, λ_R, is equal to the product of the individual locus λ_R values, that is,

$$\lambda_R = \lambda_{R1}\,\lambda_{R2} \cdots \lambda_{RN}$$

The details of the formulation of these multilocus models are given in papers by Hodge (1981) and Risch (1990a). The important principle is that these different models produce different expectations in observed familial risks as a function of degree of relationship. For the additive model, equation (13.1) (for the single locus case) also applied to an arbitrary number of additively acting loci. However, under a multiplicative model, there is a faster decrease in λ_R values as a function of degree of relationship; for example, the difference in relative risk of disease in cousins compared to parent–offspring is larger for multiplicative models than for additive multilocus models. For a model of N loci acting multiplicatively, each with a common λ_{11} value, the predicted overall λ_1 is $\lambda_{11}{}^N$ and the overall $\lambda_2 = (1/2)^N(\lambda_{11} + 1)^N$. For example, consider a disease with an overall λ_1 of 8.0 caused by three loci, which contribute equally to the familial risk. Under an additive model, the relationship between λ_2 and λ_1 does not depend on the number of loci involved and can be directly determined from equation (13.1) as follows:

$$(8 - 1) = 2(\lambda_2 - 1)$$

$$7 = 2\lambda_2 - 2$$

$$\lambda_2 = \frac{9}{2} = 4.5$$

Under a multiplicative model, however, it is necessary to first calculate the λ_1 value for each locus; since the three loci are of equal effect, this is simply $8^{1/3} = 2.0$. The expected value for second-degree relationships for each locus is calculated from equation (13.1) as 1.5 and the overall λ_2 under the multiplicative model can now be calculated as $(1.5)^3 = 3.375$.

If estimates of recurrence rates or λ_R values for a number of degrees of relationships are available, these data can be used to get some idea of the number of loci involved, and whether it is more likely that these loci are acting in an additive or multiplicative fashion. For example, Risch (1990a) used published data on recurrence risks for relatives of probands diagnosed with schizophrenia to compare a number of models of gene action. Because these studies did not find an increased recurrence risk in siblings compared to offspring, it could be assumed that $\lambda_S = \lambda_1$ and therefore all models could be expressed in terms of a single quantity. All models were constrained to fit the overall value of the observed offspring relative risk of $\lambda_o = 10.0$. These data are reproduced here in Table 13.2. If one assumes that a single locus is responsible for the observed offspring relative risk of 10.0, the predicted risk of monozygotic twins is 19.0:

Table 13.2 Multilocus Multiplicative Models for Schizophrenia

		Model Prediction[a]						
Risk Ratio[b]	Observed	I	II	III	IV	V	VI	VII
λ_O	10.0	10.0	10.0	10.0	10.0	10.0	10.0	10.0
λ_S	8.6							
λ_M	52.1	19.0	100.0	75.0	55.6	43.8	56.3	42.2
λ_D	14.2							
λ_H	3.5							
λ_N	3.1							
λ_G	3.3							
(λ_2)	3.2	5.50	3.16	3.35	3.65	3.95	3.56	3.77
λ_C	1.8	3.25	1.78	1.87	2.03	2.20	1.96	2.07

[a]Definitions of models: I, one locus ~λ10 = 10.0; II, infinite loci, each with small effects, III, ~λ10 = 2.0, infinite other loci; IV, ~λ10 = 3.0, infinite other loci; V, ~λ10 = 4.0, infinite other loci, VI, ~λ20 = 2.0, infinite other loci; VII, ~λ10 = ~λ20 = ~λ30 = 2.0, infinite other loci.
[b]Subscripts: S, sibling; O, offspring; D, DZ twins; H, half-sibs; N, niece/nephew; G, grandchilg; 2, pooled value for all second-degree relationships; C, first cousins.
Source: Risch (1990a). Courtesy of *American Journal of Human Genetics.*

$$\lambda_M - 1 = 2(\lambda_1 - 1)$$
$$\lambda_M - 1 = 2(10.0 - 1)$$
$$\lambda_M = 19$$

far less than the observed MZ concordance of 52.1. In addition, we see that the observed values for both second- and third-degree relationships are substantially lower than that predicted under the single locus model. Thus it is unlikely that susceptibility to schizophrenia is governed by a single locus. Rather, based on these results, the observed familial risks would be consistent with a model of several loci acting epistatically in a multiplicative manner to confer susceptibility to schizophrenia. From Table 13.2, we see that the best fit for the observed data is a multiplicative model consisting of a single major locus with $\lambda_1 = 3$ amid a polygenic background of a large number of loci each with small effect (model IV) or two major loci each with $\lambda_1 = 2$ with a similar polygenic background (model VI). These results are of practical importance, inasmuch as the ability to map loci for a disease will depend on the magnitude of the effect associated with the locus that contributes most to the overall familial association, and on the true way in which the multiple loci interact in causing disease. These concepts are addressed formally in the sections that follow.

Tests for Linkage Using Sib Pairs

Test Based on Identity by State The first test proposed for using identity-by-state relationships in sib pairs is that by Lange (1986) in which the observed distribution of pairs sharing 0, 1, or 2 alleles identical by state is compared to that expect-

Table 13.3 Analysis of Identity by State Sharing Probabilities for Two Sibs at a Marker with *n* Equally Frequent Alleles

Number of Alleles	Probability of Sibs Sharing Alleles Identical by State		
	0	1	2
2	0.03	0.38	0.59
3	0.06	0.48	0.46
4	0.08	0.52	0.40
5	0.10	0.53	0.37
6	0.12	0.53	0.35
8	0.15	0.53	0.32
10	0.16	0.53	0.30
20	0.20	0.52	0.28
Infinite	0.25	0.50	0.25

ed under the hypothesis of free recombination between disease loci and the marker locus. The test for linkage therefore is a simple chi-square goodness-of-fit test comparing the observed IBS distribution to the expected. The expectations for IBS sharing under the null hypothesis are, of course, a function of the allele frequencies at the marker locus which take into account the distributions of mating types compatible with the observed pair genotypes. Table 13.3 shows the expected distribution of sharing 0, 1, and 2 alleles identical by state as a function of the informativeness of the marker (indexed as the number of equally frequent alleles).

Because alleles can be IBS without being IBD, there are more pairs than expected sharing 1 and 2 alleles than the 1/2 and 1/4 expected for IBD sharing; however, as the number of marker alleles increases, it can be seen that the IBS distribution (at least in nuclear families) approaches the 1/4, 1/2, 1/4 expected for identical-by-descent relationships. Intuitively, as discussed earlier, this is because for highly polymorphic markers, the probability that one or both parents are homozygous, or both are identical heterozygotes (which makes IBD status ambiguous) becomes increasingly small. The power of these methods for sib and other relative pairs is considered in a paper by Bishop and Williamson (1990), who examined the effect on power of the strength of the genetic component, marker informativeness, and type of relative pair. Although we have denoted these methods "identity by state," it is important to recognize that here one is simply using IBS as a surrogate for the unobservable IBD status. Many of the more sophisticated IBD methods presented in the next section are, in effect, estimating the unknown IBD distribution and thus are similar to the approach presented in this section.

Tests Based on Identity by Descent

Simple Tests In the simplest case, where the numbers of pairs sharing 0, 1, or 2 alleles identical by descent can be unambiguously determined (i.e., parents are typed and informative), several simple tests have been developed to test for linkage between a marker locus and a disease. The first such proposed test was a chi-square goodness-of-fit test comparing the observed proportion of sib pairs sharing 0, 1, or 2 alleles identical by descent with the expected proportions of 1/4, 1/2, and 1/4 under the hypothesis of no linkage between the marker and disease locus. A related test proposed by Suarez et al. (1978) is to compare only the proportion of sib pairs who share exactly 2 alleles IBD with the expected proportion of 1/4. Only differences in which the observed proportion of shared alleles *exceeds* the expected are considered significant; thus a one-sided test is used. The third test, called the means test, compares the average IBD sharing in the sample of pairs with the expected value of $0.5[0 \times \frac{1}{4} + 1 \times \frac{1}{2} + 2 \times \frac{1}{4})/2$ chromosomes]. Each of these tests can be written in terms of the observed numbers of pairs with each IBD sharing status:

0	1	2	Total
n_0	n_1	n_2	N

Statistics

1. *Goodness of Fit*

$$\chi^2 = \frac{4(n_0 - N/4)^2 + 2(n_1 - N/2)^2 + 4(n_2 - N/4)^2}{N}$$

Reject H_0 if χ^2 is greater than $\chi^2_{1-\alpha}$ with 2 degrees of freedom for size α test.

2. *Proportion*

$$T_1 = 4\left(\frac{n_2}{N} - \frac{1}{4}\right)\left(\frac{N}{3}\right)^{1/2}$$ one-sided t test with $N - 1$ degrees of freedom

3. *Means*

$$T_2 = (2n_2 + n_1 - N)\left(\frac{2}{N}\right)^{1/2}$$ one-sided t test with $N - 1$ degrees of freedom

As an example of the application of these three methods, we use data taken from Cox and Spielman (1989) for HLA sharing in sib pairs affected with type I diabetes (IDDM).

	Alleles or Haplotypes IBD		
Total pairs: 137	0	1	2
Observed	10	46	81
Expected	34	69	34

The results of these analyses are shown in Table 13.4.

Table 13.4 Results of Simple Sib Pair Tests on IDDM Data

Test	Statistic[a]	Degrees of Freedom
Goodness of fit	$\chi^2 = 88.4$	2
Proportions	$T_1 = 9.22$	136
Means	$T_2 = 8.58$	136

[a] T_1, is the proportions test statistic, T_2, is the means test statistic.

Given the overwhelming evidence for the involvement of HLA in IDDM as shown by the data above, all three of the tests described give a highly significant P value ($P < 10^{-15}$), but in other situations these tests may provide different conclusions regarding the significance of observed IBD data. The choice of the optimal test from the set presented depends on the true underlying genetic model, which is usually unknown. However, studies have shown that the means test performs best under a fairly wide range of genetic models (Blackwelder and Elston, 1985).

Tests Applicable When IBD Status Cannot Be Determined The tests described above are useful when the IBD status of the entire data set is known with certainty, or when there is a sufficient subset of the entire data set to permit the subset, where it is unknown, to simply be discarded. In many situations, however, marker genotype data is unavailable on one or both parents of the affected sib pair. This is frequently the case, for example, for disorders with late age of onset, such as Alzheimer disease or prostate cancer. In these situations, however, information on additional family members often provides some information about the sharing status of the sibs. For this reason, methods that can use all the available data to make inferences about the IBD status of the affected pairs are desirable. The first such method was proposed in a seminal series of papers published by Neil Risch (1990b, 1990c) in which the author considered the basic models for general sib pair analysis and the power of these methods. The method, denoted MLS for **m**aximum **l**od **s**core method, compares the probability of the observed genotypic data in the affected sib (or other relative) pairs as a function of the proportion sharing 0, 1, or 2 alleles IBD (Z_0, Z_1, Z_2) with the corresponding probability of the observed marker data under the null hypothesis of no linkage ($Z_0 = 0.25$, $Z_1 = 0.5$, $Z_2 = 0.25$). Numerical search methods are used to find the values of Z_0, Z_1, Z_2 which make the numerator of this ratio (hence the ratio itself, since the denominator is constant) the largest. Put another way, we are finding the values of (Z_0, Z_1, Z_2) that are most consistent with the marker genotypes of the affected pairs. This method is designed to use information from affected sibs or other relatives in determining the optimal IBD proportions. To be analogous to parametric analysis (see Chapters 4 and 12), the $\log_{10}$ of the likelihood ratio is taken to produce a lod score. For example, for the HLA data above, the optimal values of the sharing probabilities (Z_0, Z_1, Z_2) are easily shown to simply be the observed proportions $10/137 = 0.073$, $46/137 = 0.336$, and $81/137 = 0.591$, respectively, since we have complete information in this sample regarding IBD status. This produces a lod score of:

$$\text{lod} = \log_{10}\left[\frac{0.073^{10}0.336^{46}0.591^{81}}{0.25^{10}0.5^{46}0.25^{81}}\right] = 16.98$$

equivalent to odds of 10^{17}:1 in favor of involvement of HLA in this disease.

However, not all values of the Z_i are compatible with genetic models. For example, the case of none of the pairs sharing any alleles ($Z_0 = 1$) is not compatible with any Mendelian model of inheritance (in large sam r es) Holmans (1993) showed that the $\{Z_i\}$ for possible genetic models must satisfy

$$2Z_0 \leq Z_1 \leq 1/2$$

which he called the "genetically possible triangle." Note that this implies that $Z_0 \leq 0.25$ and that $Z_2 \geq 0.25$. In this same paper, Holmans (1993) showed that restricting the estimation of the $\{Z_i\}$ to satisfy this criterion resulted in a more powerful test of linkage than the test based on unrestricted maximization.

Several other methods, though principally designed for quantitative traits (see below), can be used to estimate the mean IBD sharing status for pairs of relatives for single markers (Amos et al., 1989) or groups of marker loci (Goldgar, 1990; Goldgar et al., 1993; Kruglyak and Lander, 1995a).

The maximum likelihood estimates of (Z_0, Z_1, Z_2) can be used to make inferences about the contribution of this locus to the overall increased familial risk, as the expected sharing at a marker locus linked to a disease locus can be written in terms of the λ_S value for the disease locus and the distance between this locus and the marker locus (Suarez et al., 1978; Hodge, 1981; Risch, 1990a). These equations can then be solved to estimate the contribution of this locus to the overall familial risk. When there is no recombination between the marker locus and the disease locus—that is, when the marker locus is a candidate locus for the disease—the following equations relate the sharing probabilities and the familial relative risks attributable to a given locus.

$$Z_0 = \frac{1}{4\lambda_S};\ \lambda_S = \frac{1}{4Z_0}$$

$$Z_1 = \frac{\lambda_O}{2\lambda_S};\ \lambda_O = \frac{Z_1}{2Z_0}$$

$$Z_2 = \frac{\lambda_M}{4\lambda_S};\ \lambda_M = \frac{Z_2}{Z_0}$$

When the hypothesized disease gene is located some distance from the marker locus, more complex relationships (see, e.g., Risch 1990b) hold, although they can be somewhat simplified by assuming that the recombination fraction between the disease locus and marker locus, θ, is not too large (e.g., $\theta < 0.10$) and that the offspring and sibling risks due to the locus are the same (i.e., $\lambda_O = \lambda_S$). If these assumptions can be justified, then the following holds:

$$Z_0 = \frac{1}{4} - \frac{(2\omega - 1)(\lambda_S - 1)}{4\lambda_S}$$

$$Z_1 = 1/2$$

$$Z_2 = \frac{1}{4} + \frac{(2\omega - 1)(\lambda_S - 1)}{4\lambda_S}$$

where $\omega = \theta^2 + (1 - \theta)^2$.

In our HLA example, assuming zero recombination between HLA and IDDM susceptibility, we would estimate the sibling risk attributable to this locus to be $\lambda_{S\text{-}HLA} = 1/(4 \times 0.076) = 3.3$. If, however, the disease gene is not HLA but is located a distance of $\theta = 0.05$ from the HLA complex, the estimate of $\lambda_{S\text{-}HLA}$ increases to 7.1, using the estimate of Z_0. This illustrates the general confounding of the strength of the effect of a given locus and the distance of this locus from the tested marker; in a genomic search with anonymous STR markers, it will be difficult to distinguish between a relatively weak disease locus located very close to a marker and a locus of larger effect located some distance away.

If we assume that HLA is, in fact, the predisposing locus (or located very close to it), we can use the estimate of 3.3 obtained above to get some idea of how much of the overall familial risk is likely to be accounted for by the HLA system. Since observed values from a variety of studies have estimated the recurrence risk to siblings for IDDM to be 0.06 and the disease has a population prevalence of about 0.004 (Spielman et al., 1980), this yields an overall λ_S value of $0.06/0.004 = 15$. Thus based on the data, it is clear that other susceptibility loci must be involved in the determination of IDDM. The estimated contribution of the HLA locus is $\log(3.3)/\log(15) = 44\%$ for a multiplicative model and $(3.3 - 1)/(15 - 1) = 16\%$ for an additive multilocus model. The results of a complete genomic search for IDDM susceptibility loci are discussed later in this chapter.

Extensions to Basic Affected Sib Pair Methods

Multipoint As we have seen, the methods discussed in this chapter depend on our ability to distinguish identity by descent from identity by state in affected sib pairs. In addition, it is easy to see that the ability to detect linkage to a disease locus will depend on the distance between the marker locus at which IBD sharing is assessed and a disease locus influencing the disease under study. The use of multiple markers in a defined genetic region addresses both these issues by maximizing the probability that complete or partial IBD information is available in the region spanned by the markers. The first person to examine IBD sharing in nuclear families as a function of marker genotypings at multiply linked loci was Goldgar (1990), who examined the estimation of IBD sharing in a region spanned by genetic markers in order to map quantitative trait loci. For qualitative traits such as disease phenotypes, the theoretical advantages of using multiple markers (albeit perfectly informative ones) was first discussed by Risch (1990b). More recently, Hauser et al. (1996) have extended the general MLS method to interval mapping in which the evidence for linkage to a disease locus located between two markers is assessed. Kruglyak and Lander (1995) and Olson (1995a) have also devised multipoint mapping methods for nonparametric analysis of affected relative pairs (see Chapter 14).

Other Sibship Configurations Although the methods described earlier are oriented toward analyzing a set of affected sib pairs, often other affected siblings are available for study. For methods designed to analyze pairs only, this requires the formation of all possible sib pairs from the sibship. If the sibship contains n affecteds, there are $n(n - 1)/2$ distinct pairs that can be formed for the analysis. The difficulty is that these pairs are not independent, and therefore significance levels based on such data will be inaccurate, resulting in increased type I error rates. For example, if sibs A and B share no alleles IBD, and the pairs B and C also share no alleles, we know that A and C must share both alleles IBD. One solution proposed by Hodge (1984) to get around this problem is to weight the contribution of a sibship by $n - 1$ (the number of independent pairs in a sibship of size n). For certain traits, consideration of discordant pairs (one affected, one unaffected) also may provide information for linkage. In this case, one would expect there to be more pairs who share 0 alleles IBD at the expense of pairs sharing both. However, the power of the former pairs under almost all circumstances will be small compared to that for affected pairs. For diseases with variable (or late) age of onset and genes of relatively small effect, discordant pairs will be almost totally uninformative. However, if they are collected and genotyped to contribute additional information to the overall estimation of IBD status in the family, including them in the analysis may be worthwhile.

Two Trait Loci In this situation, we assume that there are two (or more) loci contributing to the trait being studied, and we want to use genetic marker information to localize these genes. Most often at least one trait locus has been localized, and we want to use this information to help us in the genomic search for one or more additional loci.

The two-locus sib pair method (Dizier and Clerget-Darpoux, 1986) and the MASC (marker association segregation chi-square) method (Clerget-Darpoux, 1988; Dizier et al., 1994) were designed to test two candidate loci or explore the interactions between two known loci, rather than to find unidentified genes. For this latter purpose, Knapp et al. (1994) devised a series of two-marker tests and compared the power of these tests based on joint consideration of the IBD distribution at two loci. They assumed that IBD status could be determined with certainty, that there was no recombination between disease and marker, and that the two loci contributed equally to the disease. Under these assumptions, they found that two-locus tests (particularly a test based on the mean two-locus sharing) could provide a substantial increase in power for a variety of models. Cordell et al. (1995) extended the work of Risch and Holmans to define an MLS test for two loci that did not require IBD to be known with certainty and incorporated two-locus restrictions on the estimated parameters. They applied this method to sib pairs with IDDM, examining the effect of two loci, while controlling for the effect of HLA at this locus. The results showed that incorporation of the HLA status into the analysis provided increased power for detecting the effect of the additional loci. Recently, in a more general context, Goldgar and Easton (1997) and Farrall (1997) showed that incorporating linkage and mutation data at known loci provided increased power for a genomic search for other loci. All the aforementioned two-locus methods are typically useful when one or more loci for a disease have already been identified and the aim is to discover additional susceptibility loci. At the present time, they are not generally useful for the simultaneous search for multiple unknown disease loci because of the extreme number of comparisons that have to be made.

Parametric Lod Score Analysis In addition to the nonparametric methods described thus far in this chapter in which there is no need to specify a genetic model, the parametric lod score method (Chapters 4 and 12) can be used in the analysis of sib pairs as well. The difficulty, of course, is that for most of the traits for which one would collect sib pairs, the mode of inheritance is unclear. However, assuming one has some knowledge of the λ_R value and an idea of the population prevalence, the following approximate equations can be used to construct a set of "reasonable" models that fit the epidemiological data:

$$\lambda_S D^2 = qP^2 \qquad \text{for dominantly acting loci}$$

$$4\lambda_S D^2 = q^2P^2 \qquad \text{for recessively acting loci}$$

where λ_S is the sibling familial risk, D is the disease prevalence, q is the frequency of the allele associated with disease susceptibility, and P is the penetrance (probability of disease) associated with the risk allele.

The risk of disease in noncarriers, s, is given by:

$$s = \frac{D - 2qP}{1 - 2q} \qquad \text{for dominant loci}$$

$$s = \frac{D - q^2P}{1 - q^2} \qquad \text{for recessive loci}$$

For any known (or assumed) values of λ_S and D, values of q, P, and s can be determined from the equations above. Then traditional lod score analysis (either two-point or multipoint) can be performed under this limited set of models, with appropriate correction of the significance threshold for the number of models tested (Hodge and Elston, 1994).

For example, for a disease in which the sibling risk due to a hypothesized locus is 4.0 and the disease population prevalence is 1%, this would be compatible with the following genetic models:

1. A rare autosomal dominant locus with an allele frequency of 0.0005 for the high-risk allele, and penetrances of 89% in individuals who carry at least one copy of the high-risk allele and 0.9% in noncarriers.

2. A common autosomal dominant locus with an allele frequency of 0.01 and a penetrance of 20% in carriers and 0.6% in noncarriers.
3. An autosomal recessive locus with a disease allele frequency of 0.05. Individuals who are homozygous for this allele have a 80% chance of disease, compared with 0.8% for heterozygous and normal homozygous individuals.

While these models all produce a familial relative risk of 4 and a disease prevalence of 1%, they differ in several respects. For example, the proportion of all cases due to the proposed risk allele will vary from 39% for model 2 to 20% for model 3 to 9% for model 1. These models will also differ with respect to the parent–offspring risk and in the proportion of families with many cases of disease. Thus, if one also has an idea of the offspring risk, it may be possible to rule out either a dominant or recessive model, and the frequency of extended families with high incidence of disease may help distinguish a rare high-penetrant allele from a more common, but lower penetrant gene. In any case, the *accurate* knowledge of the basic epidemiological profile of the disease in question can aid in choosing a small number of consistent models to use in parametric lod score analysis of affected sib pairs or other affected relative sets.

The lod score approach has the advantages of producing an estimate of the recombination fraction as well as permitting the estimation of other parameters of the genetic model. If the model is even approximately correct, the use of the lod score method should provide higher power than nonparametric methods (Goldin and Weeks, 1993). Also, under a wrong model, the type I error rate, (i.e., the chance of finding significant evidence of linkage when it is not present) is not inflated (Williamson and Amos, 1990). Given that the lod score method has been shown to be reasonably robust to deviations from the true model, has considerable flexibility in analyzing data from any pedigree structure with any number of affected and unaffected individuals (i.e., does not need to reduce the data to pairs), and makes complete use of available marker information, the use of this method should not be ruled out, even for studies dealing with sib pair data. For further discussion of these issues, the reader is advised to consult Greenberg et al. (1996).

Power Analysis and Experimental Design Considerations

Factors Influencing Power of Sib Pair Methods The power of the affected sib pair method—that is, the probability that a true disease-predisposing locus will be detected by examining a linked marker locus—was first considered in detail in the papers by Risch (1990b, 1990c) for the MLS identity-by-descent method, and by Bishop and Williamson (1990) for methods based on identity by state. As demonstrated by Risch (1990b), the power to detect linkage of a trait locus and a marker locus can be expressed solely in terms of the single parameter of sibling relative risk, λ_S, regardless of the underlying genetic model. The power is calculated as a function of the difference in the expected values of the IBD sharing probabilities (Z_0, Z_1, Z_2) if the disease is linked to the marker locus and the values expected under the null hypothesis of no linkage ($Z_0 = 0.25$, $Z_1 = 0.5$, $Z_2 = 0.25$). As we saw above, the expected distribution of (Z_0, Z_1, Z_2) can be written as simple functions of λ_S and the disease–marker locus recombination fraction θ.

For reference, Figure 13.2, reproduced from Risch's work (1990b), shows the power to detect linkage between a perfectly informative marker locus and a single disease locus that accounts for all the observed familial risk, as a function of sibling relative risk λ_S and the recombination fraction between the trait and marker locus for three sample sizes. The dramatic effect of recombination on power is easily seen, particularly for moderate (N = 100) and small (N = 40) sample sizes, even when λ_S is quite large.

Holmans (1993) showed higher power for the restricted-triangle method described above compared with the unrestricted MLS test for both dominant and recessive models. This gain in power was especially marked for cases of low marker informativeness and/or high recombination fraction between the marker and disease locus (θ = 0.1). Holmans also examined the utility of genotyping parents in sib pair analyses as a function of marker heterozygosity. The results indicated a ratio of required sample size of as much as 1.5 (i.e., not genotyping parents required 50% more pairs) when the marker polymorphism is low. If efficiency is measured by the number of individuals needing to be genotyped, then genotyping parents (which requires twice as much genotyping) could be considered to be inefficient, particularly with markers with heterozygosities of over 0.5. This of course assumes that the desired number of sib pairs needed without typing parents can be easily ascertained and that they are genotyped accurately. Power for the case of interval mapping of a disease locus has been considered by Hauser et al. (1996), who derived conclusions regarding the genotyping of parents and other

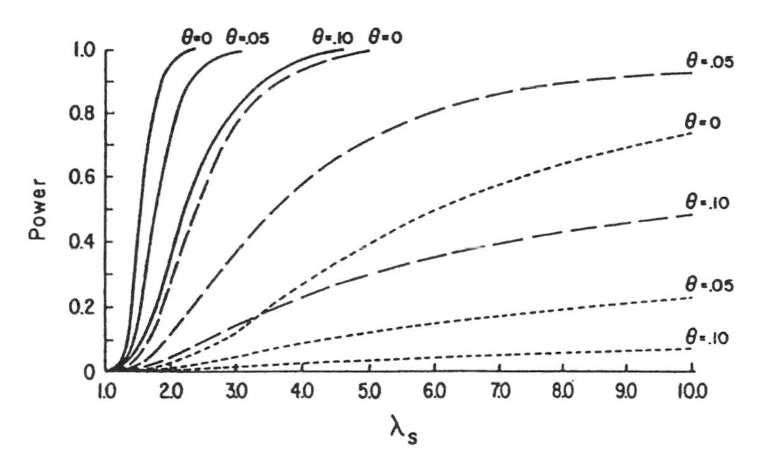

Figure 13.2 *Power calculations for MLS sib-pair analysis. (From N. Risch, Linkage strategies for genetically complex traits. II. The power of affected relative pairs. American Journal of Human Genetics, 46:229–241 (1990), © 1990. Reproduced with permission of the publisher, The University of Chicago Press.)*

relatives similar to those of Holmans but with the increased power resulting from a multipoint analysis. It is clear that in designing a study to map a complex disease gene, one should consider the estimated magnitude of the effect of interest (as measured by λ_S), the marker density, the availability of parents or other relatives for genotyping, and the relative importance of type I and type II errors. To aid in the design of such studies, the DESPAIR (design of sib pair) program, a part of the SAGE package (1994), can be used.

In some situations, however, sib pairs may not be the most efficient sampling strategy for mapping genes involved in complex disease. For example, Goldgar and Easton (1997) compared a variety of nuclear family sampling strategies in an analytical study of a number of two-locus disease models. For the majority of situations examined, affected sib trios were found to be a more efficient design for mapping disease loci than sib pairs. This was especially true when one of the two loci had been localized/identified and linkage or mutation data were available and incorporated in the analysis. Even accounting for the increased difficulty in ascertaining affected sib trios, this design (or one more extreme) still proved superior to sib pairs and should be considered for any gene localization effort.

The Example of Testicular Cancer As part of an international effort to map susceptibility loci for testicular cancer, Easton (unpublished) calculated the expected lod scores for the available sample under a variety of different models for the number and type of loci influencing the trait. All models were constrained to fit the observed sibling risk (λ_S) of 8.0 (e.g., Goldgar et al., 1994) and included one, two, or four loci acting additively or multiplicatively. These models were chosen to span a reasonable range of possibilities, although there are few data that a priori favor one model over another. The effects of intermarker distance and heterozygosity were also examined. The data consisted of 77 affected sib pairs, 2 affected half-sibs, 15 affected uncle–nephew pairs, and 6 affected cousin pairs. Table 13.5 shows the expected lod scores for each situation. While these analyses were done using a parametric analysis that assumed a rare autosomal dominant model with low penetrance, similar conclusions would apply with any of the methods discussed so far in this chapter.

Table 13.5 shows immediately that there is a dramatic decrease in the expected lod score as a function of the number of loci involved. This be expected, since the effect of any single locus is reduced. We note however that in general, if the loci are acting multiplicatively, there is higher power (as measured by the expected lod score) than if they are acting in an additive fashion, even though in the former case, the sibling relative risk due to an individual locus will be reduced (although the relative contribution will be the same). The reason for this was alluded to earlier, when it was observed that in multiplicative models, a higher fraction of pairs would be expected to be segregating risk alleles at multiple disease loci, while in additive models each pair may be due to only one of the set of disease susceptibility loci.

It is also apparent that moving from a completely informative marker completely linked to the disease locus ($\theta = 0$) to the more realistic genomic search situation of

Table 13.5 Expected Lod Scores for Mapping Testicular Cancer Susceptibility Loci Using Affected Pairs Under a Variety of Genetic Models and Genome Search Conditions

Marker Density (cM)[a]	Marker Heterozygosity (%)[b]	Single Gene	Heterogeneity[c]		Multiplicative[d]	
			2 Genes	4 Genes	2 Genes	4 Genes
0	100	14.4	3.3	0.8	6.8	1.0
10	90	7.8	2.1	0.4	3.3	1.2
(Multipoint)	80	5.8	1.4	0.3	2.6	0.9
20	90	5.0	1.2	0.3	2.2	0.8
(Multipoint)	80	3.6	0.9	0.2	1.7	0.6
20	90	3.1	0.7	0.2	1.4	0.5
(two-point)	80	2.1	0.5	0.1	1.0	0.4

[a]Spacing between genetic markers: 0 cM refers to the situation of a disease gene that is a known candidate locus. For the other situations, the disease locus is assumed to lie midway between the two markers. Either a multipoint analysis using the two flanking markers simultaneously is performed or a two-point analysis between the disease locus and each marker.

[b]Informativeness of the marker (i.e., the proportion of individuals heterozygous at a given marker locus).

[c]In the heterogeneity models, most affected pairs are caused by only one of the two loci. The locus specific risks are 4.5 and 2.75 for the two- and four-locus models, respectively.

[d]In the multiplicative models, most pairs are the result of high-risk alleles at two or more loci.

less than perfect markers with a marker density of 10–20 cM also reduces the expected lod score considerably. Assuming the usual criteria of a lod score of 3.0 for detection of linkage (note that an expected lod score of 3.0 then corresponds to 50% power), one can see that under a realistic genomic search situation, for two or more loci contributing to disease susceptibility, it is unlikely that this sample is of sufficient size to detect linkage to any locus.

QUANTITATIVE TRAITS

For quantitative traits, we no longer characterize an individual as affected or not affected, but by the specific value of some quantitative measurement made on that individual. The values of this measurement will naturally (we hope) vary from individual to individual. Our goal in analyzing such traits is to "explain" this variation in terms of specific components, both genetic and environmental. Before discussing the role of specific genes, or specific environmental effects, it is useful to get an overall idea of how much of the observed variation in the trait is due to genetic factors, how much is due to nongenetic familial factors that are shared among family members (common environment), and how much is due to random environmental effects, which are unique to each individual. To do this, we look at the degree to which different categories of related and unrelated individuals have similar values for the quantitative trait of interest. If there is significant evidence of a genetic component, we can begin the process of localizing specific genes that may be responsible for the genetic variation. The methods available for both these tasks are described in the next several sections.

Measuring Genetic Effects in Quantitative Traits

The strength of the genetic component is traditionally measured by a quantity called heritability, denoted h^2. Heritability is defined as the proportion of the total variation in the trait that is due to genetic factors. Usually, we distinguish between two different types of genetic variation, additive effects and dominance effects. The difference in these two types of variation can be illustrated by considering a locus with two alleles A and B, which is associated with a particular quantitative trait. For this locus, individuals will be one of three genotypes: AA, AB, or BB. if the mean of the quantitative trait among the AA group is, for example, 80, and that among BB is 120, the difference between the two homozygotes is responsible for the additive variance because it shows the independent effects of each allele. The other source of variation in the trait due to this locus is reflected by the mean quantitative trait value associated with the AB heterozygote. If the alleles are acting in an additive manner, then we would expect the mean of the heterozygote individuals to be midway between the two homozygotes, in this case a value of 100.

Any difference between the expected value of 100 and the observed value of the mean for the heterozygotes indicates that one of the two alleles is exerting a stronger influence than the other in the heterozygous state (i.e., is "*dominant*"). This represents additional variance due to the locus, the *dominance variance.* In addition to this difference in genotype-specific means, the magnitude of these variance components is influenced by the allele frequencies (hence the genotype frequencies) at the locus of interest. These concepts can be extended to any number of loci. One general case that is often considered is that of an infinite number of loci, each with small effect, which form the basis of the polygenic inheritance model of quantitative traits.

Just as in the case with multilocus disease models, the number of loci will not, in general, be known. However, the magnitude of these effects as a proportion of the total phenotypic variance can be estimated from similarities among relatives, analogous to recurrence risks in discrete traits. For our purpose here the important consideration is that two siblings contribute to half the additive variance and one-fourth of the dominance variance, while parents and offspring contribute to only the additive component of variance. Only bilineal relatives who can potentially share two alleles identical by descent will include dominance variation; in fact, the proportion of this shared variance is equal to their probability of sharing two alleles identical by descent.

Heritability (in the broad sense) is defined as $h^2 = (V_A + V_D)/V_T$, where V_T is the total phenotypic variance of the trait. For quantitative traits, the heritability can be estimated from the observed correlations between relatives. For nuclear family data, these consist of the intraclass correlation between siblings ρ_{SS} (estimated from the between-and within-sibship mean squares from analysis of variance) and the interclass correlation (ordinary Pearson product–moment correlation) between parents and offspring, ρ_{PO}. Thus V_A/V_T is estimated as $2\rho_{PO}$, while V_D/V_T is estimated as $4(\rho_{SS} - \rho_{PO})$.

Table 13.6 shows the sib–sib, and parent–offspring correlations and correspond-

Table 13.6 Familial Correlations in Blood Pressure

Trait	σ_{PO}	σ_{SS}	V_A/V_T	V_D/V_T	h^2 (broad sense)
Blood pressure					
Systolic	0.237	0.333	0.47	0.38	0.85
Diastolic	0.183	0.265	0.37	0.33	0.70

Source: Cavalli-Sforza and Bodmer (1971), based on work of Miall and Oldham (1963).

ing estimates of the proportion of total variance due to additive and dominance genetic factors for systolic and diastolic blood pressure.

The estimate of V_D assumes that the increased similarity of sibs compared to parents and children is due to dominance variation; but, depending on the phenotype, it could be the result of higher shared environment among sibs than parent–offspring pairs. Without detailed studies of other relatives, particularly, half-sibs, or adoption studies, it is impossible to distinguish true genetic dominance from common environment. Since we are primarily interested in this chapter in additive genetic variance, the parent–offspring correlation is usually sufficient to give a rough idea of the magnitude of the genetic component of the trait.

Early Methods The first person to address the problem of linkage with a quantitative trait in humans was Penrose (1938), who developed a method of testing by means of the interaction between the marker genotype and the quantitative phenotype in sib pairs from particular parental matings. Penrose's linkage test was designed for ordinal-level phenotypes as opposed to truly continuous ones. Lowry and Schultz (1959) looked at methods for detecting associations between continuous traits and marker loci, while Hill (1975) used a nested analysis of variance design (with marker genotype nested within sibship) for detecting and estimating linkage with a quantitative trait. One of the first nonparametric approaches for sib pair mapping of quantitative traits was proposed by Smith (1975).

A method quite different from those discussed above, proposed a generation ago by Haseman and Elston (1972), is still among the most commonly applied methods for examining linkage between a quantitative trait and a marker locus. Haseman and Elston derived the joint distribution of the number of genes IBD at a marker locus and the number of genes IBD at a hypothesized locus determining a quantitative trait in terms of the recombination fraction between the two loci. They used these results to develop a method based on the regression of the squared sib pair difference for the trait on their estimated genetic correlation at the marker locus. Equation (13.2) is the basic relation underlying the Haseman–Elston method. Assuming that the dominance variance is negligible, we have

$$E[Y_j|I_{mj}] = \alpha + \pi_{mj}\beta \qquad (13.2)$$

where $Y_j = (X_{1j} - X_{2j})^2$ is the squared sib pair difference of the jth sib pair, I_{mj} is the marker information on the jth family, π_{mj} is the estimated proportion of alleles shared identical by descent by the jth sib pair based on the marker information, and α and β are the coefficients of the linear regression and are equal to:

$$\alpha = \sigma_e^2 + 2[\theta^2 + (1 - \theta^2)]\sigma_g^2$$

$$\beta = -2(1 - 2\theta)^2 \sigma_g^2 \qquad (13.3)$$

where σ_e^2 and σ_g^2 are the respective environmental and genetic variances of the quantitative trait under study, and θ is the recombination fraction between the marker locus and hypothesized quantitative trait loci (QTL).

A significantly negative estimate of β, the slope of the regression line, indicates linkage between the marker locus and the locus that influences the trait. The basic idea underlying the method is that sib pairs who are identical by descent at a marker locus will be phenotypically similar for traits influenced by a nearby linked gene. Conversely, sib pairs who do not share genes IBD at the linked marker will tend to be phenotypically different. Originally, the method required IBD status at the marker locus to be determined with certainty, but now the method is generalized to estimate the proportion of alleles shared IBD using maximum likelihood methods. Amos et al. (1990) have derived computational algorithms for calculating the proportion of alleles at a marker locus shared IBD for an arbitrary relative pair, conditional on the observed marker data in all typed individuals in a pedigree. Amos et al. (1989) also proposed use of a weighted-least-squares (WLS) approach to compensate for the failure of the variance of squared trait values to be constant for each value of IBD sharing when linkage between the trait and marker locus is present, a circumstance that violates an assumption implicit in the use of the regression approach of Haseman and Elston. Amos showed that the use of the WLS approach resulted in higher power for the detection of linkage than the ordinary Haseman–Elston test. This general regression-based approach is implemented in the SIBPAL program of the SAGE package (1994). Figure 13.3 shows a sample graph showing the relationship between IBD status and the sib pair difference squared, together with the estimated regression line and the output from the SIBPAL program to illustrate this method.

Although this method was primarily designed for use with quantitative traits, it can also be adapted for use with disease phenotypes by one of two means: transforming disease status to a quantitative susceptibility measure based on sex, age at onset, or other factors (Dawson et al., 1990) or testing for estimated IBD proportion = 0.5 in affected sib pairs, but using estimated IBD sharing rather than relying solely on data from pairs with unambiguous IBD status, as done in the simple means test described earlier.

Multipoint Methods Another method for analyzing quantitative trait loci that uses the concept of IBD is the multipoint IBD method (MIM) proposed by Goldgar (1981, 1990). This method differs from the Haseman–Elston procedure in several ways. Most importantly, the method uses sibships, rather than sib pairs, thus avoid-

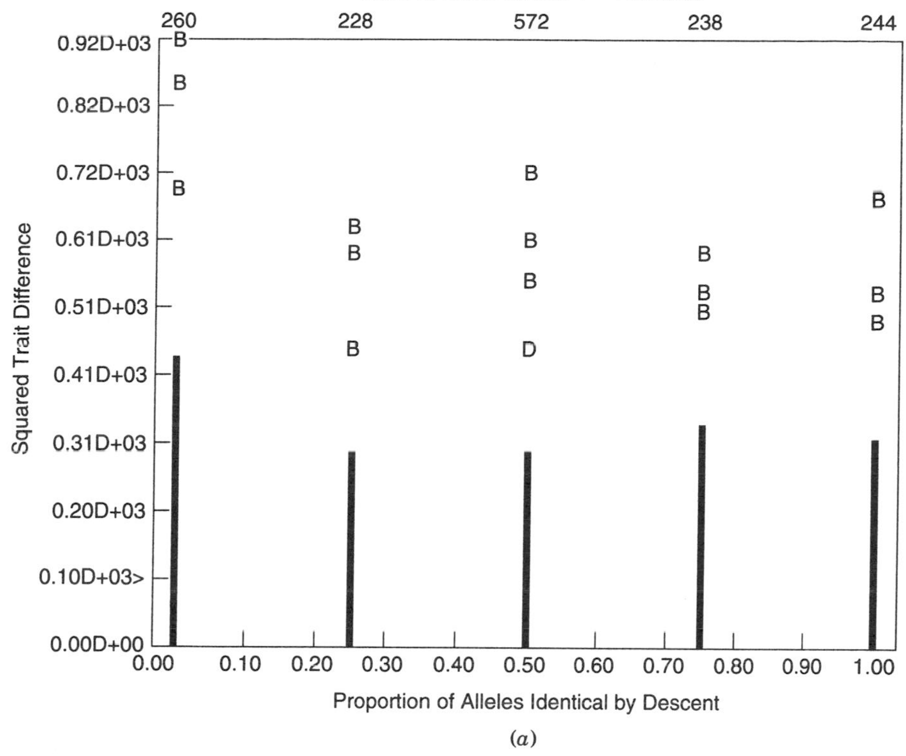

(*a*)

```
1S.A.G.E. RELEASE 2.2 -- SIBPAL VERSION 2.7          FEB 1996
 COPYRIGHT (C) 1996 BY R. C. ELSTON.


 OPTIONS:  QUANTITATIVE TRAIT,  PARENTAL DATA ARE USED,  FULL AND HALF SIBS POOLED
           NO WEIGHT IS USED WITH THE LINEAR REGRESSION

 ************************************************
 *                                              *
 * ANALYSIS FOR TRAIT NUMBER 01 (A            ) *
 *                                              *
 ************************************************

 L I N E A R   R E G R E S S I O N   A N A L Y S I S

                        Effective  Full Sibs  Half Sibs  Regress Y on Pi
 Trait       Locus      D.F.       Pi  Mean   Pi  Mean   T-values     P-values        Intercept     Slope
 ----------  ---------- ---------  --------   --------   --------     --------        ---------     --------
 A           Marker1        891    0.496433   --------   -2.7282      0.003247 **     104.3727      -27.4869

 # = THE REGRESSION LINE IS REFITTED WITH  ALPHA + BETA = 0.
```

(*b*)

Figure 13.3 *Haseman–Elston sib pair analysis graph* (a) *and output* (b).

ing the problems of using multiple, nonindependent sib pairs from the same sibship. Second, as the name implies, the method considers the IBD sharing in a chromosomal region defined by a set of linked genetic markers rather than IBD at a single locus. This method uses a variance partitioning approach in which the parameter of interest is the proportion of the total genetic variance of the trait that is due to a locus (or loci) in the region spanned by the genetic markers studies. This parameter, P, is estimated by means of a grid search of possible values and is tested by comparing the likelihood of the data under the "best" value of P compared to the likelihood under the null hypothesis P = 0. Like all such likelihood ratio tests, the test statistic defined by minus twice the natural logarithm of this likelihood ratio has an approximately chi-squared distribution with 1 degree of freedom. The method assumes that the dominance variance of the trait V_D is small relative to the additive variance. A more important assumption is that the investigator has some knowledge of the magnitude of V_A, the total additive genetic variance of the quantitative trait under study; thus it is not a "true" nonparametric method. However as we indicated above, this quantity is easily estimable from the quantitative trait data on parents and offspring via the parent–offspring correlation, or if parental data is unavailable, through the sib–sib correlation. Simulation studies (Goldgar, 1990) have shown that this method has higher power than the Haseman–Elston procedure and that the increase in power resulted from both the use of the entire sibship in a statistically meaningful way and from the use of multiply linked markers. An application of this method to a simulated problem involving the elucidation of major genes for a series of related quantitative traits can be found in Lewis and Goldgar (1995).

More recently other investigators have derived interval or multipoint mapping approaches in sib pairs, most of which represent multipoint extensions of the Haseman–Elston method (Fulker and Cardon, 1994; Fulker et al., 1995; Kruglyak and Lander, 1995b; Olson, 1995b). As expected, all these efforts demonstrate the higher power for detecting linkage that one obtains from multipoint analysis by making all pairs (or nearly all pairs) informative for determination of IBD status through construction of multilocus haplotypes. Another advantage of the multipoint approach is that it provides some information on the location of the QTL relative to the marker map; this information cannot be estimated using a series of single marker analyses.

Power and Sampling Considerations for Mapping Quantitative Trait Loci

Just as the power to detect disease loci in affected sib pairs depends on the sibling relative risk λ_S, the power for detecting quantitative trait loci depends on the degree to which the trait is genetically determined (i.e., the trait heritability h^2) and the proportion of this heritability that is due to the specific trait locus to be mapped. Studies have shown that using a random sampling approach and multipoint analysis, these methods can detect, at best, trait loci that account for about 30% of the total phenotypic variance of the trait (Goldgar, 1990; Goldgar and Oniki, 1992). Given a single trait locus that accounts for a significant fraction of the overall trait variance, it has been shown that the power to detect this locus depends to some extent on the nature of the remaining sources of variation of the trait. For example, Risch and Zhang (1996); see below) showed that the power to detect quantitative trait loci is increased when there is polygenic genetic variation in addition to the major trait locus. That is, it is easier to map a quantitative trait locus that is responsible for 40% of the total phenotypic variance of the trait when most of the remaining 60% arises from other genetic loci rather than when all the remaining variation is due to random environmental effects. Moreover, it has been shown (Goldgar and Oniki, 1992) that any factor that increases the sibling similarity, whether it is additional genetic effects or common environmental effects, results in higher power for QTL detection.

While the methods described thus far, by and large, assume random sampling of sib pairs (or sibships) from the population under study, several investigators have noted that this may not be the most powerful sampling strategy for the identification of loci influencing a quantitative trait. Just as in the case of disease mapping, where the strategy is to sample pairs of affected individuals rather than unaffected, it seems reasonable, in certain cases, to expect to derive advantages from a similar strategy of sampling sib pairs or sibships in which at least one member of the pair has an extreme trait value. The first to address this issue for quantitative traits were Carey and Williamson (1991), who proposed sampling a proband with an extreme value of the trait under study and performing a regression analysis using the other sibs' trait value as the dependent variable and the sibs' IBD sharing status with the proband at the marker locus as the independent variable.

Another approach for increasing the power of these methods is to select sib pairs who are both concordant for extreme values of the trait (e.g., both in the upper or lower 10%) (Cardon and Fulker, 1994), or as an alternative, to select sib pairs who are "extremely discordant," that is, at opposite ends of the quantitative trait distribution. Risch and Zhang (1995, 1996) showed that among these three choices (single proband with extreme value; extremely concordant pairs, extremely discordant pairs), the extremely discordant sib pair approach (EDSP) was the most powerful and was significantly better than random or single selection for a variety of genetic models. They concluded that mapping of a trait locus that controlled only 10% of the total variation was possible with a reasonable number of extremely discordant sib pairs using the EDSP method.

The major hurdle in this selected sib pair approach is in the difficulty of obtaining such pairs. For example, in their analysis, Risch and Zhang (1996) considered the case of a trait in which 60% of the variance is due to genetic factors composed of 20% due to a dominant major locus with a frequency of the dominant allele of 0.1, with the residual genetic component of 40% due to many genes of smaller effect. Assuming that the major quantitative trait locus is midway between two markers located 20 cM apart (i.e., 10 cM from both markers) and that these markers are informative 80% of the time, it can be shown that a sample size of 212 extremely discordant sib pairs would be required to detect this locus at a significance level of 0.00001 with 80% power. The EDSPs were defined as one member of the pair in the upper 10% and the other in the lower 10% of the overall trait distribution. However, to obtain 212 sib pairs that meet this criterion, an expected total of 12,366 sib pairs would have to be screened for the quantitative trait of interest. If one loosens the criterion to EDSPs in the upper 10% and lower 30%, the sample size required increases to 333, but the number of pairs to be screened is now reduced to 3719. The efficiency of any selection-based approach will depend on the relative costs of sample ascertainment and phenotyping versus the cost of genotyping the selected individuals. If getting the family material and making the phenotypic measurement are difficult and expensive, it might be wasteful to screen such families to obtain the required number of EDSPs when a smaller number of families selected at random or through a single proband would suffice.

SOFTWARE AVAILABLE FOR ANALYZING SIB PAIR AND SIBSHIP DATA

Many of the methods mentioned above are simple enough to be analyzed using standard statistical packages, or would be if the IBD distribution could be determined with certainty. For others, no user-friendly programs that implement the method are generally available (see volume appendix).

The list provided in the Appendix should not be taken as exhaustive, but rather as a resource of programs designed primarily for nonparametric analysis of complex traits (either quantitative or qualitative) in sib pairs or nuclear families, for which documented software adapted to several platforms is available on the Internet. Programs for parametric linkage analysis, nonparametric analysis in extended pedigrees, or association/linkage disequilibrium analysis, are mentioned elsewhere in this volume.

EXAMPLES OF APPLICATION OF SIB PAIR METHODS FOR MAPPING COMPLEX TRAITS

So far in this chapter, we have concentrated on presenting an overview of the methods developed to utilize the basic sib pair/sibship design in mapping genes contributing to complex human qualitative and quantitative traits. However, the reader might well ask whether these methods and sampling schemes, for all their theoretical advantages, have successfully identified genes for any complex human traits through screening of random genomic markers. There has been great progress in several disorders using this approach, primarily in insulin-dependent (juvenile) diabetes, for which in addition to the candidate loci of HLA and insulin, a genomic search for additional IDDM susceptibility locus revealed evidence for at least three other genes (Davies et al., 1994). Of course, many of these linkages have not yet been replicated in independent data sets. All these loci were characterized by relatively small effects (locus-specific λ_S values 1.3–1.5). Given that IDDM is characterized by an overall sibling relative risk of 15, even under a multiplicative model these loci account for only about 60–65% of the overall familial aggregation.

An interesting example of the use of affected sib pair methodology for a disease with a smaller overall familial component but a higher prevalence in the population is provided by the recent analysis of adult onset diabetes (NIDDM) by Hanis et al. (1996). This study utilized a primary sample of 330 Mexican-American affected sib pairs typed for 490 genetic markers distributed throughout the genome and found evidence [maximum lod score (MLS) = 3.2] of a susceptibility locus on the terminal portion of the long arm of chromosome 2, designated *NIDDM1*. The maximum effect of this locus (assuming no recombination between the linked marker and the disease) was a locus-specific λ_S value of 1.37. Since the overall sibling relative risk for *NIDDM* is 2.8, the authors estimate that this locus accounts for 30% of the overall familial risk, assuming a multiplicative model for the interaction between all the loci involved.

Another recent example of a genome-wide search for susceptibility loci for a complex disease involves several parallel genomic searches for susceptibility loci for multiple sclerosis (MS), the results of which were published in *Nature Genetics* (Ebers et al., 1996; Haines et al., 1996; Sawcer et al., 1996). Although all three studies found evidence for a susceptibility locus in (or near) the HLA complex (which had been suggested by earlier association studies and biological plausibility), no additional locus was found by more than one of the three studies. Risch (1987) has estimated that less than 30% of the familial aggregation in MS is due to HLA, so it is somewhat surprising that no consistent additional susceptibility loci were identified. This could be because all additional loci are of too small an effect to be consistently detected, or alternatively, because the familial aggregation has nongenetic (e.g., viral) etiology.

For quantitative traits, the success stories for human mapping are somewhat harder to come by, although Cardon et al. (1994) reported linkage to chromosome 6 markers for test scores underlying reading ability in a sample of sibships ascertained for a proband with dyslexia, and this result was recently replicated in an independent sample by Grigorenko et al. (1997).

SUMMARY

We have introduced the reader to both the theoretical and practical aspects of the use of sib pairs (or sibships) in the search for genes that either confer increased susceptibility to a qualitative trait (e.g., a disease) or contribute to variation in human quantitative traits. In addition, we have provided a basic understanding of the variety of models of familial aggregation and have shown how these models can be distinguished by means of available epidemiological and statistical data. Other analysis strategies and sampling designs, as alluded to here, may be preferred over sib pair designs and the corresponding nonparametric methods, depending on the characteristics of the phenotype being studied. In particular, methods analogous to those presented here but designed for use in extended pedigrees are presented in Chapter 4. Because, in terms of both theory and implementation, this area is moving forward at a rapid pace, we have concentrated on the underlying principles involved in all these methods rather than specific details of any single approach. Moreover, we have pointed out some of the practical limitations of sib pair methodology and provided the reader wishing to map genes for complex traits some idea of when these methods would be appropriate. These cautions aside, sib pair and related methods are today the most commonly used vehicles for mapping complex human traits. Given the rapid advances in human genome technology and the continuing development of powerful statistical methods, it is uncertain how long this will continue to be the case.

REFERENCES

Amos C, Elston RC, Wilson AF, Bailey-Wilson JE (1989): A more powerful robust sib pair test of linkage for quantitative traits. *Genet Epidemiol* 6:435–449.

Amos C, Dawson D, Elston RC (1990): The probabilistic determination of identity-by-descent sharing for pairs of relatives from pedigrees. *Am J Hum Genet* 47:842–853.

Bishop DT, Williamson JA (1990): The power of identity-by-state methods for linkage analysis. *Am J Hum Genet* 46:254–265.

Blackwelder WC, Elston RC (1985): A comparison of sib-pair linkage tests for disease susceptibility loci. *Genet Epidemiol* 2:85–97.

Cardon LR, Fulker DW (1994): The power of interval mapping quantitative trait loci, using selected sib pairs. *Am J Hum Genet* 55:825–833.

Cardon LR, Smith SD, Fulker DW, et al. (1994): Quantitative trait locus for reading disability on chromosome 6. *Science* 266:276–279.

Carey G, Williamson J (1991): Linkage analysis of quantitative traits: Increased power using selected samples. *Am J Hum Genet* 49:786–796.

Cavalli-Sforza LL, Bodmer WF (1971): *The Genetics of Human Populations.* San Francisco: Freeman.

Clerget-Darpoux F, Babron MC, Prum B, Lathrop GM, Deschamps I, Hors J (1988): A new method to test genetic models in HLA associated diseases: The MASC method. *Ann Hum Genet* 52:247–258.

Clerget-Darpoux F, Babron M-C, Prum B, Lathrop GM, Deschamps I, Hors J (1988): A new method to test genetic models in HLA associated diseases: The MASC method. *Ann Hum Genet* 52:247–258.

Cordell HJ, Todd JA, Bennett ST, Kawaguchi Y, Farrall M (1995): Two-locus maximum lod score analysis of a multifactorial trait: Joint consideration of *IDDM2* and *IDDM4* with *IDDMI* in type I diabetes. *Am J Hum Genet* 57:920–934.

Cox NJ, Spielman R (1989): The insulin gene and susceptibility to IDDM. *Genet Epidemiol* 6:65–69.

Davies JL, Kawaguchi Y, Bennett ST, et al. (1994): A genome-wide search for human type 1 diabetes susceptibility genes. *Nature* 371:130–136.

Dawson DV, Kaplan EB, Elston RC (1990): Extensions to sib-pair linkage tests applicable to disorders characterized by delayed onset. *Genet Epidemiol* 7:453–466.

Dib C, Fauré S, Fizames C, Samson D, Drouot N, Vignal A, et al. (1996): A comprehensive genetic map of the human genome based on 5,264 microsatellites. *Nature* 380:152–154.

Dizier MH, Clerget-Darpoux F (1986): Two disease locus model: Sib pair method using information on both HLA and Gm. *Genet Epidemiol* 5:343–356.

Dizier MH, Babron M-C, Clerget-Darpoux F (1994): Interactive effects of two candidate genes in a disease: Extension of the marker–association–segregation chi-square method. *Am J Hum Genet* 55:1042–1049.

Ebers GC, Kukay K, Bulman DE, et al. (1996): A full genome search in multiple sclerosis. *Nat Genet* 13:472–476.

Fisher RA (1935): The detection of linkage with recessive abnormalities. *Ann Eugen* 6:339–351.

Fulker DW, Cardon LR (1994): A sib-pair approach to interval mapping of quantitative trait loci. *Am J Hum Genet* 54:1092–1103.

Fulker DW, Cherry SS, Cardon LR (1995): Multipoint interval mapping of quantitative trait loci using sib pairs. *Am J Hum Genet* 56:1224–1233.

Goldgar DE (1981): Partitioning the variance of a quantitative trait to specific chromosomal regions. Unpublished doctoral dissertation, University of Colorado.

Goldgar DE (1990): Multipoint analysis of human quantitative genetic variation. *Am J Hum Genet* 47:957–967.

Goldgar DE, Easton DF (1997): Optimal strategies for mapping complex diseases in the presence of multiple loci. *Am J Hum Genet* 60:1222–1232.

Goldgar DE, Oniki R (1992): Comparison of a multipoint IBD method with parametric multipoint linkage analysis for mapping quantitative traits. *Am J Hum Genet* 50:598–606.

Goldgar DE, Lewis CM, Gholami K (1993): Analysis of discrete phenotypes using a multipoint identity by descent method: Application to Alzheimer disease. Proceedings of Genetic Analysis Workshop 8. *Genet Epidemiol* 10(6):383–388.

Goldgar DE, Easton DF, Cannon-Albright LA, Skolnick MH (1994): A systematic population-based assessment of cancer risk in first degree relative of cancer probands. *J Natl Cancer Inst* 86:1600–1608.

Goldin LR, Weeks DE (1993): Two locus models of disease: Comparison of likelihood and nonparametric linkage methods. *Am J Hum Genet* 53:908–915.

Greenberg DA, Hodge SE, Vieland VJ, Spence MA (1996): Affected-only linkage methods are not a panacea. *Am J Hum Genet* (letter) 58:892–895.

Gyapay G, Morisette J, Vignal A, Dib C, Fizames C, Millasseau P, Marc S, et al. (1994): The 1993–94 Généthon human genetic linkage map. *Nat Genet* 7:246–339.

Haldane JBS (1934): Methods for the detection of autosomal linkage in man. *Ann Eugen* 6:26–55.

Haldane JBS, Smith CAB (1947): A new estimate of the linkage between the genes for colorblindness and haemophilia in man. *Ann Eugen* 14:10–31.

Haines JL, Ter-Minassian M, Basyk A, et al. (1996): A complete genomic screen for multiple sclerosis underscores a role for the major histocompatibility complex. *Nat Genet* 13:469–471.

Hanis CL, Boerwinkle E, Chakraborty R, et al. (1996): A genome-wide search for human non-insulin-dependent (type 2): Diabetes genes reveals a major susceptibility locus on chromosome 2. *Nat Genet* 13:161–166.

Haseman JK, Elston RC (1972): The investigation of linkage between a quantitative trait and a marker locus. *Behav Genet* 2:3–19.

Hauser ER, Boehnke M, Guo S-W, Risch N (1996): Affected-sib-pair interval mapping and exclusion for complex genetic traits: Sampling considerations. *Genet Epidemiol* 13:117–137.

Hill AP (1975): Quantitative linkage; a statistical procedure for its detection and estimation. *Ann Hum Genet* 38:439–449.

Hodge SE (1981): Some epistatic two-locus models of disease. I. Relative risks and identity by descent distributions in affected sib pairs. *Am J Hum Genet* 33:381–395.

Hodge SE (1984): The information contained in multiple sibling pairs. *Genet Epidemiol* 1:109–122.

Hodge SE, Elston RC (1994): Lods, wrods, and mods: The interpretation of lod scores calculated under different models. *Genet Epidemiol* 11:329–342.

Holmans P (1993): Asymptotic properties of affected-sib-pair linkage analysis. *Am J Hum Genet* 52:362–374.

Knapp M, Seuchter SA, Bauer MP (1994): Two-locus disease models with two marker loci: The power of affected-sib-pair tests. *Am J Hum Genet* 55:1030–1041.

Kruglyak L, Lander ES (1995a): Complete multipoint sib-pair analysis of qualitative and quantitative traits. *Am J Hum Genet* 57:439–454.

Kruglyak L, Lander ES (1995b): A non-parametric approach for mapping quantitative trait loci. *Genetics* 139:1421–1428.

Kruglyak L, Daly MJ, Lander ES (1995): Rapid multipoint linkage analysis of recessive traits in nuclear families, including homozygosity mapping. *Am J Hum Genet* 56:519–527.

Lange K (1986): The affected sib-pair method using identity by state relations. *Am J Hum Genet* 39:148–150.

Lewis CM, Goldgar DE (1995): Screening for linkage using a multipoint identity-by-descent method. *Genet Epidemiol* 12:777–782.

Lowry DC, Schultz FT (1959): Testing association of metric traits and marker genes. *Ann Hum Genet* 23:83–90.

Miall WE, Oldham PD (1963): The hereditary factor in arterial blood pressure. *Br Med J* 19:75–80.

Morton NE (1955): Sequential tests for the detection and linkage. *Am J Hum Genet* 7:277–318.

Olson JM (1995a): Multipoint linkage analysis using sib pairs: An interval mapping approach for dichotomous outcomes. *Am J Hum Genet* 56:788–798.

Olson JM (1995b): Robust multipoint linkage analysis: An extension of the Haseman–Elston method. *Genet Epidemiol* 12:177–193.

Penrose LS (1935): The detection of autosomal linkage in data which consists of pairs of brothers and sisters of unspecified parentage. *Ann Eugen* 6:133–138.

Penrose LS (1938): Genetic linkage in graded human characters. *Ann Eugen* 9:133–138.

Penrose LS (1953): The general purpose sibpair linkage test. *Ann Eugen* 18:120–124.

Risch N (1987): Assessing the role of HLA-linked and unlinked determinants of disease. *Am J Hum Genet* 40:1–14.

Risch N (1990a): Linkage strategies for genetically complex traits. I. Multilocus models. *Am J Hum Genet* 46:222–228.

Risch N (1990b): Linkage strategies for genetically complex traits. II. The power of affected relative pairs. *Am J Hum Genet* 46:229–241.

Risch N (1990c): Linkage strategies for genetically complex traits. III. The effect of marker polymorphism on analysis of affected relative pairs. *Am J Hum Genet* 46:242–253.

Risch N, Zhang H (1995): Extreme discordant sib pairs for mapping quantitative trait loci in humans. *Science* 268:1584–1589.

Risch N, Zhang H (1996): Mapping quantitative trait loci with extreme discordant sib pairs: Sampling considerations. *Am J Hum Genet* 58:836–843.

SAGE (1994): Statistical Analysis for Genetic Epidemiology, Release 2.2. Computer program package available from the Department of Epidemiology and Biostatistics, Case Western Reserve University, Cleveland.

Sawcer S, Jones HB, Feakes R, et al. (1996): A genome screen in multiple sclerosis reveals susceptibility loci on chromosome 6p21 and 17q22. *Nat Genet* 13:464–468.

Smith CAB (1975): A nonparametric test for linkage with a quantitative character. *Ann Hum Genet* 38:451–460.

Spielman RS, Baker L, Zmijewski CM (1980): Gene dosage and susceptibility to insulin-dependent diabetes. *Ann Hum Genet* 4:135–150.

Suarez BK, Rice J, Reich T (1978): The generalized sib pair IBD distribution: Its use in the detection of linkage. *Ann Hum Genet* 42:87–94.

Thomson G (1986): Determining the mode of inheritance of RFLP-associated diseases using the affected sib-pair method. *Am J Hum Genet* 39:207–221.

Williamson JA, Amos CI (1990): On the asymptotic behavior of the estimate of the estimate of the recombination fraction under the null hypothesis of no linkage when the model is misspecified. *Genet Epidemiol* 7:309–318.

14

Affected Relative Pair Analysis

Jonathan L. Haines
Program in Human Genetics
Department of Molecular Physiology and Biophysics
Vanderbilt University School of Medicine
Nashville, Tennessee

INTRODUCTION

Traditional lod score analysis has been highly successful in the mapping of single gene disorders. However, when the underlying genetic model cannot be specified with any confidence, as is the case for more genetically complex traits, traditional, parametric lod score analysis loses much of its power, and the results can be misleading (Chapter 12). Even when a single gene is known to be acting, if there are many ungenotyped individuals in a pedigree, the accurate specification of disease and marker allele frequencies may become critical; even small discrepancies could produce misleading results. In several pedigrees, for example, early-onset Alzheimer disease has been traced back seven generations. Since DNA is available only for individuals in the most recent two generations, however, most affected individuals are connected through many ungenotyped individuals. Thus while there may be many genotyped affected individuals, it is difficult to determine whether they have inherited the same marker allele from a common ancestor. That is, it is difficult to determine the number of alleles identical by descent (IBD, Chapters 4, 13). Since a large amount of information needs to be inferred in this case, every specification

Approaches to Gene Mapping in Complex Human Diseases, Edited by Jonathan L. Haines and Margaret A. Pericak-Vance. ISBN 0-471-17195-6

(however inaccurate) in the genetic model carries a greater weight in the lod score calculation.

The primary solution to the problem just stated is to utilize procedures that rely less completely on the genetic model specification. Such procedures, less powerful than parametric methods if the model is specified accurately, offer a robust approach when model assumptions are less certain. Sib pair analysis (Chapter 13) is one such nonparametric method. It has the advantages of simplicity of family structure and several well-developed methodologies. The major disadvantage of sib pair analysis is its inability to use additional information that can be extracted from other affected relatives, such as aunts, uncles, cousins, grandparents, or more distantly related individuals.

Methods that use information from such affected relatives have been developed to take greater advantage of these extended families. These affected relative pair (ARP) methods have proven very useful for linkage studies (St. George-Hyslop et al., 1990; Pericak-Vance et al., 1991). We discuss four common ARP methods, as well as an approach using the traditional parametric lod score method that reduces its dependence on the genetic model specification. Some of these methods examine the data by means of comparisons of each possible pair of affected individuals, while others examine all affected individuals simultaneously.

USE OF MAXIMUM LIKELIHOOD (PARAMETRIC LOD SCORE) ANALYSIS

It may seem contradictory to suggest that parametric lod score calculations could be used as a "nonparametric" method. However, by appropriately modifying the penetrance parameters, it is possible to approximate a nonparametric method by minimizing the effect of some parameters. This was the first, and still perhaps the most widely practiced ARP method. It is often called the "low-penetrance" or "affecteds-only" lod score analysis (see Chapter 12).

Method

For genes that moderately increase risk, the penetrance (i.e., the probability of expressing the trait, given the gene) is low. If the penetrance parameter is set to zero or to a very small value (e.g., 0.001), the phenotypic information from unaffected individuals is not used, and only phenotype information from affected individuals is used for the trait. Driving this approach is the underlying assumption that the affected phenotype is much more certainly associated with the presence of the trait allele, while the unaffected phenotype is far less certainly associated with the absence of the trait allele.

As an example of how to set up an affecteds-only analysis, we will return to the Alzheimer example (Table 14.1). Using the LINKAGE program format, the penetrance matrix on the left is a standard age-at-onset-curve matrix with variable pene-

Table 14.1 Penetrance Values for Age-at-Onset-Curve Versus Affecteds-Only Analysis

	Age-at-Onset-Curve			Affecteds-Only		
Liability Class[a]	NN	NA	AA	NN	NA	AA
N,A	0.00	1.00	1.00	0.00	1.00	1.00
N40	0.00	0.01	0.01	0.00	0.00	0.00
N50	0.00	0.05	0.05	0.00	0.00	0.00
N60	0.00	0.10	0.10	0.00	0.00	0.00
N70	0.00	0.30	0.30	0.00	0.00	0.00
N80	0.00	0.75	0.75	0.00	0.00	0.00
N90	0.00	0.95	0.95	0.00	0.00	0.00

[a]"Liability class" refers to the phenotype classification of each individual: N, normal; A, affected; NXX, normal with an age-at-exam in the XX decade.

trance defined by an age-at-onset distribution (Pericak-Vance et al., 1991). The penetrance matrix on the right mimics this matrix, but provides for no penetrance for those individuals who are at risk for Alzheimer disease (e.g., those who had been assigned a penetrance other than 0 or 1).

How are the results from such an analysis interpreted? The offhand answer is, *carefully*. It is not clear whether the usual significance levels (Chapter 11) for a genomic screen are valid when such assumptions are made, nor what kind of power such an analysis has if the inheritance pattern is misspecified (although it certainly will be lower than the power obtained when inheritance patterns are correctly specified).

Advantages

The major advantage of this method is that marker genotype information is used from all individuals, and the maximum likelihood approach captures all the possible marker information and thus generates the best estimates of IBD status. These estimates are used only in conjunction with the most certain of the trait information (i.e., phenotypes of affected individuals). This method can also be used on large and arbitrarily complex pedigrees, and multipoint analysis can be performed. It also considers all affected individuals simultaneously, not just in a pairwise fashion. Finally, standard programs such as LINKAGE, VITESSE, and MENDEL can be used for these calculations.

Disadvantages

The major disadvantage of the affecteds-only analysis is that the other parameters of a genetic model must still be specified. Even if penetrance is set to zero, the inheritance pattern (e.g., autosomal dominant, autosomal recessive, or X-linked) remain to be specified, and misspecification will still generate misleading results. In addition, the amount of computer time needed for the calculations can become excessive if the pedigrees are large or have marriage or consanguinity loops, or if multipoint analysis is undertaken.

Usefulness

Although widely used, affecteds-only analysis only compensates for poor knowledge of penetrance; it is not a satisfying solution to the problems extant in complex traits. Nevertheless, it can be used in several situations when other methods cannot be implemented:

1. In single gene disorders when the age at onset, or other penetrance function, cannot be well specified.
2. When a major gene is suspected, but cannot be proven.
3. When large or complex pedigrees are being studied, even if the mode of inheritance is unknown.
4. When multipoint analysis is being contemplated, since other ARP methods (below) may be limited to two-point analysis.

AFFECTED PEDIGREE MEMBER (APM) ANALYSIS

In 1988 Weeks and Lange described the affected pedigree member (APM) method, an extension of earlier work in this field, which was nonparametric because it required no knowledge of the underlying genetic model, which can then be arbitrarily complex. Unlike most sib pair methods, the basis of the APM tests for deviation from the expected distribution of identity-by-state (IBS), not identity-by-descent (IBD) relationships. For alleles to be IBS they must simply match in state, most often meaning that each allele is the same size (e.g., 117 bp; Chapters 4, 13). However, it is not known whether the allele has been inherited from the same common ancestor (i.e., IBD). To accurately estimate IBD, genotypes on all the individuals connecting the affected relative pairs must be available (or completely inferable). The use of IBS instead of IBD eliminated the need for such complete genotyping.

Method

While the APM is usually described as nonparametric, or model free, this is not strictly true. More accurately, it is trait-genetic-model free, since no assumptions about the underlying genetic etiology of the trait are made. However, the method does depend on three factors: the genotypes of the affected individuals, the marker allele frequencies, and the pedigree relationships of these individuals. The APM statistic for any pair of relatives is:

$$Z_{ij} = \frac{1}{4}\delta(G_{ix}, G_{jx})f(\rho_{Gix}) + \frac{1}{4}\delta(G_{ix}, G_{jy})f(\rho_{Gix}) + \frac{1}{4}\delta(G_{iy}, G_{jx})f(\rho_{Giy})$$

$$+\frac{1}{4}\delta(G_{iy}, G_{jy})f(\rho_{Giy}) \quad (14.1)$$

where

i and j represent the two affected individuals being compared

G_{ix}, G_{iy}, G_{jx}, and G_{jy} represent the maternal (x) and paternal (y) alleles at the marker locus, respectively, of individuals i and j

$\delta(G_{ix}, G_{jx})$, $\delta(G_{ix}, G_{jy})$, $\delta(G_{iy}, G_{jx})$, and $\delta(G_{iy}, G_{jy})$ are 1 if the two alleles (G) match in state, or 0 if they do not match in state

$f(\rho_{Gix}$ and $f(\rho_{Giy})$ represent an arbitrary weighting function of the marker allele frequencies

Prior to the application of marker allele frequencies, the four possible allele comparisons between these two individuals are equally weighted, so that the maximal value of Z_{ij} is 1, and the minimal value is 0. For example, assuming 4 equally frequent alleles, if two individuals both have genotype 1/1, then $Z_{ij} = 1$; if both have genotype 1/2, then $Z_{ij} = 1/2$; if one has genotype 1/2 and the other has genotype 1/3, then $Z_{ij} = 1/4$; and if one has genotype 1/3 and the other has genotype 2/4, then $Z_{ij} = 0$. A weighting function based on the marker allele frequencies may be introduced to allow for the greater likelihood that rare alleles shared by a pair of relatives indicates linkage than the sharing of a common allele. In this case each of the four possible allele comparisons contributes $1/4f(p)$. Three weighting functions are commonly used: $f(p) = 1$, $f(p) = 1/\sqrt{p}, f(p) = 1/p$, where p is the frequency of an arbitrary allele.

The overall APM statistic for a given family is simply the sum over all pairs:

$$Z = \sum_{i<j} Z_{ij} \quad (14.2)$$

The mean and variance of this statistic are more complicated to calculate, and the reader is referred to Weeks and Lange (1988) for more detail. It should be noted that equation (14.2) uses only marker genotypes and marker allele frequencies. The pedigree relationship between the two individuals is used only in the calculation of the mean and variance.

One special situation arises for parent–child pairs. In the initial formulation of the APM statistic, parent–child pairs were included with all other pairs. However, parent–child pairs by definition share exactly one allele IBD, are thus not informative for linkage, and cannot have a Z_{ij} that is less than ½. Since inclusion of parent–child pairs adds only random variation (Schroeder et al., 1994), parent–child pairs are no longer used in the current version of APM.

The overall statistic across families is simply the weighted sum from each family:

$$T = \frac{\sum_m \{w_m[Z_m - E(Z_m)]\}}{\left\{\sum [w_m^2 \operatorname{var}(Z_m)]\right\}^{1/2}}$$

Where T = total APM statistic across families
w_m = $(r_m - 1)^{1/2}/[\text{var}(Z_m)]^{1/2}$, the weights used for each family, m
r_m = number of genotyped affected individuals in each family, m

If the number of pedigrees is large, then the APM statistic T should follow a normal distribution. Since linkage should increase the sharing of alleles IBS, a one-sided test is appropriate. Because T is a weighted statistic, it is not possible to simply add T scores across families or APM runs. This is in contrast to lod scores, which can be added. Thus multiple runs must be performed to test any specific subsets of families.

A multilocus extension to the APM statistic was described in 1992 (Weeks and Lange, 1992). To implement this extension, an additional set of parameters, the intermarker recombination fractions (θ_i) between marker loci ($m_1, m_2, \ldots, m_i$), needs to be specified.

It should be noted that while the APM was constructed and is used primarily as a test of linkage, it can also be influenced by association, most likely arising from linkage disequilibrium. Linkage results in an increased sharing of specific alleles within a pedigree. However, if most pedigrees are linked to the same locus (although the allele shared in each pedigree will not be the same), the statistic will produce a positive sum across pedigrees. In association (Chapter 15), there will be increased sharing of a specific allele not only within pedigrees but across pedigrees, also resulting in a positive sum across pedigrees. Thus a significant result for the APM can indicate linkage, association, or both.

Three marker allele frequency weighting functions are provided with the APM programs. These are $f(p) = 1$, $f(p) = 1/p$, and $f(p) = 1/\sqrt{p}$, where p represents the frequency of an arbitrary allele. The first of these provides no actual weighting by allele frequency; thus sharing of rare alleles is given no more weight than sharing of common alleles. The second function gives much greater weighting to rare alleles, while the third provides a balance between the first two. The use of $f(p) = 1/\sqrt{p}$ is recommended because it seems to be the most robust in both simulated (Weeks and Lange, 1988) and real data sets (Pericak-Vance et al., 1991). Of course, it should be decided *before* the analysis which of these weighting functions will be used. Otherwise multiple tests are being performed and corrections must be made to the P values. Even so, some estimate of the sensitivity of the data set to assumptions in marker allele frequencies can be gained by comparing the results of the three weighting functions. For example, if the P value generated using $f(p) = 1$ is 0.04, while the P value generated using $f(p) = 1/p = 0.0001$, then it is likely that the allele being shared has been indicated as being rare and that the allele frequencies being used must be estimated with care (Chapters 4, 12).

Advantages

There are several advantages to the APM method. First and most important for a genetically complex trait, APM allows the testing of linkage without having to specify an underlying trait genetic model. The second advantage is that APM incorporates data from affected relatives other than just siblings, so that this additional information is not thrown away.

The third advantage is speed. While lod score analysis calculations can be extremely slow for large and/or complex pedigrees, especially when there are many ungenotyped individuals, APM calculations are much faster and are not hindered by the level of genotyping. Although the calculations are somewhat slower for larger or more complex pedigrees, reasonable speed can be achieved even for pedigrees with marriage loops and/or consanguinity. Thus pedigrees for which lod score calculations are essentially impossible can be tested using APM. The fourth advantage is that multipoint analysis can be performed. Information from closely linked adjacent markers can help to more uniquely define a haplotype, increasing the similarity between IBS and IBD estimates, and thus increasing the amount of segregation information being extracted from the pedigree.

Disadvantages

Perhaps the primary disadvantage is the reliance on IBS rather than IBD statistics. As described in Chapters 4 and 12, individuals who happen to be IBS without being IBD will provide false evidence of linkage. This effect is somewhat mitigated by the use of highly polymorphic microsatellite markers and multipoint analysis.

A second disadvantage is the sensitivity of the APM statistic to marker allele frequencies. While the incorporation of weighting functions helps in this respect, the estimates of allele frequencies are still critical and must be done accurately. Fluctuations in allele frequencies can have a major effect on the resulting *P* values (Wijsman, 1993). In this respect, at least 50 additional unrelated individuals drawn from the same populations as the trait pedigrees should be genotyped and used for marker allele frequency estimates.

A third disadvantage is that the APM throws away potential information. If genotypes are available on connecting relatives (which would help determine the IBD status of alleles in affected individuals), they are not used. In well-genotyped pedigrees this can cause a substantial loss of information.

A fourth disadvantage is that an inflated false positive (type I) error rate has been consistently observed for APM. While theoretically the APM statistic follows a normal distribution, this is true only for large samples. In general, "large" means more than approximately 30 pedigrees. If the number of pedigrees is smaller than 30, there is an ancillary program that should be used to calculate an empiric *P* value based on the given pedigree structures and marker allele frequencies. Even using this empiric *P* value, however, most of the distributions for both actual and simulated data sets seem to have a positive tail (i.e., the distributions do not follow a normal distribution). While the cause of this skewing is unknown, the result is a higher false positive rate. This has been noted in several studies (MSGG, 1996; Terwedow et al., 1996), with about 10% of markers generating a nominal asymptotic *P* value of 0.05 or less. Thus any positive result has to be considered carefully.

A fifth disadvantage is that APM considers the data only one pair of affected individuals at a time. Thus sharing of alleles among three or more affected individuals in the same family is not weighted any more than sharing of alleles in three pairs from separate families.

Usefulness

The APM can be used in several situations:

1. When a complex pedigree structure precludes standard lod score calculations for a single gene disorder.
2. When many affected relatives other than siblings are available for a genetically complex trait. Any single pedigree may have only one or a few affected relative pairs, but in aggregate there are many such pairs.
3. When multipoint analysis is desired.

See Table 14.2 for a comparison of the properties, advantages, and disadvantages of APM and other ARP methods.

SimIBD ANALYSIS

In 1996 Davis et al. described several new ARP statistics: SimAPM, SimKIN, SimIBD, and SimISO. The SimAPM statistic, an improved version of the APM statistic, still relies on IBS relationships. The latter three attempt to estimate IBD relationships. In the comparison of these three statistics, SimIBD and SimISO were shown to provide the more accurate estimates of IBD sharing, with SimIBD examining sharing between affected relatives, and SimISO examining, in addition, sharing among unaffected relatives. While SimISO may have some applications, the greatest power to detect linkage in genetically complex traits will arise from examining affected individuals, and so the SimIBD measure is generally preferred.

Method

The general approach in SimIBD is to calculate an observed statistic given all the marker and pedigree information, and then compare it to a simulated distribution. The simulated distribution is generated by performing many replicates in which the only data generated randomly (within the confines of Mendelian inheritance) are the marker genotypes for the affected individuals conditional on the marker genotypes of the unaffected individuals. The assigned P value is simply the empirical P value obtained by comparing the observed statistic to the distribution generated from the replicates of the simulated data. SimIBD depends on three factors: the genotypes of the affected and unaffected individuals, the marker allele frequencies, and the pedigree relationships of these individuals.

The SimIBD statistic for any pair of relatives (parent–child pairs excluded) is:

$$Z_{ij} = \sum_{a=1}^{2} \sum_{b=1}^{2} \left(\frac{1}{\sqrt{p}} \right) \alpha_{ia,jb} \tag{14.3}$$

where $1/\sqrt{p}$ = weighting function based on the marker allele frequency p
$\alpha_{ia,jb}$ = probability that the two alleles being compared are IBD

In equation 14.3 $\alpha_{ia,jb}$ equals 1 if the two alleles can be absolutely determined to be IBD and 0 if the two alleles can be absolutely determined not to be IBD. If IBD status cannot be absolutely determined, $\alpha_{ia,jb}$ is estimated given the marker genotypes in the pedigree and will fall between 0 and 1. The estimation procedure uses a recursive algorithm. The reader is referred to Davis et al. (1996) for more detail.

The SimIBD statistic for any pedigree is:

$$Z_p = \frac{1}{(r-1)^{1/2}} \left[\sum_{i=1}^{r} \sum_{j=1}^{r} Z_{ij} \right]$$

where r is the number of affected individuals in the pedigree.

The overall statistic across pedigrees is simply the sum of each pedigree statistic:

$$Z = \sum_{p=1}^{m} Z_p$$

Advantages

The primary advantage of the SimIBD statistic is that it uses IBD information when available, and estimates it otherwise. This convention minimizes the amount of potentially misleading information and would make its way into a purely IBS approach, such as the APM. As with APM, SimIBD does not require any specification of the trait genetic model, making it a desirable method for genetically complex traits. In most situations, it outperforms the APM, having a lower false positive rate and higher power. Therefore, SimIBD is recommended over APM.

Disadvantages

There are three disadvantages to the SimIBD method. The first is that arbitrarily complex pedigrees cannot be analyzed. The current version of this program can handle simple pedigrees as well as those with one or two marriage loops or consanguineous loops. While this is not a major limitation for most studies, it may severely restrict application of SimIBD in investigations of very rare recessive conditions or inbred populations. The second limitation is that currently no multipoint extension is available. Thus the added information gained by examining nearby markers cannot be utilized. Finally, the program can take substantial amounts of computer time to run. This is due primarily to the simulation aspect of the statistic. Although the observed statistic is calculated only once, the simulated distribution requires hundreds of replicates to achieve reasonable accuracy. We recommend that at least 1000 replicates be performed.

Usefulness

The SimIBD method can be used in several situations:

1. When an extended pedigree structure precludes standard parametric lod score calculations in single gene disorders. However, the reduction in computer time may not be substantial.
2. When many affected relatives other than siblings are available in genetically complex traits. Any single pedigree may have only one or a few affected relative pairs, but in aggregate there are many such pairs.
3. When genotypes for unaffected individuals who connect affected individuals are available.

See Table 14.2 for a comparison of the properties, advantages, and disadvantages of SimIBD and other ARP methods.

NPL ANALYSIS

In 1996 Kruglyak et al. introduced an ARP approach they termed NPL (nonparametric linkage) as part of the GENEHUNTER computer program. Like the other ARP methods, the NPL statistic measures allele sharing among affected individuals within a pedigree. The NPL method can analyze the data using a pairwise approach (using the NPL_{pairs} statistic), but it has been extended to provide a simultaneous comparison of alleles in all affected individuals in a pedigree (the NPL_{all} statistic). The NPL approach is inheritently multipoint, since it calculates the IBD probability for any given point along the chromosome (the "inheritance distribution"), using all available marker data on that chromosome.

Method

The calculation of the NPL statistic can be broken into two parts. The first part, the calculation of the inheritance distribution, will be used to estimate the allele sharing among a set of affected individuals, be they just pairs or complete family groupings. This is done as outlined in Lander and Green (1987) and Kruglyak et al. (1996), and the reader is referred to these sources for a detailed description. The second part of the calculation is the individual evaluation of the scoring function that determines whether the inheritance information is indicative of linkage. As mentioned above, two scoring functions are described, the NPL_{pairs} and the NPL_{all}.

If the inheritance pattern can be determined unambiguously in a pedigree, the resulting NPL_{pairs} statistic is:

$$S_{\text{pairs}} = \Sigma S_{ij}$$

where i and j are the two individuals being compared and $S_{ij} = 0$, 1, or 2, depending on how many alleles are shared IBD.

In most genetically complex traits, however, the inheritance pattern cannot be determined unambiguously. In this case, the S_{pairs} statistic is an average taken across all possibilities.

To compare S_{pairs} to a statistical distribution, it is normalized:

$$Z_{\text{pairs}} = \frac{S_{\text{pairs}} - E(S_{\text{pairs}})}{[\text{var}(S_{\text{pairs}})]^{1/2}}$$

where $E(S_{\text{pairs}})$ and $\text{var}(S_{\text{pairs}})$ are the expectation of the mean and variance of S_{pairs} under the null hypothesis.

The overall statistic across pedigree is:

$$Z = \Sigma\left(\frac{1}{\sqrt{m}}\right)(Z_{\text{pairs}})_i$$

where $(Z_{\text{pairs}})_i$ = normalized score for a pedigree
m = number of pedigrees

The NPL_{all} statistic is defined as

$$S_{\text{all}} = 2^{-a} \sum_h \left[\prod_{i=1}^{2f} b_i(h)!\right] \tag{14.4}$$

where a = number of affected individuals in the pedigree
h = collection of alleles generated by taking one allele from each affected individual (there are 2^a possible collections)
$2f$ = total number of founder alleles in the pedigree (i.e., the total number of different alleles of distinct origin)
$b_i(h)$ = total number of a specific founder allele (i) in the collection (h)

This statistic is averaged over all feasible inheritance patterns, normalized, and weighted across pedigrees in the same manner as the $\text{NPL}_{\text{pairs}}$ statistic.

Statistical significance is determined by comparing the Z score to the standard normal distribution. The use of the standard normal distribution is an approximation, usually a conservative one. In other words, the true P value is often smaller than the P value obtained by using a standard normal table.

Advantages

There are several advantages to the NPL approach. Like most linkage analysis programs developed at the Whitehead Institute, GENEHUNTER, the program that calculates NPL, has a friendly and easy-to-use interface.

The NPL_{all} statistic is the only ARP that can consider all affected relatives simultaneously, rather than as a combination of all possible comparisons of pairs. This implements the intuitively attractive idea that if, for example, five affected individuals in a pedigree *all* share the *same* allele IBD, this information should carry more weight than if *each* of the 15 possible pairs in the same pedigree shared *some* allele IBD, not necessarily the *same* allele. This increased weighting is seen in equation (14.4) by including the factorial of $b_i(h)$, rather than $b_i(h)$ itself. However, the NPL_{all} statistic does not perform as well as the NPL_{pairs} statistic in some situations, particularly those involving very common, recessive-acting alleles.

Perhaps the greatest advantage of the NPL statistic is that the data for all markers on a chromosome can be evaluated simultaneously using a multipoint approach. Because GENEHUNTER uses the Lander–Green algorithm for calculating the IBD distribution, the amount of computer time for an analysis increases only linearly with the number of markers included in the analysis. This is in contrast to the Elston–Stewart algorithm used for most parametric lod score analyses, which causes computer time to increase exponentially with the number of markers.

Disadvantages

There is one potential disadvantage to the NPL statistic, related to its inherently multipoint approach. The Lander–Green algorithm can incorporate many markers because computer time increases only linearly with the addition of new markers. However, computer time increases exponentially with the number of individuals in the pedigree. This contrasts with the Elston–Stewart algorithm, where computer time increases only linearly with the number of individuals in the pedigree, but exponentially with the number of markers. Thus NPL is limited to smaller pedigrees. This is dictated by the rule that $2n - f \leq 16$, where n is the number of nonfounders and f is the number of founders. More complicated or extended pedigrees cannot be handled with any confidence, even on very fast workstations. The NPL statistic also appears to be quite conservative, meaning that P values of 0.05 occur less often than expected. While this reduces the chance of a false positive result, it also reduces the power to detect true linkages.

Usefulness

The NPL statistic can be used in several situations:

1. When pedigrees of moderate size are available. Even pedigrees with only a single affected sib pair can be useful, since a large number of markers can be examined simultaneously. Large pedigrees cannot be examined.
2. When many relatives other than siblings are available for a genetically complex trait. Each pedigree needs only a single affected relative pair.
3. When a large number of linked markers are being (or could be) examined. This is particularly useful if pedigrees have somewhat “sparse” genotyping

(i.e., many individuals providing the biological links between affected individuals are not available for genotyping). This situation generally leads to difficulty in generating the IBD status for a single marker, and the problem may be reduced substantially if many linked markers are genotyped.

WPC ANALYSIS

In 1994 Commenges described a new nonparametric statistic he termed the weighted pairwise correlation (WPC) statistic, based on a score test of homogeneity among strata (Liang, 1987). In this particular case, the strata are defined by the marker genotypes within each family. If a marker is unlinked to a trait locus, the phenotypes should be homogeneous across these strata. However, if the marker is linked, such homogeneity will no longer exist because certain phenotypes will associate with certain genotypes. By evaluating the IBS status for all pairs of family members, the WPC statistic evaluates the null hypothesis of homogeneity, and thus the null hypothesis of no linkage.

It should be noted that unlike the other methods described in this chapter, WPC considers all pairs of individuals, not just affected individuals. In fact, WPC cannot be applied to data on just affected individuals unless they have an associated quantitative trait that varies among affected individuals.

Method

The derivation of the WPC statistic is quite complex, and the reader is referred to Commenges (1994) for more detail. Unlike the ARP statistics, the WPC method was derived specifically for quantitative traits. Qualitative traits can also be evaluated using a rank or proportional hazard version of the statistic, assuming that some sort of ranking (e.g., by age) can be applied across all individuals. The statistic depends on three factors: the phenotype of each individual, the genotype of each individual, and the pedigree structure (e.g., relationships between individuals).

The WPC statistic for each family is:

$$S_L = \sum_{i=1}^{n-1} \sum_{j=i+1}^{n} W_{ij} U_i U_j$$

where W_{ij} is a weight given to the relative pair consisting of individuals i and j. The weight is either 0, 1, or 2, depending on the number of alleles shared IBS, and is centered by subtracting the average number of alleles shared by all pairs of the same relationship (e.g., all sib pairs, all cousin pairs, etc.).

The phenotype scores for individuals i and j (U_i, U_j) are the residuals after the expected phenotype has been subtracted under the null hypothesis of no linkage.

When linkage exists, the phenotypes will tend to be similar (e.g., both positive or both negative) if marker sharing exceeds the average ($w_{ij} > 0$). Thus S_L will be positive, and large values of S_L are indicative of linkage. The phenotypes will tend to be dissimilar (e.g., one positive, the other negative) if marker sharing is less than average ($w_{ij} < 0$), making S_L negative. To simplify the score and make it more robust, Commenges replaced U_i and U_j with R_i and R_j, the ranks of the phenotype scores. This resulting weighted rank pairwise correlation statistic (WRPC) is:

$$\mathrm{WRPC} = \frac{S_R - E(S_R)}{[\mathrm{var}(S_R)]^{1/2}}$$

where S_R is analogous to S_L except that the ranks of the phenotype scores are used, $E(S_R)$ is the expectation of S_R, and $\mathrm{var}(S_R)$ is the variance of S_R.

The overall WRPC statistic across families is simply the sum across all families (p):

$$\sum_{p=1}^{n} \mathrm{WRPC}_p$$

Advantages

The WRPC method has several advantages. As with the ARP methods, it does not require any specification of the trait genetic model, making it a desirable method for genetically complex traits. Unlike all the ARP methods, it is derived for quantitative traits and has the greatest power when it is used for this purpose. The use of the ranks, however, allows an easy conversion for qualitative traits (assuming the phenotype can be easily ranked), highlighting the flexibility of WPC. In addition, this is the only method that inherently allows the incorporation of modifying environmental factors. This is accomplished in forming the phenotype residuals. This approach is quite similar to the method of incorporating covariates in the Haseman–Elston sib pair method (Chapter 13). In fact, if only pairs of siblings are considered, the WPC method simplifies to very nearly the Haseman–Elston sib pair method. The strength of the WPC method is that it uses all relatives, not just those who are affected, thus gaining power by increasing the number of pairs examined. Finally, the calculation of the WRPC statistic is relatively fast, similar to that for APM.

Disadvantages

Some of the strengths of the WPC method are also its weaknesses. For example, the use of both affected and unaffected individuals, while increasing the number of pairs examined, also increases the chance that heterogeneity or misdiagnosis will introduce unwanted variation. In addition, the reliance on ranks requires a ranking scheme. The most commonly used scheme is to rank by age within each trait class (e.g., affected and unaffected). However, this requires both an assumption that age is important and some ranking function (usually a simpler linear function), data for which may not be available on many members of the pedigree.

Because WPC uses centered residuals, more than a single pair of relatives is required. With only a single pair, there is no variation, thus no residual, and therefore the statistic is zero. Consequently families with only a single sib pair cannot be used. Additionally the current implementation lumps more distant relationships together into a single category, making the analysis less accurate. As a result, very large pedigrees may produce unexpected and uninterpretable results.

The WPC statistic appears to be anticonservative. That is, when no linkage exists, WPC exceeds the value corresponding to a nominal *P* value of 0.05 more often than 5% of the time. Numerous simulation studies (Rogus, 1996) indicate that the inflation of this type I error can be as high as 10%.

The WRPC statistic currently can be calculated only for two-point analysis; no multipoint extension is available. It is also restricted to examining pairs of individuals at one time.

Usefulness

The WRPC method can be used in several situations:

1. When pedigrees of moderate size are available. Pedigrees with only a single sib pair, or very large pedigrees, cannot be examined with any confidence.
2. When many relatives other than siblings are available in a data set for a genetically complex trait. Each pedigree must have several relative pairs.
3. When a quantitative trait is being examined. WPC allows the use of trait information on all individuals without having to arbitrarily define "affected" or "unaffected."
4. When covariates must be included in the analysis. Qualitative covariates can be included by converting them to a linear scale, while quantitative measure can be used directly.

See Table 14.2 for a comparison of the properties, advantages, and disadvantages of WRPC and other ARP methods.

SUMMARY

Table 14.2 presents a summary of the traits of the five methods discussed. The choice of a method to use for a nonparametric analysis depends on many factors, including the type of pedigrees being examined and the type of data being generated. The following guidelines may be useful:

1. If the data set consists mostly or completely of sib pairs, then sib pair methods (Chapter 13) should be used, and ARP methods should not be considered.
2. If the data set consists of pedigrees segregating a Mendelian trait, then traditional parametric lod scores should be used unless the computer time requirements would be prohibitive. The alternative is then APM if the pedigrees are

Table 14.2 Comparison of ARP Methods

Factor	Affecteds-Only Lod Score	APM: Affected Pedigree Member	SimIBD	NPL: Nonparametric Linkage Analysis	WRPC: Weighted Rank Pairwise Correlation
Parameters					
Trait genetic model	Yes (no penetrance	No	No	No	No
Recombination fraction	Yes	Marker map only	No	Marker map only	No
Marker allele frequencies	Needed	Needed	Needed	Needed	No
Arbitrary pedigrees?	Yes	Yes	Yes (limited size)	Yes (limited size)	Yes
Multipoint analysis	Yes	Yes	No	Yes	No
Uses phenotypes from unaffected individuals	No	No	No	No	Yes
Can consider all affected individuals simultaneously	Yes	No	No	Yes	No

very large, and NPL if the pedigrees are only of moderate size. Each allows both two-point and multipoint calculations.

3. If the data set consists of small to moderate-sized pedigrees and the trait has a complex genetic etiology, NPL should be used if multiple linked markers are going to be genotyped and the pedigrees have 12 or fewer nonfounders. Otherwise SimIBD can be used.
4. For a genetically complex trait that is quantitative or has a quantitative risk factor, and the pedigrees are of moderate size, the WRPC method may be useful, especially if modifying covariates are available.

REFERENCES

Commenges D (1994): Robust genetic linkage analysis based on a socre test of homogeneity: The weight pairwise correlation statistic. *Genet Epidemiol* 11:189–200.

Davis S, Schroeder M, Goldin LR, Weeks DE (1996): Nonparametric simulation-based statistics for detecting linkage in general pedigrees. *Am J Hum Genet* 58:867–880.

Kruglyak L, Daly MJ, Reeve-Daly M, Lander ES (1996): Parametric and nonparametric linkage analysis: A unified multipoint approach. *Am J Hum Genet* 58:1347–1363.

Lander ES, Green P (1987): Construction of multilocus genetic maps in humans. *Proc Natl Acad Sci USA* 84:2363–2367.

Liang KY (1987): A locally most powerful test for homogeneity with many strata. *Biometrika* 74:259–264.

The Multiple Sclerosis Genetics Group (1996): A complete genomic screen for multiple sclerosis underscores a role for the major histocompatibility complex. *Nat Genet* 469–471.

Pericak-Vance MA, Bebout JL, Gaskell PC, et al. (1991): Linkage studies in familial Alzheimer's disease: Evidence for chromosome 19 linkage. *Am J Hum Genet* 48:1034–1050.

Rogus JJ (1996): A gene-environment interaction model for linkage analysis. Ph.D. Thesis. Harvard School of Public Health, Department of Biostatistics.

Schroeder M, Brown DL, Weeks DE (1994): Improved programs for the affected-pedigree-member method linkage analysis. *Genet Epidemiol* 11:69–74.

St George-Hyslop PH, Haines JL, Farrer LA, et al. (1990): Genetic linkage studies suggest that Alzheimer's disease is not a single homogeneous disorder. *Nature* 347:194–197.

Terwedow H, Rimmler JB, Ter-Minassian M, Pritchard M, Weeks DE, Pericak-Vance MA, Haines JL (1996): A comparison of APM and SimIBD, two statistical methods of linkage analysis. *Am J Hum Genet* 59:A238.

Weeks DE, Lange K (1988): The affected-pedigree-member method of linkage analysis. *Am J Hum Genet* 42:315–326.

Weeks DE, Lange K (1992): A multilocus extension of the affected-pedigree-member method of linkage analysis. *Am J Hum Genet* 50:859–868.

Wijsman E (1993): Genetic Analysis Workshop 8 summary paper on the Alzheimer's disease data. *Genet Epidemiol* 10:348–360.

15

Linkage Disequilibrium and Allelic Association

Margaret A. Pericak-Vance
Center for Human Genetics
Section of Medical Genetics
Duke University Medical Center
Durham, North Carolina

INTRODUCTION

There are two primary approaches for mapping genes that either cause or increase susceptibility to human disease. The first is the linkage analysis approach, either a parametric lod score analysis when the genetic model is known or model-independent affected relative pair analysis when the genetic model is unknown. The linkage approach has been discussed in detail elsewhere in this book (Chapters 12, 13, 14). The second approach is the application of allelic association studies. Allelic association is another nonparametric approach to mapping disease genes. It is a useful and often necessary tool in identifying disease gene loci, particularly susceptibility genes in genetically complex diseases. Allelic association can be explained either by direct biological action of the polymorphism (e.g., the *APOE-4* allele in Alzheimer disease), or by linkage disequilibrium with a nearby susceptibility gene. Association studies can play a critical role in the analysis of genetically complex traits, in the evaluation of candidate gene loci as well as in the fine-mapping of a region once linkage studies have indicated a region of interest for follow-up analysis. Certainly, as the Human Genome Initiative identifies and characterizes each gene in

Approaches to Gene Mapping in Complex Human Diseases, Edited by Jonathan L. Haines and Margaret A. Pericak-Vance. ISBN 0-471-17195-6

the human genome, this approach will become more widespread, particularly for localization and gene identification.

Allelic association refers to a significantly increased or decreased frequency of a marker allele with a disease trait and represents deviations from the random occurrence of the alleles with respect to disease phenotype. Allelic association can be due to either linkage or association. We use linkage disequilibrium to mean allelic association maintained by tight linkage. Linkage disequilibrium occurs when a particular marker allele lies so close to the disease susceptibility allele that these alleles will be inherited together over many generations. Thus the same allele will be detected in affected individuals in multiple apparently unrelated families. Conceptually this is the same as standard linkage analysis, except that the recombination distances being measured are now very small (generally < 1 cM), and the recombination events can only be inferred based on the level of sharing of the same allele. Population substructure most often occurs with the recent admixture of populations. In the case of population substructure, alleles may show a statistical association simply by chance due to differences in allele frequencies in the two mixing populations. This can occur even when there is no biological association or true genetic linkage.

LINKAGE DISEQUILIBRIUM

To understand one way in which linkage disequilibrium can come about, consider this hypothetical example. A new mutation occurs in a gene that results in a disease-causing phenotype. At the time of the initial mutation, every marker allele for every marker on the chromosome is "associated" with the disease mutation. The chromosome with the disease mutation is then transmitted to the offspring of the original individual in whom the mutation occurred. Transmission over several generations gives the opportunity for recombination to occur and thus for the rearrangement of the alleles at the marker loci. Alleles at marker loci that are further away from the disease mutation will exchange faster than markers that are closer to the disease mutation. The closer to the marker is to the disease gene, the longer the marker allele/disease association will persist.

When marker and disease loci are very close together on a chromosome, genetic crossing over will have occurred at such a low rate that the marker will appear to cosegregate with the gene regardless of the family studied. This is in contrast to the situation of two loci further apart but still linked, in which case repeated crossing over will allow all possible combinations of chromosomal haplotypes to appear with frequencies as predicted by the equation for Hardy–Weinberg equilibrium (Chapter 2). Thus, linkage disequilibrium can be very useful in defining the ancestral haplotype of a disease gene in relation to several marker loci; it can be used for fine-mapping of the disease gene even when complete linkage ($\theta = 0.0$) is established in the families being studied.

There are many measurements of linkage disequilibrium (Devlin and Risch, 1995). The most commonly used is the disequilibrium coefficient D.

$$D = P_{11} - p_1 q_1$$

where P_{11} is the observed frequency of the 1/1 haplotype, p_1 is the frequency of the "1" allele at locus 1 in the general population and q_1 is the population frequency of the "1" allele at locus 2. Generally, the "1" allele at each locus is defined as the most common of the alleles at that locus. Because we assign "1" as the most common allele, the coefficient D ranges from 0.25 to 0.25. Positive values of D indicate that the common alleles at each locus segregate together. Negative values indicate that the common allele at one locus segregates with the rare allele at the other locus. The rate of decay of linkage disequilibrium is dependent on the distance between loci:

$$D_t = D_0(1 - \theta)^t$$

where t is the current generation number, D_t is the current amount of disequilibrium, D_0 is the disequilibrium at generation 0, and θ is the recombination fraction between loci.

Allelic association due to population admixture, selection, or genetic drift between unlinked loci will decay fairly rapidly in comparison to linkage disequilibrium between tightly linked genetic loci, and thus is a short-term phenomenon that will be almost impossible to detect in the typical study. However linkage disequilibrium will decay rather slowly, dependent primarily on the recombination distance between the markers and the number of generations that has passed since the initial event (Fig. 15.1). The slowness of linkage disequilibrium decay makes this a useful mapping tool.

The general rule of thumb is that the stronger the disequilibrium, the closer the marker is to the disease locus. This is not always the case, however, for several reasons. First, the frequencies of the marker alleles have an impact on the power to detect linkage disequilibrium. For example, if the disease susceptibility allele is associated with a marker allele whose general population allele frequency is 0.50, an

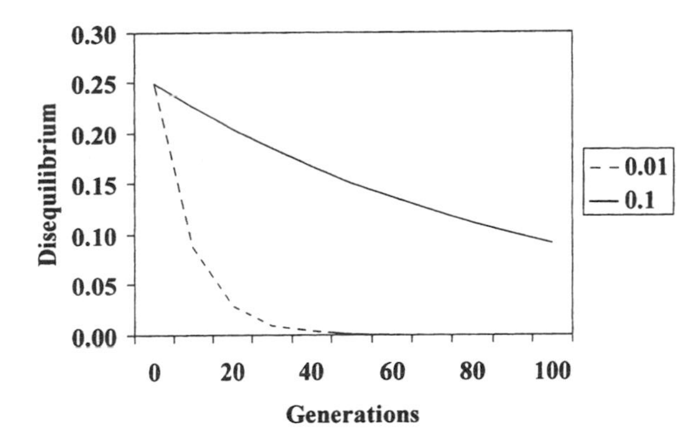

Figure 15.1 Decay in linkage disequilibrium for recombination fractions of 0.01 and 0.10.

observation that this marker allele frequency is 0.80 in affected individuals represents only a 60% increase in the frequency. However, if the marker allele population frequency is only 0.20, an observed 0.80 frequency is a 400% increase. Mutation rates at the marker locus also affect disequilibrium by increasing the chance that the associated marker allele will change and so seem to be representing a different chromosome.

Population bottlenecks, where the effective population size is reduced to a very small number for a period of time before the population size increases again, can create or reinforce an existing association. This is done by the random loss to the genome pool of most chromosomes carrying the susceptibility allele; what remains may have existed in only one individual who survived the bottleneck. Chance loss of susceptibility-allele-bearing chromosomes (random genetic drift) can also generate linkage disequilibrium. Two phenomena that can complicate the analysis of allelic association are selection in favor of a particular phenotype and new mutations (at either the disease or marker loci) arising in the population.

MAPPING GENES USING LINKAGE DISEQUILIBRIUM AND SPECIAL POPULATIONS

In most cases, allelic association will result from linkage disequilibrium, unless a specific polymorphism in the actual susceptibility gene is being studied. The power of linkage disequilibrium is best exploited in its use in fine-mapping. Because linkage disequilibrium rarely extends more than 1 cM from the susceptibility locus, its detection signals a significant decrease in the minimum candidate region. However, this great strength is also its great drawback. Because the effect is so localized, it will be very hard to find against the background of the entire human genome. In genetically complex diseases, the further complications of genetic heterogeneity and/or gene–gene interaction may make detection even more difficult.

Mapping and/or gene identification using linkage disequilibrium is especially powerful in genetically unique or isolated populations (so-called special populations, such as Amish or Finnish populations). These populations have already been used successfully for Mendelian diseases, since they are often homogeneous in disease origin. In other words, there are likely to be only a few founding individuals who carried specific chromosomal haplotypes on which the original mutation occurred. If the population is relatively new (i.e., the result of recent admixture of two populations), this approach can also be useful in the general mapping of disease loci as well, as has been shown in rare recessive disorders (Hastbacka et al., 1992). This is possible because the linkage disequilibrium is likely to extend over larger areas (several centimorgans) of the chromosome, since the number of generations available to allow decay of the linkage disequilibrium by recombination is small.

These populations can be equally useful in mapping complex traits. The basic premise is that genetically isolated populations will have fewer genes contributing toward a disease trait, and therefore the effect of each remaining susceptibility gene will be easier to detect. The value of these populations in genetic mapping studies has long been realized. Homozygosity mapping was described early on in the molecular revolution (Lander and Botstein, 1987). Recently these advances have been expanded to include the use of pooling strategies (Sheffield et al., 1994) and the exploitation of the phenomena of linkage disequilibrium together with the isolated inbred nature of these groups through the approach of "shared segment" mapping of a complex phenotype (Houwen et al., 1994; Durham and Feingold, 1997). Thus the great advantage of the special population is its power to detect linkage. However, it must be pointed out that this power comes at the potential cost of specificity. Only one or a few of the entire suite of susceptibility genes may be found, and the effect of this gene or genes may be limited to the special population being studied.

ASSOCIATION STUDIES: IMPLEMENTATION

There are two types of association studies. *Case–control* studies compare allele frequencies in a set of unrelated affected individuals to a set of matched controls. The control populations should be matched with respect to ethnicity as well as other factors such as age. Spurious associations can result because of population stratification (i.e., the existence of multiple population subtypes in what is assumed to be a relatively homogeneous population). Such stratification can represent either recent admixture or the incorrect matching of cases and controls. The existence of these confounding factors can lead to a significant result even in unlinked loci (Table 15.1) or unassociated loci within stratum.

An example of a case–control linkage study is the identification of the *APOE-4* allele as the susceptibility gene in late-onset familial and sporadic AD. Table 15.2 presents data on a large study of over 500 Alzheimer disease patients and age and ethnically matched controls. Using standard chi-square analysis, a significant association was found [$p < 0.001$]. As indicated in Table 15.2, there appears to be an in-

Table 15.1 Example of Population Stratification[a]

Population A			Population B			Population C (mixed)		
	A (0.80)	*a* (0.20)		*A* (0.20)	*a* (0.80)		*A* (0.50)	*a* (0.50)
B (0.80)	0.64	0.16	*B* (0.20)	0.04	0.16	*B* (0.50)	0.25	0.25
b (0.20)	0.16	0.04	*b* (0.80)	0.16	0.64	*b* (0.50)	0.25	0.25

[a]In this example populations A and B have very different allele frequencies for the disease gene *A* and the unlinked marker *B*. If the populations mix evenly, then the overall allele frequencies are as seen in population C. If comparisons of the haplotype frequencies are made from population C to either population A or B, the results will be significant even if no linkage exists. For example, if the affected individuals are drawn from population C, they will have an *A/B* haplotype frequency of 0.25, assuming no assiciation with the disease. If the controls are drawn from population B the *A/B* frequency will be 0.04. The erroneous conclusion is that there is an association of the disease with the *B* allele (0.25 vs. 0.04), since they occur together more often than in population B. In most data sets, it is not possible to determine whether the sample was drawn from one, two or more populations.

Table 15.2 Case-Control Association Studies: APOE-4 allele and Alzheimer Disease

a. Observed Counts

APOE-4 allele	Cases	Controls	Total
APOE-4	240	60	300
Not APOE-4	360	340	700
Total	600	400	1000

b. Expected Counts

APOE-4 allele	Cases	Controls	Total
APOE-4	180	120	300
Not APOE-4	420	280	700
Total	600	400	1000

χ^2 = (observed-expected)^2/expected = {(240 – 180)^2/180} + {(60 – 120)^2/120} + {(360 – 420)^2/420} + {340 – 280)^2/280)} = 71.5. P < 0.0001.

crease in the *APOE-4* allele, and a concomitant decrease in the *APOE-3* allele, in Alzheimer patients.

Family-based studies control for the possibility of genetic differences between the case and control populations by comparing the frequencies of alleles transmitted to the affected child to the alleles not transmitted. The only samples necessary are those from the affected individual and his or her two parents (the TDT triad). This approach eliminates the concern that population substructure may be the cause of the association. These studies include the transmission disequilibrium test (TDT) (Spielman et al., 1993), the haplotype relative risk test (HRR) (Falk and Rubenstein, 1987), and the AFBAC method (Thomson, 1995). The HRR and AFBAC approaches were developed as family-based tests for association, and the AFBAC method is designed to detect association in the presence of linkage. The TDT approach tests for linkage in the presence of association. Both AFBAC and TDT have little power unless linkage and association coexist. The difference between these two methods is that the TDT can also function as a test of association in the presence of population admixture and can be used as a valid test of linkage. The statistical differences between these methods are subtle and are not described here.

The TDT is the most widely used of all the tests. It can be a more powerful test to detect linkage than the affected sib pair linkage tests, especially when the genetic effect is small (Spielman et al., 1993; Risch and Merikangas, 1996), as is often the case with genetically complex traits. The disadvantage of this approach is that the TDT has no power to detect linkage if association is not present. The TDT test examines the number of transmissions of allele 1 (*A1*) or allele 2 (*A2*) from a heterozygous parent to an affected offspring. An example of the TDT is given in Figure 15.2. As a test of linkage the counts needed in Figure 15.2 can come from simplex, multiplex or even multigenerational family data. The statistical significance is tested by standard chi-square (McNamar's test). The data used in con-

Example of the TDT

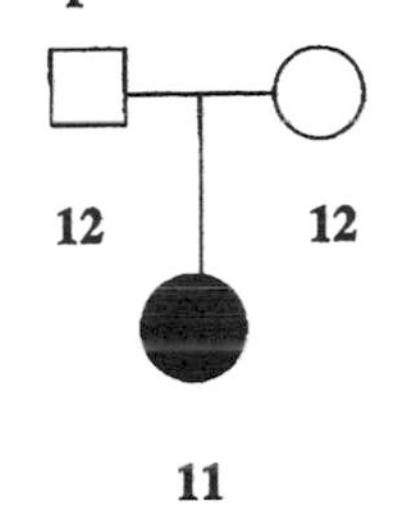

Transmitted	Not Transmitted	
	Allele 1	Allele 2
Allele 1	0	2
Allele 2	0	0

Figure 15.2 *Example of scoring a transmission disequilibrium test (TDT) family.*

structing the counts comes only from heterozygous parents. The TDT statistic is as follows:

$$\chi_1^2 = \frac{(b-c)^2}{b+c}$$

where b is the number of times an $A1/A2$ parent transmits an $A1$ to an affected offspring and c is the number of times an $A1/A2$ parent transmits an $A2$ to an affected offspring.

The test statistic then tests for deviations from the expected equal transmission rate into the two categories from the heterozygous parents. Homozygous parents do not need to be scored. This is different from the AFBAC method, which uses both heterozygous and homozygous parents. A significant result indicates that the marker is linked to the disease locus. The TDT can find linkage only in the presence of association. If there were only linkage and no linkage disequilibrium, then across families there would be no difference between b and c, since the allele in coupling with the disease gene in each family is random, preventing the detection of linkage. The TDT can also serve as a test of linkage disequilibrium if only simplex families are used or if only one affected individual and his or her parents are included per family. This use of the TDT is critical for narrowing a broad region of interest identified by linkage analysis. Analysis of association could potentially identify the markers that are closest to the actual disease susceptibility locus.

The TDT approach was a useful approach in identifying the relationship of the

Table 15.3 Transmission Disequilibrium Test (TDT) and Diabetes[a]

	Nontransmitted[b]	
Transmitted	*A1*	*A2*
A1	NI	78 (62)
A2	46 (62)	NI

[a]In this example, the number of *A1* alleles transmitted from a heterozygous parent to the affected (IDDM) child is 78, while the number of *A2* alleles transmitted is 46. If no association exists, the expectation is that each allele would be transmitted 62 times. This result is highly significant.
[b]NI, not informative; expected number in parentheses.
Source: Spielman et al. (1993).

insulin gene in insulin-dependent diabetes mellitus (IDDM) (Spielman et al., 1993) (Table 15.3). The TDT method does not require researchers to go to the expense and effort of recruiting families with multiple affected individuals, indeed, the clinical status of the parents does not have to be known. In some cases, however, these advantages may be outweighed by practical problems. In the diabetes situation, both parents of the affected family members were available for study, making this approach ideally suited to IDDM. However, in the Alzheimer/*APOE-4* example, since AD is a late-onset disease, parental DNA is almost always unavailable, and thus the traditional TDT approach was impossible. A novel approach that circumvents this difficulty is to use unaffected siblings as controls rather than relying on parental controls (Curtis, 1997; Boehnke and Langefeld, 1998; Spielman and Ewens, 1998). This sib-TDT (S-TDT) approach compares marker allele frequencies in affected and unaffected siblings. The test requires only a simple affected/unaffected sibling pair, although power can be increased somewhat if additional siblings are available. The S-TDT retains the advantages of the TDT in that it provides a test for linkage and association and is immune to the effects of sampling bias.

The TDT test was originally developed to look at biallelic marker systems or situations of alleles that could be readily collapsed because of prior information regarding a known association. With the availability of a multitude of multiallelic markers, several new statistics have been proposed. These included the symmetry statistic (T_c), the marginal statistic (T_m) of Bickeboller and Clerget-Darpoux (1995), the likelihood ratio statistic (T_l) of Sham and Curtis (1995), and the marginal statistic with only heterozygous parents (T_{mhet}) of Spielman and Ewens (1996). Kaplan et al. (1997), investigated the properties of these four statistics and determined that the T_{mhet} was the most appropriate and efficient method. It had equivalent power to the other tests and it gave a (approximately) valid chi-square test of linkage. Their recommendation for multiallelic markers was to implement the T_{mhet} using critical values of χ^2 with ($m - 1$; m = number of alleles at the marker locus) df and including all available affected individuals and their heterozygote parents in the analysis. Table 15.4 provides the definition of the T_{mhet} statistic.

Table 15.4 Multiallelic TDT: T_{mhet}[a]

Nontransmitted allele	Transmitted allele			
	1	2	3	Total
1	n_{11}	n_{12}	n_{13}	$n_{1.}$
2	n_{21}	n_{22}	n_{23}	$n_{2.}$
3	n_{31}	n_{32}	n_{33}	$n_{3.}$
Total	$n_{.1}$	$n_{.2}$	$n_{.3}$	$n_{..}$

[a] $T_{mhet} = \frac{m-1}{m} \sum_{i=1}^{m} \frac{(n_{i.} - n_{.i})^2}{n_{i.} + n_{.i} - 2n_{ii}}$ where m is the number of alleles.

THE USE OF THE TDT IN GENOMIC SCREENS

The TDT was originally proposed as a test for linkage to specific candidate loci. More recently discussions have centered on the use of the TDT approach for entire genome scans (Risch and Merikangas, 1996). The ability to replicate in a comparable data set is as critically important when one is using the TDT as it is in the performance of linkage studies using affected relative pair analysis. With the advent of microchip technology on the horizon, and the interest in the development of ancestral single nucleotide repeat polymorphisms (SNPs), the expanded use of this methodology to quickly map and identify genes is a certainty.

However, there are several potential problems with the TDT in genomic scanning. First, the problem of multiple comparisons arises in this situation. That is, when so many statistical tests are performed false positive results are likely by chance alone unless the usual significance value of 0.05 or 0.01 is modified. Thus use of a critical value that is greater than the nominal P value is warranted. The usual Bonferroni correction approach (simply dividing the desired nominal significance level by the number of tests performed) will be too conservative because it assumes each test to be independent of the others. This will not be the case, since many of the markers will be linked and associated with each other. Unfortunately, it is not clear what the appropriate correction needs to be, although simulation-based statistics are being explored.

The second problem is simply the number of polymorphic markers necessary. Even at one marker per centimorgan, over 3000 well-mapped markers are needed. In addition, these markers must actually be located at 1 cM distances, not just scattered with a 1 cM average distance. While many more than 3000 markers exist, there are still many regions of the genome up to 20 cM in length with no known polymorphisms.

The third problem is that the TDT approach rests completely on the assumption that some level of linkage disequilibrium exists. While this may be true in many cases, susceptibility alleles arising from frequent mutation events, existing as extremely old mutations, or arising in regions with very high recombination rates, will have little or any detectable linkage disequilibrium.

SUMMARY

Allelic association may arise for several reasons, but association due to linkage disequilibrium can be exploited to aid in the mapping of genetically complex diseases. Both case–control and family-based methods can be used. The latter have several advantages, especially when tested using the TDT or its variant, the sib-TDT. The TDT is both simple and powerful, and it will have substantial power for detection of susceptibility alleles as better and more finely spaced markers are available for study.

REFERENCES

Bickeboller H, Clerget-Darpoux F (1995): Statistical properties of the allelic and genotypic transmission/disequilibrium test for multiallelic markers. *Genet Epidemiol* 12:865–870.

Boehnke M, Langefeld CD (1998): Genetic association mapping based on discordant sib-pairs: The discordant alleles test (DAT). Submitted.

Curtis D (1997): Use of siblings as controls in case control association studies. *Ann Hum Genet* 61:319–333.

Devlin B, Risch N (1995): A comparison of linkage disequilibrium measures for fine-scale mapping. *Genomics* 29:311–322.

Durham LK and Feingold D (1997). Genome scanning for segments shared identical by descent among distant relatives in isolated populations. *Am J Hum Genet* 61:830–842.

Falk CT, Rubinstein P (1987): Haplotype relative risks: An easy way to construct a control sample for risk calculations. *Ann Hum Genet* 51:227–233.

Hastbacka J, de la Chappelle A, Kaitila I, Sistonen P, Weaver A, Lander E (1992): Linkage disequilibrium mapping in isolated founder populations: Diastrophic dysplasia in Finland. *Nat Genet* 2:204–211.

Houwen R, Baharloo S, Blankenship K, Raeymaekers P, Juyn J, Sandkuijl LA, Freimer NB (1994): Genome screening by searching for shared segments: Mapping a gene for benign recurrent intrahepatic cholestasis. *Nat Genet* 8:380–386.

Kaplan NL, Martin ER, Weir BS (1997): Power studies for transmission/disequilibrium tests with multiple alleles. *Am J Hum Genet* 60:691–702.

Lander ES, Botstein D (1987): Homozygosity mapping: A way to map human recessive traits with the DNA of inbred children. *Science* 236:1567–1570.

Risch N, Merikangas K (1996): The future of genetic studies of complex human disorders. *Science* 273:1516–1617.

Sham PC, Curtis D (1995): An extended transmission/disequilibrium test (TDT) for multiallele marker loci. *Ann Hum Genet* 59:323–336.

Sheffield V, Carmi R, Kwitek-Black A, Rokhlina T, Nishimura D, Duyk GM, Elbedour K, Sunden SL, Stone EM (1994): Identification of a Bardet-Biedl syndrome locus on chromosome 3 and evaluation of an efficient approach to homozygosity mapping. *Hum Mol Genet* 3:1331–1335.

Spielman RS, Ewens, WJ (1996): The TDT and other family-based tests for linkage disequilibrium and association. *Am J Hum Genet* 59:983–989.

Spielman RS, Ewens WJ (1998): A sibship test for linkage in the presence of association: The insulin gene region and insulin-dependent diabetes mellitus (IDDM). *Am J Hum Genet* (in press).

Spielman RS, McGinnis RE, Ewens WJ (1993): Transmission test for linkage disequilibrium: The insulin gene region and insulin-dependent diabetes mellitus (IDDM). *Am J Hum Genet* 52:506–516.

Terwilliger JD, Ott J (1992): A haplotype-based 'haplotype relative risk' approach to detecting allelic associations. *Hum Hered* 42:337–346.

Thomson G (1995): Analysis of complex human genetic traits: An ordered-notation method and new tests for mode of inheritance. *Am J Hum Genet* 57:474–486.

16

Using Public Databases

Jonathan L. Haines
Department of Molecular Physiology and Biophysics
Program in Human Genetics
Vanderbilt University School of Medicine
Nashville, Tennessee

INTRODUCTION

Computers have played a central role in statistical analysis of human gene mapping data since the LIPED program was first distributed to the research community in 1974 (Ott, 1974). The suite of available programs grew rapidly throughout the 1980s, taking advantage of the exponential increase in the power, ease of use, and availability of computers. At the same time, the role of computers has expanded past computation and into communication and data storage and retrieval. It is interesting to note that one of the first uses of computers to compile genetic data actually predates the widespread use of computers for linkage analysis. McKusick's seminal compendium of human genetic disorders, *Mendelian Inheritance in Man* (McKusick, 1994), has been compiled using computer systems since its inception in the mid-1960s. The volume has grown from the initial description of 1487 genetic disorders in the first print version (McKusick, 1966) to describing over 9400 genetic disorders in OMIM, the online version of *Mendelian Inheritance in Man* (*http://www.ncbi.nlm.nih.gov/Omim/*).

The goal of this chapter is to introduce the reader to some of the most useful resources now available through the World Wide Web, the most common method of access and retrieval of data now in use. We cover the access and use of genetic marker databases, physical mapping databases, sequence databases, model organism databases, and other potentially useful sources of information. The very nature

Approaches to Gene Mapping in Complex Human Diseases, Edited by Jonathan L. Haines and Margaret A. Pericak-Vance. ISBN 0-471-17195-6

of the Web is dynamic change. New sites appear and old ones disappear with somewhat disruptive regularity. We have tried to limit ourselves to sites that are most likely to remain stable, and thus useful for the human gene mapper.

GENETIC MAP AND MARKER DATABASES

There are two very commonly used databases of genetic maps and markers. The Cooperative Human Linkage Center (CHLC, or "Chelsea"), encompassing researchers in Iowa, Wisconsin, and Pennsylvania (Murray et al., 1994), has developed extensive genetic maps of every chromosome using primarily tri- and tetranucleotide repeat microsatellite markers. As discussed in Chapters 9 and 11, these markers have several technical advantages over the more common dinucleotide microsatellite markers. Généthon, centered in Evry, outside of Paris, France, has developed a set of over 5000 dinucleotide repeat markers and generated a highly detailed genetic map of the entire human genome (Dib et al., 1996). The CHLC has attempted to integrate marker data from several different sources, including Généthon and another global mapping effort in Utah and provides several different maps for comparison.

The subsections that follow list these two sites plus several others that offer data on genetic markers, maps, or both. There are many human-chromosome-specific databases that also have genetic maps associated with them. While the quality and depth of the genetic map data varies greatly among these databases, they should be examined along with the global sites listed below. We discuss the chromosome-specific databases in detail later in this chapter.

Sites

CHLC The primary universal resource locator (URL), or home page, for CHLC is *http://www.chlc.org*. This site provides descriptions of CHLC activities and resources, genetic maps of several different types, information (primer sequence, allele size, allele sizes for several CEPH individuals, heterozygosity, repeat type, etc.) on markers, lists of publications, and links to other useful sites.

Généthon The primary URL for Généthon is *http://www.genethon.fr/genethon_en.html*. This site provides descriptions of Généthon activities, and access to the genetic, physical, and transcript maps. The information on the genetic maps includes the final Généthon map, as well as detailed information (primer sequence, allele sizes, allele sizes for CEPH 1347-02, heterozygosity) on each marker, along with links to other sites.

CEPH The primary URL for CEPH (Centre d'Étude du Polymorphisme Humain) is *http://www.cephb.fr*. This site provides information about CEPH and its activities and access to the CEPH genotype database. The CEPH genotype database contains data submitted by over 100 laboratories around the world and includes RFLP and

VNTR data along with extensive microsatellite marker data. The database provides information on the marker including allele sizes and frequencies, and heterozygosity (when known). It can also provide the actual genotypes for individuals in the CEPH pedigrees (Chapters 2 and 11), extremely useful information for promoting consistency of allele calling between gels, projects, and laboratories (Chapter 9). Note that CEPH does not provide primer sequence data.

Marshfield The primary URL for Marshfield Medical Research Foundation (MMRF), Center for Medical Genetics, is *http://www.marshmed.org/genetics/.* This site provides information on the activities of the Center for Medical Genetics and access to maps and markers generated and used by the center. This includes detailed information on the genotyped markers, the maps, the multiplex screening sets, and the laboratory methods used. Individual genotypes for the CEPH families genotyped at the MMRF are available.

Utah The primary URL for the Eccles Institute of Human Genetics in Salt Lake City is *http://www.genetics.utah.edu/home.html.* The genetic maps are located at *http://www.genetics.utah.edu/totalmap/index.html.* This site provides information about the Eccles Institute, along with the maps generated using markers developed at the institute. Marker data includes primer, sequence, and number of alleles. Also available are the LINKAGE format files used to generate the maps.

LDB The primary URL for LDB, the location database, is *http://cedar.genetics.soton.ac.uk/public_html/.* This site provides summary maps of all the chromosomes. These maps are developed and updated using a statistical algorithm that attempts to integrate data from many different maps and sources including genetic maps, radiation hybrid maps, cytogenetic maps, and physical maps. Links from the summary maps can provide specific marker information from other databases, such as the Genome Database (GDB).

GDB The primary URL for the GDB, or **G**enome **D**atabase, is *http://gdbwww.gdb.org.* This site provides very detailed information on genetic markers, genetic maps, cytogenetic maps, and clones. It is searchable in several formats.

GENLINK The primary URL for this site is *http://www.genlink.wustl.edu.* This site provides access to an extensive genotype database of telomere information along with access to information from the CEPH database and the Généthon genetic marker database. It contains a graphical output of the CEPH genotypes by family, which can be downloaded. It also contains links to some genetics software.

To Find Information on Genetic Markers

Several of the sites above have detailed information on genetic markers, and much of the marker information is available at multiple sites. Thus the choice of site depends somewhat on user preferences. GDB, perhaps the most complete source, can provide information on alternate names, flanking regions, primer sequences, allele sizes, allele frequencies, and both genetic and physical mapping information. However, many individuals find the GDB interface hard to use, and five levels of screens are required to get some of this information. In addition, since GDB no longer requires curation of the data before it is entered into the database, the quality of the data cannot be guaranteed. Finally, as with all the databases, data that might be found in other databases may be missing from GDB. As of July 1, 1998 no additional data will be entered into GDB. The status of future updates to GDB is currently uncertain as is the prospect of any replacement database.

Another good source of marker information is CHLC. The marker information at this site is limited to markers generated by the CHLC collaborators. Thus it is restricted to tri- and tetranucleotide repeat CHLC markers and dinucleotide repeat markers generated by the Marshfield group. The data available at this site include primer sequence, clone sequence, allele sizes, and allele frequencies, and actual genotypes for CEPH individuals 1331-01 and 1331-02 (see Chapters 2 and 11). While this site is quite intuitive to use, the data are incomplete. Many markers list only the name, or only the allele sizes, with no frequencies or primer sequences. If these data are not available here, they are also not likely to be available at GDB. For markers being used in genomic screening, the MMRF site is now the most complete, up-to-date, and accurate.

The Généthon site provides detailed information on the markers generated there. Thus primer sequence, allele sizes, number of alleles, and the genotype of CEPH individual 1347-02 are available on every marker. GENLINK provides an alternate interface to the Généthon database. The Utah site covers only Utah-generated markers and data are limited to the name and primer sequences for each. Unlike the other databases, it is not keyword searchable, further restricting its utility.

To Find Information on Allele Frequency Estimates

As with genetic marker information, there are several sources of marker allele frequency estimates. Many of these allele frequency estimates, however, are based on small numbers of chromosomes (perhaps as few as 36–40) and should be used with caution, as discussed in Chapters 4 and 12.

Along with genetic marker information, GDB often provides allele frequency estimates, perhaps on several different populations. This data is found by moving down through the marker, polymorphism, allele sets, and allele frequency links. Another source of allele frequency estimates exists at Duke University (*http://www2.mc.duke.edu/depts/medicine/medgen/allele_freqs/index.html*). While the number of markers listed is not as great as in GDB, the population sizes tend to be larger, leading to more accurate estimates. The CHLC Web site allele frequency information is obtained on the marker information page, but not all markers have this information. The Marshfield Web site allele frequency information is available from the parameter sheets in the marker data field.

To Find or Integrate Genetic Maps

Although there are several good sources of genetic maps, the first place to look is the *CHLC site.* This site contains several genetic maps and has attempted to integrate into them from different sources. The maps are easy to use, and the user can simply click on the marker of interest for specific information. These maps have been generated using the CHLC and the Généthon data. While the Généthon maps rely solely on their own microsatellites, the CHLC maps integrate markers from Généthon, CHLC, and Marshfield. This integration by including markers from various maps provides common landmarks between maps, allowing more direct comparison. In general these maps are highly accurate and of high quality because they include only markers with strong support for a unique location (e.g., lod scores > 3.00 against any other position in the map). While the CHLC-specific maps use mostly tri- and tetranucleotide markers, they tend to be sparse, in that the marker spacing averages 5–10 cM, with many larger gaps. The Généthon maps tend to be more dense but rely on the less desirable dinucleotide repeat markers.

The following types of map are available from the CHLC:

1. *Skeletal maps.* In these maps, which use only microsatellite marker data, the ordering is based on fulfilling several statistical diagnostic criteria, thus using only the best of the available marker data. Each marker included in the map must fit into a single location with a lod difference of 3 or more. Each marker must also not expand the overall length of the map (unless placed at one of the ends) by more than 2 cM (sex-equal maps). In addition, only markers in which all families show similar recombination fractions with nearby markers are included (e.g., markers for cases of significant between-family heterogeneity are excluded).
2. *Framework maps.* These maps use skeletal maps as their starting point. Two criteria are relaxed. First, RFLPs and other nonmicrosatellite markers are included. Second, markers with significant between family heterogeneity may also be included. The lod criterion of 3 is still used, however. These maps include more markers than skeletal maps, and tend to be longer.
3. *Recombination minimization maps.* The ordering in these maps is based on reducing to the smallest possible number the number of overall recombinants in the resulting map. Thus the number of double (or higher) recombinants is reduced as well. This map format uses a skeletal map as the starting point and adds all microsatellite markers than can be uniquely placed into a single interval within a given map. For each interval, all possible orders of these markers are tested, and the order with the fewest recombinants is chosen as the best order. The data must come from a subset of 15 of the CEPH families.
4. *Integrated maps.* These maps use the skeletal, framework, and recombination minimization maps as the starting point, and call on all publicly available CEPH data to generate the final map.

In addition to the markers placed into the maps, the remaining markers generated by CHLC, as well as many Généthon and other markers, have been “binned” into specific regions along the high-quality integrated maps. Thus these markers may fall within a 5–40 cM region, without any further localization. Such binning, although helpful, is far from ideal, since knowing the relative order of all markers in a region is of critical importance to the researcher contemplating high-resolution genetic mapping. Other sources of order information, such as radiation hybrid maps and physical maps, may help in this regard (see below).

One of the difficulties with using the CHLC site is that the data for all CHLC markers is referred to by their clone name (e.g., GATA12B06) rather than by their more standard D-number nomenclature (e.g., DXXX). This can make easy referencing of markers that appear in different maps next to impossible.

The *Généthon site* handles the data somewhat differently. Each map is presented as a complete list of all markers on that chromosome, p arm to q arm, with each marker placed in its most likely position. Support for order is not considered, but the advantage is a very dense map (Dib et al., 1996). This is an excellent source of a very large number of markers in almost any region.

The *Marshfield maps* are based on microsatellite markers generated by Marshfield, Utah, Généthon, and CHLC. They are comprehensive maps based on genotypic data for the eight most commonly used CEPH families. All markers are placed in the map in their best possible location. Recent efforts have been made to remove errors in the data and markers that are most prone to error. Thus these maps tend to include more markers than CHLC maps. They are good places to look if the markers or regions you are interested in are not represented in the CHLC maps.

The *Utah maps* are based on markers generated by the Utah group and genotyped on four CEPH families. Because of this limitation on the data set, the maps tend to be sparsely populated by uniquely placed markers. Many additional markers are regionally placed. This is a good place to look for a specific Utah marker or to search a region that is not well represented in any other map.

The *LDB maps* are based on data integrated from many different sources and mapping approaches, including genetic, physical, radiation hybrid, and cytogenetic. Thus many loci included in these maps are not polymorphic. The mapping information is integrated using a statistical algorithm (Collins et al., 1996) that provides a measure of support for order and attempts to correct for errors and interference. While these maps are perhaps the most comprehensive of all statistical maps, local support for order can be small (while the support for order is given, no attempt is made to limit the maps to only the most strongly ordered loci).

Genlink provides a compilation of several published genetic maps, including those generated in 1992 by a U.S.-French collaborative group (NIH/CEPH, 1993) and by Dib et al. (1996). Some additional data, including telomeric markers for some chromosomes and two-dimensional recombination minimization maps, are available as well.

Accuracy of the Genetic Maps

One underappreciated attribute of all the genetic maps is the accuracy of the estimated map distances. For cost and efficiency, most of the global maps have been generated using only a subset of the CEPH pedigrees (see Chapters 2 and 11). In the case of the Généthon maps, 8 families were used; CHLC and Marshfield used 8 or 15 families, and Utah used 4 families. Thus the number of potential recombination events useful for ordering the markers is very limited. This leads to estimates that may vary by as much as 5 cM, making interpolation of order strictly from distance estimates in different maps highly suspect.

Standardized Genotype Data

Particularly in large collaborative, multi-institutional projects, it is essential that all alleles be consistently labeled in each laboratory. This is most easily done using the actual allele sizes (see Chapter 9). Since alleles may vary by as little as two base pairs, even automated systems require control samples for standardization. This is usually done by using the widely available DNA samples from the CEPH pedigrees. The most commonly used samples are 1331-01, 1331-02, 1347-02. Allele sizes for these individuals are available from several sources, including CHLC, Marshfield, Généthon, and Duke. A more complete listing of genotypes for all CEPH individuals actually genotyped is available from CEPH. It should be noted that none of these databases is complete, and a search of several may be necessary to find the requisite information.

PHYSICAL MAPPING DATABASES

One of the first steps toward the goal of generating the entire sequence of the human genome is to have complete physical maps of each chromosome. This goal has had several intermediate steps, and these have included yeast artificial chromosome (YAC) contigs (Cohen et al., 1993; Chumakov et al., 1995), cosmid contigs (Ashworth et al., 1995), radiation hybrid maps (*http://shgc-www.stanford.edu/RH/index.html*), and most recently sequence-ready contigs. Not only are these maps useful for large-scale sequencing, identifying specific genes, and mapping nonpolymorphic sequence-tagged sites (STS), and expressed sequence tags (ESTs), they are also useful to the genetic mapper who is trying to identify disease genes through linkage analysis. Microsatellite markers are often placed on these maps, and the order of markers from these maps is often more readily available (particularly if they are closely linked to each other), than from genetic maps. The paragraphs that follow describe various resources and tell how they may be used to simplify gene mapping. There are three major sources of YAC and cosmid contigs: two maps and a series of databases.

The *CEPH YAC map* (*http://www.cephb.fr/ceph-genethon-map.html*) provides all the data on the STSs mapped to YACs used to generate the first whole-genome physical map (Cohen et al., 1993; Chumakov et al., 1995). Data include information on each YAC, the STSs mapped to them, Alu-PCR hybridization data, fingerprint and sizing data, and fluorescence in situ hybridization (FISH) data, and the confidence in both the STS content and ordering of the markers.

The *MIT Whitehead Institute map* (*http://www-genome.wi.mit.edu/cgi-bin/con-*

tig/phys_map) provides detailed information on over 24,000 total STSs, including over 6500 microsatellite markers, 5500 STSs, and 12,000 ESTs mapped against either or both of the CEPH mega-YAC library or in a radiation hybrid map (see below). The data for each STS includes chromosomal assignments, primer sequences, product sizes, PCR conditions, sequence information, and YACs that are positive for the STS.

Another potentially useful source of physical mapping information comes from the chromosome-specific databases (Table 16.1). The information content is highly variable from chromosome to chromosome.

There are several sources of radiation hybrid (RH) mapping information, maintained by academic, governmental, and private institutions.

The *Stanford Radiation Hybrid Map* (*http://shgc-www.stanford.edu/RH/index.html*) provides mapping information on the G3 mapping panel for approximately 11,000 unique STSs, approximately 2500 of which are polymorphic. SHGC presents the data as a series of bins. Bins are ordered with respect to each other with odds of 1000:1 (lod 3) against all markers outside the bin. However, within each bin, the order is less certain; the most likely order is given. Other database references (e.g., GenBank accession numbers, GDB locus designations, dbSTS accession numbers) are provided where known.

The *Whitehead Institute Radiation Hybrid map* (*http://www-genome.wi.mit.edu/cgi-bin/contig/phys_map*) provides mapping information on over 14,000 STSs mapped to the Genebridge 4 panel. These are a subset of the total of over 24,000 STSs mapped mentioned above. The framework of approximately 1600 markers are ordered with odds of approximately 300:1 (lod 2.5); the order of the over 10,000 additional markers is less certain.

Another integrated physical map, primarily using radiation hybrid information, is the *Science 96 gene map* (*http://www.ncbi.nlm.nih.gov/genemap*). This map provides the locations of over 16,000 unique ESTs across the entire human genome. Detailed information on location, adjacent markers, and access to other database information (names, primer sequences, mRNAs sequences, etc.) is available.

The *Sanger Centre* (*http://www.sanger.ac.uk/HGP/rhmap/*) is using both the G3 and Genebridge 4 radiation hybrid panels to generate RH maps. Maps are available on all chromosomes, and detailed information on the markers (names, primer sequences, references to sequence information, etc.) is available.

Finding a Marker and Its Relative Position

Any of the resources above can be accessed for a specific marker; many of these databases provide links, or directly access, other databases such as GDB to provide detailed information. For example, if the CEPH YAC map is queried for a genetic marker—say "D11S922" (the query is case sensitive)—data will be returned on allele sizes, primer sequences, heterozygosity, the YACs positive for this marker, and the neighboring YACs. The YAC information can be used to place this marker between other markers mapped to neighboring YACs. The same marker queried in the

Table 16.1 List of Chromosome-Specific Databases

Chromosome	Address (URL): *http://*	Workshop Reports	Genetic Map	Radiation Hybrid Map	Cytogenetic Map	Physical Map	Integrated Map	STSs	ESTs	Sequence Data	Notes
1	*linkage.rockefeller.edu/chr1/*	X	X							X	
1	*www.med.upenn.edu/~poncol/chr1 /resources.html*	X	X	X	X	X	X	X	X		
1	*www.sanger.ac.uk./HGP/chrl/*		X						X		
3	*mars.uthscsa.edu*	X	X	X	X	X		X			
3	*www-eri.uchsc.edu/Chrom3.html*					X					
5	*chrom5.hsis.uci.edu*			X	X	X					
6	*www.sanger.ac.uk/HGP/Chr6/*	X		X						X	
7	*www.genet.sickkids.on.ca/chromosome7*		X		X	X		X	X	X	Disease gene references
7	*www.genlink.wustl.edu/workshops/ chr7/index.html*	X									
7	*www.nhgri.nih.gov/DIR/GTB/CHR7/*					X		X	X		
8	*gc.bcm.tmc.edu:8088/chr8/home.html*	X									
8	mars.uthscsa.edu	X			X						
9	*www.gene.ucl.ac.uk/chr9/*	X	X			X	X				
10	*www.cric.com/genesequences/ index.html*	X		X	X	X	X		X		
11	*mcdermott.swmed.edu*	X		X		X	X	X	X	X	
11	*shows.med.buffalo.edu/home.html*	X				X		X			
12	*paella.med.yale.edu/chr12/Home.html*	X			X	X	X	X		X	
13	*genome1.ccc.columbia.edu/~genome/*	X				X		X			
16	*www-ls.lanl.gov/TOC.html*				X		X	X			
16	*www.tigr.org/tdb/humgen.html*									X	

(continued)

Table 16.1 List of Chromosome-Specific Databases *(continued)*

Chromosome	Address (URL): *http://*	Workshop Reports	Genetic Map	Radiation Hybrid Map	Cytogenetic Map	Physical Map	Integrated Map	STSs	ESTs	Sequence Data	Notes
17	*bioinformatics.weizmann.ac.il/chr17/*						X				
18	*www.childrenshospital.org/chromosome18*	X									
19	*www-bio.llnl.gov/bbrp/genome/genome.html*				X	X				X	
20	*www.sanger.ac.uk/HGP/chr20/*	X		X						X	
21	*www.cephb.fr/chromosome21.html*				X	X		X		X	
21	*www-eri.uchsc.edu/*							X	X	X	
22	*www.cbil.upenn.edu/cbil_home/Databases.html*					X	X	X		X	
22	*www.genome.ou.edu/human.htm*									X	
22	*www.sanger.ac.uk/HGP/Chr22*					X		X	X	X	
X	*www.sanger.ac.uk/HGP/ChrX*	X								X	
X	*www.mpimg-berlin-dahlem.mpg.de/~xteam/*					X				X	
X	*Gc.bcm.tmc.edu:8088/chrx/home.html*	X	X			X		X			
X	*Genome.wustl.edu/cgm/cgm.html*						X	X		X	
Mitochondria	*www.gen.emory.edu/mitomap.html*									X	Functional maps; mutation maps

Whitehead database will return the primer sequences, radiation hybrid mapping information, links to complete sequence information, PCR conditions, GenBank and GDB accession numbers, the Généthon genetic map, and the CEPH YAC map. By accessing the complete chromosomal map, placement of this marker against other markers is found. The same marker has not been mapped on the Stanford radiation hybrid panel, but in general would return the chromosomal location, the 1000:1 bin, its position compared to other markers, the GenBank accession number, the GDB locus, and the lod and distance information. This marker queried against the SCIENCE96 map will return cDNAs/ESTs mapped within the same interval as this marker.

As with the genetic maps, integration across the various physical maps can be difficult. Fortunately, there has been a greater effort to use many of the same markers (primarily the Généthon microsatellite markers) in the different maps, making cross-referencing somewhat easier than with genetic maps. However, as the foregoing example indicates, not all maps will contain a particular marker.

Building a Virtual Physical Map

The arduous process of building a physical map (Chapter 17) in the laboratory can be aided substantially by simple database queries. Whereas massive screening of large YAC or cosmid (now often P1 or BAC) libraries was formerly required, now a much smaller set of confirmatory experiments may suffice. The process is perhaps best seen through a simple example using the Whitehead integrated map, although similar approaches can be used in other databases. The first query (for D12S373) identifies several YACs containing this marker, and a large contig already developed (WC12.1). If we examine one of these YACs, 772-D-5, we discover that it has several other STS hits, including a number of STSs on chromosomes 9 and 12. If we examine another YAC (844-G-10), we discover it also contains marker D12S1630, and upon following this link, we will find that D12S1630 is contained in several other YACs, including 847-F-2, which also contains D12S373 and D12S363, confirming that D12S1630 and D12S373 are near each other. If we look at D12S373, we find that it is contained in 799-H-1, which also contains D12S1630 and D12S363, as well as a new marker, D12S1715. if we examine D12S1715, we can see that it is contained in YAC 880-D-11, which contains only this marker and D12S1630. This leads us to infer the order of the markers as D12S1715, D12S1630, D12S373, D12S363. The contig is shown in Figure 16.1.

There are two major difficulties with building physical maps in this manner. This first arises because many YACs are chimeric (i.e., they have pieces of more than one chromosome joined together). Thus markers from chromosomes 9 and 12 can appear on the same YAC. If the chromosomal location of a marker is not known in advance, this can lead to jumping from one chromosome to another, and a false contig will result. Thus verification is absolutely essential. This can be done by finding additional YACs that contain the markers in question, or by using alternative mapping procedures (e.g., testing against single chromosome hybrids; radiation hybrid mapping) to be sure that the markers are all on the same chromosome. The second prob-

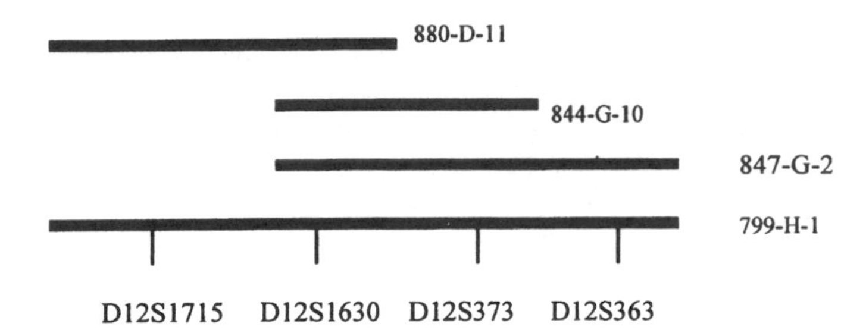

Figure 16.1 *Diagram of an overlapping YAC contig. Details are discussed in the text.*

lem is that map makers often must give up upon encountering a YAC that contains only one marker. Without a second marker to use as a starting point, it is impossible to continue building the contig. There are methods for making such markers if they do not exist (Chapter 17), but not without substantial effort in the laboratory. As the number of markers increases and the mapping reagents improve, the difficulty of providing a second marker should become less daunting, and virtual contigs should become the norm.

SEQUENCE DATABASES

The genomics researcher has available sequence data of two different types. The first contains primer information useful for making PCR products. These sequences are short usually (15–30 base pairs long) and are given as information for specific loci (be they microsatellite markers, STSs, or ESTs); they are available from most databases, depending on the type of marker being examined. The second type of sequence data is the longer sequence. This can range from 100 bp of sequence from a clone containing the microsatellite marker, STS, or EST, to kilobases or even megabases of sequence generated as part of whole-chromosomal sequencing efforts. These latter sequences are available from several sources, including chromosome-specific databases and the global repositories.

The two largest global sequence databases are GenBank and the European Molecular Biology Laboratory (EMBL) in Heidelberg. These tools can be accessed directly or through various servers. Most institutions provide remote access to these data through institution-specific servers, and thus data access and analysis using such approaches as FASTA and BLAST are not discussed here. However, direct access to the data can be obtained most easily via the Web page of the National Center for Biotechnology Information (NCBI) (*http://www.ncbi.nlm.nih.gov*). This page is constantly being updated, but contains general information on NCBI activities, access to GenBank (including search and data submission utilities), access to several gene databases, including the human gene map, the Unigene database of unique genes, and databases of ESTs and STSs. OMIM (Chapter 3) can also be accessed through this site.

The Entrez program provides a convenient method of examining several different maps simultaneously for the markers of interest. It provides a graphical view of maps from the Whitehead Institute, Stanford, CHLC, Généthon, and GDB, along with sequence data from GenBank. Markers can be identified via query, and nearby markers can be found by examining each of the maps. What these summary views cannot provide is confidence in order of the markers being examined. Thus any information on order should be examined carefully and, if possible, confirmed from the primary source.

CHROMOSOME-SPECIFIC DATABASES

Most human chromosomes have their own dedicated Web sites. Table 16.1 provides a listing of these chromosome-specific sites. These may be good places to turn if other sources do not have all the information being sought. The sites vary tremendously in their content and in their frequency of updating. Some contain primarily summaries of chromosome-specific workshops, while others contain detailed genetic, physical, and radiation hybrid maps, databases of clones and markers, STSs, ESTs, and sequence information.

MODEL ORGANISM DATABASES

The human genome, with approximately 3 billion base pairs of DNA, is still virtually unexplored. Thus any shortcut to localization should be investigated. One such shortcut may arise by examining similar phenotypes in a model organism, such as the mouse. Thousands of mutant phenotypes have been described in mice, and many have been mapped to narrow stretches of a chromosome. If the human phenotype is similar to a mouse phenotype, the location of the mouse gene will suggest a candidate region in the human. This is possible, since most genes in mice and man are conserved, as are many of the gene orders (e.g., synteny), at least over short distances, are also conserved. Thus maps and data from other, similar, but smaller genomes can be examined, and this ability may make finding trait susceptibility genes in the human much easier. There are numerous other reasons to examine model organisms. While each model organism has its own advantages (and sometimes disadvantages) as a system for study, they all share one strength: they offer the ability to more directly manipulate the genetic architecture of the organism and its population. This manipulation most often manifests itself in designing specific mating crosses based on desired phenotypes. It is beyond the scope of this chapter to discuss the uses of model organisms in detail. We include a short discussion of the mouse database as just one example of the types of database available.

The most commonly used model organism relating to human studies is the mouse. The primary mouse database is the Mouse Genome Database (MGD; *http://www.informatics.jax.org/mgd.html*). MGD contains detailed information on mouse genes and markers, including their probes and PCR primers. It also provides maps and mapping data, including cytogenetic, linkage, recombinant inbred, and recombinant congenic maps. Of particular usefulness are the homology mapping data (e.g., similar genes mapped in both species) and the listings of the mouse mutant phenotypes. The latter are analogous to the listing of human mutant genes in OMIM. By querying the phenotypes database, an investigator may be able to identify a phenotype similar to the human trait under study. If genes for this phenotype have been mapped and/or cloned, they become logical candidate genes for study in humans. All such data is readily available in MGD.

Many other model organism databases are available, including those for dog, sheep, cattle, swine, chicken, puffer fish, zebra fish, *Caenorhabditis elegans, Drosophila melanogaster*, yeast, *E. coli,* mycobacteria, maize, fungi, Archaea, Eubacteria, rice, *Arabidopsis thaliana,* and conifers. An excellent and comprehensive list of these databases, with links, is available at *http://gdbwww.gdb.org/gdb/hgpResources.html.*

SUMMARY

There are a large number of public databases brimming with genetic data useful for mapping genes involved in complex traits. Although most databases of human data are organized around methodologies (e.g., genetic mapping, physical mapping, radiation hybrid mapping, sequence data), other databases have attempted to integrate these diverse data types using biologically relevant groupings (e.g., model organisms, specific human chromosomes). These sites are constantly evolving as technology and experience, and the sheer volume of data, allow more and better ways of organizing and accessing information. As the amount and quality of information grows, virtual experiments, and even virtual cloning of genes, will be possible.

REFERENCES

Ashworth K, Batzer MA, Bandriff B, Branscomb E, de Jong P, Garcia E, Garnes J, Gordon LA, Lameerdin JE, Lennon G, Mohrenweiser H, Olsen AS, Slezak T, Carrano V (1995): An integrated metric physical map of human chromosome 19. *Nat Genet* 11:422–427.

Chumakov IM, Rigault P, Le Gall I, Bellane-Chantelot C, Billault A, Guillou S, Soularue P, Guasconi G, Poullier E, Gros I, Belova M, Sambucy JL, Susini L, Gervy P, Gilbert F, Beaufils S, Bui H, Perrot V, Saumier M, Soravito C, Bahouayila R, Cohen-Akenine A, Barillot E, Bertrand S, Codani JJ, Caterina D, Georges I, Lacroix B, Lucotte G, Sahbatou M, Schmit C, Sangouard M, Tubacher E, Dib C, Fauré S, Fizames C, Gyapay G, Millasseau P, Mguyen S, Muselet D, Vignal A, Morissette J, Menninger J, Lieman J, Desai T, Banks A, Bray-Ward P, Ward D, Hudson T, Gerety S, Foote S, Stein L, Page DC, Lander ES, Weissenbach J, Le Paslier D, Cohen D (1995): A YAC contig map of the human genome. *Nature* 377:175–183.

Cohen D, Chumakov I, Weissenbach J (1993): A first-generation physical map of the human genome. *Nature* 366:698–701.

Collins A, Frezal J, Teague J, Morton NE (1996): A metric map of humans: 23,500 loci in 850 bands. *Proc Natl Acad Sci USA* 93:14771–14775.

Dib C, Fauré S, Fizames C, Sampson D, Drouot N, Vignal A, Millasseau P, Marc S, Hazan J, Seboun E, Lathrop M, Gyapay G, Morissette J, Weissenbach J (1996): A comprehensive genetic map of the human genome based on 5,264 microsatellites. *Nature* 380:152–154.

McKusick V (1966): *Mendelian Inheritance in Man.* Baltimore, London: Johns Hopkins University Press.

McKusick V (1994): *Mendelian Inheritance in Man,* 11th ed. Baltimore, London: Johns Hopkins University Press.

Murray JC, Beutow KH, Weber JL, Ludwigsen S, Scherpbier-Heddma TS, Manion F, Quillen J, Sheffield VC, Duyk GM, Weissenback J, Lathrop GM, White R, Ward D, Dausset J, Cohen D (1994): A comprehensive human linkage map with centimorgan density. Cooperative Human Linkage Center (CHLC). *Science* 265:2049–2055.

NIH/CEPH Collaborative Mapping Group (1992): A comprehensive genetic linkage map of the human genome. *Science* 256:67–86.

Ott J (1974): Estimation of the recombination fraction in human pedigrees: Efficient computation of the likelihood for human linkage studies. *Am J Hum Genet* 26:588–597.

17

Laboratory Approaches Toward Gene Identification

Douglas A. Marchuk
Department of Genetics
Duke University Medical Center
Durham, North Carolina

INTRODUCTION

Once linkage has been established to a genomic region, and the critical interval narrowed sufficiently, the work of positional cloning enters the laboratory for a full-scale analysis of the DNA in the critical region. In terms of the ultimate goal of identifying the gene(s) behind a given phenotype, there is yet much more work required. The progression from the establishment of genetic linkage to the eventual identification of causative mutations within a gene requires a number of intermediate goals. These include the construction of a physical map of the critical region where the entire region is cloned in a contiguous set of overlapping clones, commonly referred to as a clone contig. Second, a complete transcript map for the region must be determined where all the genes encoded within this interval are placed onto the physical map. In addition to approximate location, it is important to determine some preliminary information about each transcript. This generally includes additional sequence information and a homology search to ascertain potential functions due to relatedness to other known genes. In addition, for each transcript an expression profile must be determined from a number of tissues, including tissues relevant to the pathophysiology of the disease. The final stages of this process involve DNA se-

Approaches to Gene Mapping in Complex Human Diseases, Edited by Jonathan L. Haines and Margaret A. Pericak-Vance. ISBN 0-471-17195-6

quence analysis of these transcripts from patients with the disease. This stage usually occurs in a scanning mode in which multiple samples are scanned for base pair changes, which are subsequently sequenced. Finally, one must include a population study of sequence variants uncovered to distinguish polymorphic base changes in the normal population from disease-causative mutations.

The goals of the Human Genome Project are such that for many genomic regions, a physical contig map of the region is already available. The reader is referred to Chapter 16 on the methods for establishing a YAC contig using public databases accessible via the Internet. The second objective, that is, a transcript map of the interval, is also becoming a reality, as evidenced by the recent placement of over 16,000 expressed sequence tags (ESTs) on a combined genetic and physical map (Schuler et al., 1996). Extensive mapping efforts are under way for most chromosomes, with markers of many different types being placed on the YAC genomic–clone contigs. Included in this list are genetic markers, sequenced tagged site markers (STSs), and particularly important for this endeavor, ESTs. These ESTs, which are being sequenced at an astonishing rate, are rapidly being assigned precise positions on physical maps. Assuming that all the transcripts are not known, we discuss a few commonly used methods to identify transcripts from a given fragment of DNA. We anticipate that in the not-so-distant future, as the transcript maps fill up, these will be increasingly less important. Nonetheless, current positional cloning efforts require some additional transcript identification to ensure that every candidate gene is identified in the critical interval.

The ultimate and final physical map is the complete sequence of the human genome, which can be thought of as a physical representation of the genome at a resolution of one base pair. At some point in time, the complete DNA sequence will be available and some of the above-mentioned techniques will be of only historical interest, at least with regard to positional cloning for human disease. However, the task of the individual investigator will not be complete. The study of inheritance requires knowledge of the different DNA sequences that lead to altered phenotypes. Thus, irrespective of the point at which one enters this process and the level of groundwork done by the Human Genome Project, it will always be necessary to determine the DNA sequence variation in the patient population under investigation. Therefore we examine a number of different mutation scanning techniques that enable rapid and preliminary analysis of the loci that vary in sequence in comparison to normal controls. We also discuss the important issue of methods to determine the approximate frequency of any base pair change in the general population. This is critical to the distinction between polymorphism and causative mutation for any base pair change that is identified.

The work described in this chapter is rapidly evolving as the goals of the Human Genome Project are advanced; indeed, certain aspects are becoming obsolete as new techniques are developed. The entire range of procedures and techniques covered in this chapter could easily fill an entire separate volume. Our goal, however, is not to provide detailed protocols for all the numerous steps involved in physical mapping, transcript identification, and mutation analysis. It is rather to give a broad overview of the remainder of the process of positional cloning. We discuss the strengths and weaknesses of many of the most commonly used protocols, with the goal of giving intelligent guidelines to developing an efficient strategy for the individual investigator who wishes to identify genes involved in common and complex human phenotypes. For detailed protocols for many of these procedures, the reader is referred to the excellent series *Current Protocols in Human Genetics* and its supplements (Dracopoli et al., 1994).

PHYSICAL MAPPING

The first stage in this process is to obtain a contiguous set of overlapping genomic clones (contig) over the critical region. This becomes the primary resource for all subsequent analysis of the region. The clones can be used to isolate and characterize additional genetic markers mapping within this interval, such as new microsatellite repeats. New polymorphisms can be identified by conventional screening by hybridization (Hudson et al., 1992) or by a selection cloning strategy (Ostrander et al., 1992). These markers can be then used on the existing distal and flanking crossovers in existing families, or typed on newly ascertained families, to further narrow the critical region. In addition, the contig becomes the primary resource for isolation and identification of genes which map within the region.

As mentioned earlier, in many cases any critical interval for a given phenotype will already have a complete or nearly complete YAC contig associated with it. This is a result of the tremendous progress in physical mapping due to efforts funded by the Human Genome Project. In most cases the high-resolution genetic maps have been used to anchor the physical maps. In other words, the closely linked genetic markers have been used as the starting points to isolate large YAC clones. In some cases merely using these anchor points as probes has allowed large stretches of contiguous DNA to be mapped. In other cases the contigs will be incomplete and additional clones will need to be isolated and fingerprinted using Southern blot hybridization with repetitive probes. These fingerprints help establish additional overlaps among the clones. Most of the work of first-level physical mapping has been accomplished, so our purpose is not to discuss in detail the various methods involved in map construction. Nonetheless, at this point in time any YAC contig is likely to harbor either real or potential gaps, and we discuss the options for bridging these gaps, beginning with a brief description of the different clone options for physical mapping.

Yeast Artificial Chromosomes (YACs)

YAC clones (Burke et al., 1987) contain human inserts in the size range of a hundred or more kilobases to over a million bases (megabases). These inserts are carried as an artificial linear chromosome within the *Saccharomyces cerevisiae* nucleus in a single copy per cell. YAC libraries, as well as PAC and BAC libraries discussed below, are difficult and quite laborious to construct, and it is unlikely that most investigators would begin a positional cloning endeavor by attempting to construct a novel human

YAC library. However, a number of good total human genomic YAC libraries exist. Perhaps the most relevant is the so-called mega-YAC library created by genome researchers at the Centre d'Étude Polymorphisme Humain (CEPH) in Paris (Chumakov et al., 1992). Individual clones from the CEPH YAC library are available through collaborations with laboratories maintaining a copy of the library or through a number of commercial repositories. Owing to the large size of inserts, and their use in some of the earliest physical mapping endeavors, YAC clones provide some of the best coverage of the human genome. Most of the first- and second-generation physical maps of chromosomes are based on this technology, and they provide long-range continuity throughout many regions of the genome. This is in fact their strength—that for most regions, a complete contig can be established by merely a computer search. Thus, despite the advent of newer cloning technologies, YAC contigs will often be the choice for an entry level complete contig of a genomic region. However, these clones are more difficult to work with because of their large size, their single copy nature, and the difficulty of purifying the YAC itself from the yeast cell except by means of pulsed field gel electrophoretic techniques.

Another of the plagues of YAC technology is the relatively high rate of chimerism in the inserts. A chimera is defined as two (or more) fragments, not contiguous in the genome, that have both inserted in the same clone (e.g., inserts derived from parts of chromosomes 6 and 18 within the same clone). Chimerism in the YAC clones is likely due to recombination within yeast between repetitive elements as well as co-ligation events during cloning. Thus in any positional cloning endeavor it is critical to keep in mind that some YAC clones will undoubtedly be chimeric.

Chimerism can wreak havoc with electronic physical mapping efforts. Many an investigator can tell the story of building a YAC contig via a computer search only to find that the contig started and ended in the critical region, but some of the clones within the contig were wholly derived from a number of different chromosomes. This result indicates that at some point in the contig construction, a chimeric portion of a YAC clone was chosen as the query for additional overlaps, whereupon the contig "moved" to a new chromosome. This phenomenon is especially prevalent among contigs containing regions based on a single overlap between two clones. With dense physical maps consisting of a number of overlapping clones, these mishaps can usually be avoided, since it would be highly unlikely for two or more independent clones to be chimeric for the exact same regions of the genome. Nonetheless, investigators must be wary of electronically derived contigs that have not been confirmed in the laboratory.

There are two ways to ensure that the contig under consideration is real. One is by ensuring that all overlaps consist of two or more YACs anchored by two or more STS sites (Arratia et al., 1991). Thus the clone and STS density of the map works in favor of identifying true overlaps. With progress on whole-chromosome physical maps, in most cases this would in fact be a reality. Where a single overlap between clones exists, it may be necessary to confirm the overlap with a combination of end rescue, fingerprinting, and even fluorescence in situ hybridization to metaphase chromosome spreads to ensure that both clones come from the same approximate genomic region. These techniques are discussed later.

Bacterial Artificial Chromosomes (BACs)

BACs clones (Shizuya et al., 1992) are based on the *Escherichia coli* fertility factor and carry the insert in a circular artificial chromosome. They are capable of harboring an insert of 300 kb, although most libraries contain inserts in the range of 120–150 kb. BAC clones exist at a copy number of one per host chromosome but are fairly easily isolated from the host strain by means of modified large-insert plasmid purification protocols. These useful characteristics tend to promote the use of BACs as the initial clones for large-scale sequencing projects. A number of good human BAC libraries exist, and screening can be done through a number of commercial repositories. Although BACs are in some ways easier to manipulate than YACs, in general, one would not create a BAC contig over a very large (many megabases) region de novo, since the endeavor would potentially take too many screenings of the library, unless of course, one has unlimited access to a library.

P1 Artificial Chromosomes (PACs)

P1 artificial chromosomes (Ioannou et al., 1994) are based on a combination of the best features of both the *E. coli* P1 bacteriophage cloning system (Sternberg et al., 1990; Sternberg, 1994), which is not discussed here, and the *E. coli* fertility factor–based BACs. P1 artificial chromosomes have an average insert size of approximately 150 kb, and a low chimerism rate. Perhaps their strongest attribute is that like BACs, they are much more easily isolated from the host DNA via modified plasmid or cosmid preparation protocols. At this date, few publicly assessible long-range contigs are available using PAC clones, however, this may change with time. The same issues concerning long-range coverage that were discussed for BACs also apply to PACs.

Cosmids

Cosmids are circular molecules that replicate as a plasmid in *E. coli.* They are created by means of vectors derived from the cohesive ends of the lambda bacteriophage. They have a maximum insert size of approximately 40 kb, and as such represent a physical map having much higher resolution than the previously described vector systems. In addition, cosmids are quite easy to purify and give much higher yields of DNA than the other systems. Because they are so small, however, it is very difficult to obtain long-range continuity with cosmid clones, although some exceptions are acknowledged. Nonetheless it is very unlikely that the critical region for any given gene hunt would be assessible in a cosmid contig constructed de novo for that particular project.

Gap-Filling

Chapter 16 discussed building YAC contigs by searching physical maps available through electronic databases. The YAC contig is established primarily by using this tool. Nevertheless it is likely that any given region under investigation will have in the region one or more gaps that need to be bridged. Thus the ends of the YACs must be isolated if these are to be used as probes for rescreening a YAC or other large insert library. There are a number of ways to obtain sequence from the ends of the YACs, each involving a PCR amplification system, and these are discussed next in order of their simplicity. Other techniques for plasmid end rescue exist (Hermanson et al., 1991; Marchuk et al., 1992), but these require yeast transformation and cloning and are not discussed here.

The easiest way to amplify the end of a YAC is to use vector-specific primers in addition to an Alu repetitive primer (Breukel et al., 1990; Nelson et al., 1991). Controls would include the vector-specific primer alone, which should give no amplification product, and the Alu repetitive primer alone, which very well may give a number of specific inter-Alu fragments. However any unique band that appears in the vector-plus-repeat primer lane can be excised from the gel and sequenced. This DNA fragment should extend from the vector arm to the first repetitive element within the DNA insert. Some such fragments consist primarily of repetitive DNA and thus are useless for rescreening a genomic library. However if unique sequence is encountered, PCR primers can be designed within this sequence and can be used for gap-filling by rescreening the library with this STS.

A slightly more complicated method for the isolation of YAC ends is via inverse PCR (Ochman et al., 1988; Silverman, 1993). In this approach PCR primers are designed to be within the vector sequence just beyond the cloning site but in opposing directions from each other. DNA from the YAC is cut with one of a number of particular restriction endonucleases. This DNA is then diluted and ligated such that intramolecular circles are the primary product. In this conformation the PCR primers can amplify across the ligated circle and create a product that extends from the first restriction endonuclease site within the vector to the first site within the insert. Specificity can be increased by using a nested set of primers for second-round amplification. This technique is more time-consuming than the Alu-vector PCR, since it works best if the YAC DNA is excised from a pulsed field gel and involves some preparative work (digestion and circularization) prior to the PCR amplification step. In addition, it is subject to the spacing of the available restriction sites in the vector arms. Unfortunately, there are only a few restriction endonucleases to choose from, and many of these cut resulting in blunt ends, which decreases the efficiency of circularization via ligation.

A third PCR amplification procedure for the isolation of YAC ends is also more complex but in principle should allow for the amplification of essentially any YAC clone. This technique involves the digestion of the YAC DNA with a restriction endonuclease and the ligation of a specific linker to all of the fragments. This creates a primer annealing site at each end (Riley et al., 1990). This technique, called bubble-linker PCR or vectorette-PCR, involves a linker that has within its sequence a large region that displays no match between the two strands, creating a bubble in the sequence. To understand this technique, it is necessary to recognize that the primer for PCR amplification cannot anneal to the sequence within the bubble because it is identical to the DNA sequence within the bubble, which has no complementary strand. The complementary sequence can be generated only if there is first-strand synthesis from a primer annealed to the opposite strand. This first-strand synthesis is provided by a vector-specific primer just outside the cloning site and directed outward toward the insert DNA. Here again specificity can be increased by using nested primers. PCR amplification will give unique band corresponding to the end of the YAC.

A caveat with respect to any of these approaches is that if this sequence is part of a chimeric clone, it might lead to building a chimeric contig. This is critical even with well-characterized contigs, where the ends of many clones may be of chimeric but uncharacterized sequence. It is critical therefore that the starting clone be shown to be nonchimeric. A chimeric YAC clone is identified when the STS generated from each end of the YAC insert is localized to determine whether it maps to the correct region of the genome. A chromosome hybrid mapping panel such as the NIGMS #2 mapping panel will show whether or not a YAC-end derives from the correct chromosome. Better yet, one can assay a radiation hybrid (RH) mapping panel to determine whether the STS derives from the specific region of interest. These panels consist of approximately 80 to 160 independently-derived rodent cell lines which contain distinct radiation-induced fragments of the human genome. Many thousands of genetic markers and other STSs have been assayed on these panels, resulting in high density RH maps. Any new STS, such as the end of a YAC insert, can be mapped in relation to these fixed loci. Two commonly used panels are the Stanford G3 panel (Stewart et al., 1997) and the Genebridge4 panel (Gyapay et al., 1996). If one has additional mapping resources such as cell hybrids which contain the chromosomal region of interest, these also can be used as DNA templates to check for chimerism.

To complete the contig, YAC end isolation may need to continue on subsequent YAC isolates (coming in from each end) until the gap has been bridged. In each case both ends of the newly isolated YAC should be isolated and tested for chimerism as well as orientation. The end furthest into the gap should be present only on the newly isolated YAC, whereas the other may be present on other YACs in the contig.

Overlaps between clones can also be identified by fingerprinting the clones (Bellanne-Chantelot et al., 1992). Since the clones are invariably large and contain numerous fingerprints, a simple restriction digest would be too complex to interpret. Thus the clones are digested with a number of different restriction enzymes but then blotted and probed with a human repetitive element such as Alu or LINE-1 (long interspersed repeat element 1: L1). The choice of the repeat will depend on the size of the inserts in the library. In principle, overlaps can be visualized by a number of hybridizing bands of the same size. In practice, this can become quite difficult to discern when multiple clones are being analyzed from gels run on different days. For large-scale genome mapping, these variables are controlled by scanning the multitude of gel images into the computer and realignment of bands to a common molecular weight standard using computer software. For the individual investigator creating a YAC contig within a critical region of the genome, however, these variables can be overcome by the smaller scale of the project if the prospective overlapping clones are rerun on the same gel to confirm the overlap.

Screening Large-Insert Clone Libraries

YAC and other large-insert clone libraries are arrayed in 96- or 384-well gridded arrays. Various pooling schemes have been devised to allow screening by PCR amplification using locus-specific PCR primers (Green and Olson, 1990). Most involve a hierarchical scheme, where the DNA from clones from individual plates are combined into a number of pools and in turn these pools are combined into a number of superpools. Thus screening involves the identification of the superpools that contain a positive signal indicating the existence of a positive clone somewhere within them. The DNA samples from the positive superpools are next screened, to identify any microtiter plate that contains a positive clone. In the last step, DNA made from rows and columns is used as template, and the intersection point between the rows and columns, if unique, defines the positive clone. This last stage can also be done by hybridization of a gridded copy of the DNA from the microtiter array on a nylon filter.

Other schemes involve more elaborate pooling in three dimensions, using PCR amplification to identify the positive clones. In addition, DNA samples from entire YAC and other large-insert clone libraries have been copied onto high density arrays on nylon filters for screening by radiolabeled hybridization (Larin et al., 1991). These are sometimes available for purchase from repositories. The individual investigator who wishes to screen a YAC library will use usually whatever scheme is available through the source of the YAC library, whether it be a research collaborator or a commercial repository. It is critical during PCR screening to have a robust PCR reaction that gives a unique band in human DNA and does not show a signal when yeast DNA is used.

Reconstruction of Physical Maps in Smaller Insert Clones

As suggested earlier, most large-scale physical mapping projects will take advantage of the large size of YACs and the long-range continuity and coverage they afford to construct the contig. In addition, for the most part, YACs may exhibit better representation of the underlying genomic DNA than clones in other vectors. Because of the difficulty of working with these clones, continued analysis of the DNA is often accomplished by first reconstructing the contig in a bacterial-based vector such as cosmids or bacteriophage. This can be accomplished in one of two ways. First, it may be possible to recreate the contig in one of these vectors by rescreening a cosmid library with the STSs within the YAC contig. This of course, will be possible only with a very dense set of STS sites within the contig. However, given the large number of expressed sequenced tags that are being mapped, this will increasingly become a reality. This approach will be aided by the high-resolution mapping efforts of the Human Genome Project, where contigs are being constructed in cosmids for preparation for high-throughput sequencing. Thus at some point it may be possible to begin the physical mapping process by querying databases for cosmid contigs in the critical region. In this case it may suffice to bridge gaps with larger insert clones such as YACs. In fact, for some chromosomes and subchromosomal regions this is already a possibility.

Another option would be to use inter-Alu PCR products generated from the YACs of the contig to reprobe a gridded cosmid library for that particular chromosome or chromosome arm. Inter-Alu PCR is a commonly used technique to isolate the unique genomic DNA between Alu (or other) dispersed repeats (Nelson et al., 1989; Tagle and Collins, 1992). Primers can be used that will generate the lowest amount of Alu repeat within the probe. The YAC DNA can be used as template using these primers in a PCR reaction and the corresponding amplified products radiolabeled as probe. This probe in turn is used to screen a cosmid (or BAC or PAC) library. Most often these experiments are done in collaboration with a group concentrating on the chromosome of interest, who may have a gridded cosmid library available for screening. A library with clones occupying individual positions in a microtiter array allows rapid screening using multiple probes, without the requirement of secondary and tertiary screens to identify positive clones.

For the near future, it may not be possible to directly recreate a cosmid contig from a YAC contig for the critical region from existing clones. Thus it is up to the individual investigator to generate subclones from the YACs. YAC DNA is purified by pulsed field gel electrophoresis so that the individual YAC migrates in a position away from the endogenous yeast chromosomes. It is then excised from the gel. Since the DNA molecules of the original YAC are large, extreme caution must be taken not to shear the DNA. Usually this involves working with the DNA embedded in an agarose matrix to reduce shear forces. This process can also be done using DNA isolated from the yeast cell itself. Most of the clones will thus be derived from the yeast DNA. Clones derived from the YAC insert can be identified by a screen for human positive clones using a human-specific repetitive probe. With either isolated YAC DNA or whole yeast DNA from the clone, the DNA is partially digested with a frequent cutting restriction endonuclease, and subcloned into the cosmid or bacteriophage vector of choice. One can then recreate a higher resolution contig of this region. Isolation and purification of the insert are much more efficient in these vectors, and give higher yields of insert. Thus they are more amenable to subsequent analysis such as transcript identification.

TRANSCRIPT IDENTIFICATION

As the Human Genome Project progresses, many of the expressed sequence tags are being mapped to their corresponding positions on YAC contigs. These data provide an initial framework transcript map. In addition, many of the known genes will be placed on these maps as well. Thus the investigator is in a strong position from the start to attempt to identify potential genes involved in complex human phenotypes. As these maps evolve, it is likely that the individual investigator will begin mutation analysis on the basis of the transcripts already known. Some of these genes will be characterized already, and knowledge of their sequence homologies or established functions may make them good candidate genes for the disease.

In most cases, ESTs that map to the critical candidate region will have a paucity of associated sequence information, usually only approximately 300 bases from the poly-A tail. Often this portion will consist primarily of the 3′ untranslated region of the message, hence will give very little useful information about the potential function of the gene product. In some cases, this EST sequence will be annotated with an additional 300 bases of sequence from the other end of the clone insert. Even so however, the investigator will have access to very little data about most of the transcripts that map to the critical region. The same will occasionally be true for any cDNA isolated via any of the other techniques. Thus additional sequences, or even the full sequence, must be determined for each of these transcripts.

Isolation of Full-Length Transcripts

Using standard plaque hybridization, a random-primed cDNA library can be probed with radiolabeled EST clone insert to isolate overlapping clones for any EST. In this way one can obtain full-length sequence for each transcript in the critical interval. Although this is a reasonable procedure for one or a few transcripts, it becomes inefficient when the goal is to isolate complete transcripts for a large number of ESTs and partial cDNAs over a large genomic region. Two procedures based on PCR amplification reduce the workload for this task: rapid amplification of cDNA ends and leapfrog PCR.

In the method of isolating full-length cDNA clones called RACE (for **r**apid **a**mplification of **c**DNA **e**nds: Frohman et al., 1988), mRNA is copied into cDNA by means of a specific primer derived from the known sequence—for example, within sequence of the EST. This first-strand synthesis using reverse transcriptase can extend to the 5′ end of the message and is tailed with dATP using terminal deoxynucleotidyltransferase. Amplification can then occur via a specific primer designed from known sequence and a poly-dT primer. Upon addition of cloning sites within both primers, the amplified products can be cloned and sequenced.

Leapfrog PCR involves repetitive screening of a random-primed cDNA library by PCR, so to reduce the workload by making it unnecessary to actually isolate individual clones (Gibbons et al., 1991). In this way even very long transcripts can be sequenced with a minimum of cloning steps. In each successive step, clones that extend outward from known sequence are amplified by means of two sets of nested primers, one set near the end of the known sequence and directed outward, and a second nested set of vector primers just beyond the cloning site of the vector. Resulting PCR products can be excised from the gel and directly sequenced, doing away with the requirement to clone the products, or even the need to screen the library by hybridization. Subsequent walking outward from the newly derived sequence continues to create new sequence. One variation of this might involve primary screening of the library but obviating the need to purify the plaques. Since it may be necessary to obtain full-length sequence for a large number of transcripts, this approach reduces the isolation of full-length clones for each of these to a series of PCR steps. A principal disadvantage of this technique is failure to yield a collection of clones that can be used in a biological study of the gene product. However, once the correct gene for the genetic disease has been identified, the PCR fragments can be used to screen a cDNA library by plaque hybridization to obtain these clones.

Isolation of Novel Transcripts

Although the available transcript maps are increasing in their density, we are currently quite far from having detailed transcript maps of most regions of the genome. Thus positional cloning efforts will require the isolation of novel transcripts from the genomic contig of the critical region. Earlier techniques for this step involved isolation of conserved sequences from the genomic contig, using so-called zoo blots. Random DNA fragments are used as a probe against a battery of DNA samples isolated from different organisms. Fragments that are conserved across a number of species often indicate coding regions of genes. This technique is not very efficient and is not discussed, although it is of historical interest (Rommens et al., 1989). Other techniques such as cloning undermethylated CpG islands, which often mark the promoter regions of some genes, can be used, but these again will not provide all the genes. Two commonly used techniques to identify transcripts within a large genomic region are exon trapping and cDNA selection. These techniques and their variations are well established and have a number of strengths and weaknesses.

Exon trapping is a method of cloning exons from genomic DNA by passaging genomic subclones through an mRNA intermediate in mammalian cells (Duyk et al., 1990). It is in essence an attempt to locate genes by the presence of functional splicing signals within the genomic DNA. This procedure, as currently practiced, involves subcloning portions of the larger genomic insert into a vector that will express the cloned DNA from a strong promoter but has splice donor and acceptor sites surrounding the site for subcloning (Buckler et al., 1991). Larger clones sometimes require a complex pool of subclones containing thousands of individual clones. The subcloned DNA is transfected into mammalian cells (COS-7 cells) that make heterogeneous nuclear RNA from the promoter within the vector. This RNA is spliced by the nuclear spliceosome so that any exons in the correct orientation will undergo the splicing reaction. Messenger RNA is isolated from the cells and copied into DNA using reverse transcriptase PCR, the amplification occurring via primers within the vector sequence on either side of the cloning site. A second round of PCR using nested primers containing dU residues follows. These trapped exons can be cloned into phagemid vector using the uracil DNA glycosylase system (Nisson et al., 1991) and sequenced.

Exon trapping can be fraught with artifacts, although the background noise has been reduced by a number of improvements in vector design. Most notable is the ability to remove clones that do not contain a complete exon (Church et al., 1994). The most recent improvement is meant to reduce cryptic splicing within the intron of the trapping vector so that essentially all the trapped inserts are derived from genomic DNA (Burn et al., 1995). This new vector also will have increased cloning efficiency owing to a new selectable marker.

Exon trapping has the great advantage that in principle, most internal exons in the entire genomic region are clonable, regardless of their pattern or level of expression. Thus the investigator is not forced to make an educated guess as to which tissue might express the gene or when during development expression might occur; in this respect, exon trapping differs from cDNA selection. However, the expected yield tends to be about 1 exon per cosmid. The system does not trap initial, terminal, and some alternatively spliced exons (Andreadis et al., 1993). Moreover, the exons that are trapped are small, with many in the range of 100–200 bp, and many exons may be isolated as single units, which will be part of the larger transcript. It is occasionally nontrivial to sort out which of the many smaller exons derive from the same transcriptional unit. To prove that the trapped DNA is part of a transcribed gene, Northern blots or RT-PCR must be done, using RNA from a number of tissue sources.

A variation of exon trapping that attempts to overcome some of these drawbacks is 3′ exon trapping (Krizman and Berget, 1993; Krizman et al., 1995). The technique is in principle similar to exon trapping as just described, except that only the 3′ exons of the transcripts are trapped. The vector contains a promoter and viral exons, but lacks a terminal exon. Thus the transcription unit will not produce polyadenylated mRNA unless a terminal exon is provided by the cloned insert. One obvious advantage of this technique is that only one exon will be trapped per terminal exon, usually translating to one exon per gene. This exon is generally larger than the average coding exon, which makes conversion to an STS much easier. However, the 3′ terminal exon often consists primarily if not exclusively of noncoding sequence, providing little information about the function of the exon.

Another common method for isolation of transcripts within a given genomic region is cDNA selection, also called direct selection. In its original form, this procedure was performed essentially as the reverse of screening a cDNA library with genomic DNA, with genomic DNA (rather than cDNA) immobilized on a nylon filter (Lovett et al., 1991; Parimoo et al., 1991). However, sensitivity can be increased by changing the process to solution hybridization rather than filter hybridization. As currently practiced, an entire population of cDNA molecules is modified so that the inserts are easily amplified. Highly repetitive DNA is suppressed by blocking with repetitive DNA. These cDNA molecules are hybridized in solution to genomic DNA, which has been tagged with biotin prior to hybridization. The genomic DNA/cDNA hybrids can be captured by Streptavidin-coated magnetic beads (Morgan et al., 1992). The beads are washed to remove nonspecific hybridizing clones, and the cDNAs are eluted. These primary-selected cDNAs are PCR-amplified and cloned or, more often, recycled through the entire process to enrich for region-encoded transcripts. With a secondary selection, enrichments of greater than 100,000-fold can be obtained.

The success of cDNA selection is highly dependent on four variables: the source of the mRNA, the abundance of the transcript in question, the size of the cDNA inserts, and the amount of repetitive DNA within the genomic target. First, one must select the correct cell type from which to isolate mRNA for cDNA synthesis. This must be guided by the proper knowledge of which human tissues exhibit the defect of the disease under investigation. Although most often such selection will be possible given some knowledge about the pathophysiology of the disease, in some cases only an educated guess may be possible as to which tissue might express the transcript of interest. Second, success is dependent on the relative abundance of the transcript of interest. Unfortunately, this cannot be known in advance. Although some transcripts will be abundantly or moderately represented in the mRNA pool, some low-abundance transcripts will be present in a copy number of less than one in a million. Since many cDNA libraries have on the order of one million clones, it will be impossible to find such a low-abundance message in these populations, especially when one is working with a tissue having a complex pattern of gene expression (e.g., brain). Therefore it is critical to begin with a large number of cDNA clones for cDNA selection. Because of this, it is recommended that the investigator first create a cDNA library from mRNA isolated from the appropriate tissue, rather than reamplifying and subcloning an existing library. With each round of reamplification, the representation of each transcript can change, to the point that some low-abundance clones will be lost entirely.

Another reason for de novo synthesis of the cDNA starting material has to do with the size of the cDNA inserts for this procedure. Most cDNA libraries are constructed to maximize the size of the inserts in the library. However, larger cDNAs will amplify poorly when they are competing with shorter clones in the pool during the amplification stages of cDNA selection. This selection against larger inserts means that transcript representation may be skewed, this time toward smaller insert clones. The problem can be minimized by constructing a primary cDNA library and attaching oligonucleotide linkers for amplification. By random priming the RT reaction that limits the size of the transcription, one can build an overlapping set of smaller clones to represent the mRNA of a given tissue.

Although repetitive DNA such as Cot 1 DNA is used to block hybridization in the procedure, artifacts related to this class of DNA can still make interpretation of the final clones difficult. All positive clones should be checked to see that they hybridize to the genomic clone(s) from which they came and do not appear as a smear on genomic DNA, indicating repetitive DNA. However, some true positive cDNA clones will contain repetitive DNA, often within the 3′ untranslated region, so that they appear as a repetitive clone. Additionally, pseudogenes, and members of gene families, and low-copy repeats can also make interpretation of clone status difficult. Another repeat-related artifact involves selection of ribosomal sequences, especially when YACs are the starting genomic material, due to contamination with yeast chromosomal ribosomal sequences in the starting genomic material. Although the YAC DNAs can be gel-purified, occasionally this contamination still occurs, and blocking with ribosomal sequences in the hybridization steps must be designed into the procedure.

Direct head-to-head comparisons of exon trapping and cDNA selection protocols have shown that both have inherent biases, such that each will identify transcripts that the other will not (Brody et al., 1995; Friedman et al., 1995; Osborne-Lawrence et al., 1995; Yaspo et al., 1995). Thus irrespective of the first choice of the individual investigator, it may be necessary to switch to an alternative approach to identify all the transcripts in the region.

One technique that can be used to identify transcripts residing on a clone from a given target tissue is the screening of arrayed bacteriophage or cosmid libraries with radiolabeled cDNA probe made from total mRNA of a target tissue (Hochgeschwender et al., 1989). This requires the construction of a large contig over the region in cosmid or bacteriophage clones (which are placed in a gridded array) and DNA made from a copy of this array on a nylon filter. In and of itself such a procedure would be a major undertaking, but when this resource is in hand, cosmids or phage inserts that contain transcribed sequences can in fact be identified. The clones that contain positive signals are digested with restriction endonucleases and reprobed with the cDNA probe. Genomic fragments carrying the transcribed sequences can then be subcloned to reduce the complexity of the clones for analysis. These can be either directly sequenced, subcloned into exon trapping vectors, or used to screen cDNA libraries by hybridization.

A final method to identify transcripts within a candidate region is to directly sequence the entire region and to use gene-finding computer algorithms to identify exons. Unless one has access to a laboratory dedicated to high-throughput DNA sequencing, this will usually not be an option. However, with the ever-increasing flood of DNA sequence available for many regions of the human genome, the annotation of the sequence by the individual investigator is becoming routine. A number of different gene finding programs are available, including GeneMark (Borodovsky and McInich, 1993), GRAIL II (Xu et al., 1994), GeneFinder (Green, 1994), xpound (Thomas and Skolnick 1994), and Genie (Kulp et al., 1996). Each uses a unique search algorithm to identify open reading frames, which correspond to individual exons of genes. Genotator (Harris, 1997) is sequence annotation workbench that runs five different gene-finding programs, three homology searches, and also searches for promoters and splice sites. The output can be viewed with an interactive graphical browser. Homology searches can identify the most promising candidate exons, and primers can be designed to amplify these for Northern blot expression analysis and subsequent mutation analysis. As the Human Genome project continues to accumulate DNA sequence, this approach represents the future of transcript mapping for positional cloning projects.

Preliminary Characterization of Transcripts

As transcripts are identified in the critical region, it is necessary to rank them in order of their promise. There are characteristics of each to be ascertained. First, one must determine something about the expression profile of each of the transcripts. Given the pathobiology of the disease in question, certain tissues might be expected to express the transcript. Second, one would like to determine any sequence homologies that might provide a clue to the function of each of the candidate transcripts.

Expression profiles can be obtained by probing Northern blots of RNA isolated from various tissues. Northern blots are less sensitive than RT-PCR and may miss low-level expression of transcripts. Nonetheless, most often one would be looking for a reasonable level of expression in the tissue(s) of interest. In addition, Northern blots are more easily quantitated for the relative level of expression of the gene among different tissues. Premade Northern blots from a variety of human tissues are commercially available, making this screening process much easier to accomplish. RT-PCR can also be performed using RNA from a variety of sources. RT-PCR is very sensitive—so much so that it is more difficult to quantitate expression levels without internal controls.

An important caveat concerning expression profiles is that ubiquitous expression (expression in all or nearly all tissues examined) can be difficult to rule out for the gene for any particular disease. Although, for example, one might expect the gene for a neurological disorder to be expressed only in the brain, it may be present in other tissues at respectable levels. A notable example is expression of the *SOD1* gene (superoxide dismutase) in amyotrophic lateral sclerosis (Rosen et al., 1993). Thus, one should not rule out a transcript that is expressed in the tissue of interest just because it is also expressed in a number of other tissues.

Sequence homologies can often provide additional information concerning the candidate transcript. These methods check the level of sequence similarity between all other known sequences in a database and rank them according to their similarity to the query sequence. With this information, one can identify genes whose function may be known or genes with which a phenotype is associated. In addition, one can identify other family members related to the transcript of interest. Occasionally, certain sequence motifs (such as a DNA binding motif) present in the transcript may give some information about a particular aspect of the candidate gene. All this can be used in conjunction with the expression profile to rank the transcripts in the critical region with regard to likelihood of involvement in the disease.

The most commonly used sequence homology programs are the BLAST (Basic Local Alignment Search Tool) family of programs (Altschul et al., 1990, 1994). These programs, available on the Internet, can be used to search homologies at the nucleotide or amino acid level. Searches using a combination of these are recommended, since each has inherent strengths and weaknesses. For example, one would want to search at both the nucleotide and translated amino acid levels to ensure that the proper reading frame has been chosen for the entire sequence, and that no frameshift sequencing errors have been made. Although the basic idea of a homology search is quite simple, the computer algorithms and statistics underlying these searches have become quite sophisticated. The reader is referred to the NCBI/BLAST Internet home pages (*http://www.ncbi.nlm.nih.gov/BLAST*) for more information on how to access and use this tool.

MUTATION IDENTIFICATION

Amplification of cDNA from Patient Samples

Once a candidate transcript for mutation analysis has been chosen on the basis of map location, expression profile, and sequence homologies, one must obtain copies of this transcript from a cohort of affected patients (and unaffected controls). This task can be quite trivial or very difficult, depending on the level of expression of the transcript in the tissues available from affected individuals. The preferred approach for mutation analysis and for initial mutation characterization, transcript (mRNA)-based amplification, will reduce the workload tremendously, since it will not be necessary to characterize the genomic structure (intron–exon borders and sequence) for each of the candidate genes. RNA-based approaches are capable of identifying most mutations within the coding region of the gene of interest. However, this approach will not enable identification of mutations of a number of different types; most notable are those that are outside the coding region and those that render the message unstable. However, initial characterization of mutations using this approach will determine whether the transcript under investigation has a role in the pathology of the disease. If so, gene structure can be determined and a combination of RNA-based and genomic DNA-based mutation analysis should identify most if not all mutations in the patient cohort.

Some candidate transcripts will be ubiquitously expressed, allowing easy access to mRNA for amplification by RT-PCR from tissue available from patients. Tissues that are commonly used include peripheral blood leukocytes isolated by venipuncture, and fibroblasts or keratinocytes isolated from skin punch biopsy. The choice of which to use will be dictated by the level of expression within each cell type and the cooperation of the affected individuals in providing samples. Peripheral blood leukocytes are the most commonly used, and B lymphoblasts are often immortalized using Epstein–Barr virus, as discussed in Chapter 8. Some investigators choose to immortalize cells from all the patients, or at least one member of each family enrolled in the study. These provide a renewable source not only of DNA but also of mRNA for mutation analysis.

Many genes are not normally expressed at significant levels in the tissues available from patients. For the most part, these transcripts can still be amplified, owing to the phenomenon of illegitimate or ectopic transcription (Sarkar and Sommer, 1989). Illegitimate transcription occurs when extremely low-level expression (often less than one copy per cell) occurs for nearly all genes in most cell types. This is true even for "tissue-specific" genes and is presumably due to the difficulty of completely silencing any gene by the transcriptional regulatory system within the nucleus. Thus, it should be possible, by means of nested primers, to amplify most (some might claim all) genes from one of these tissues with two successive rounds of PCR. Messenger RNA is isolated from lymphocytes and reverse transcribed using an oligo-dT or sequence-specific primer. Primers for first-round synthesis can be designed to amplify up to a 1 kb product. Primers for second-round PCR should be designed to amplify a product in the size range required for mutation analysis (see below).

RT-PCR from illegitimate transcripts can be subject to a number of artifacts. These include contamination from other sources of the transcript in the laboratory, amplification of Taq polymerase errors in replication, and aberrant PCR products due to deletion or alternative splicing. With proper controls and procedures, these can be minimized, and RT-PCR from mRNA is always the procedure of first choice when genomic DNA sequence is not available.

Amplification of Genomic DNA from Patient Samples

In certain cases the transcript will not routinely amplify even after two rounds of nested-primer PCR. In addition, alternative splicing may give rise to a different transcript that does not correspond to the sequence expressed in the tissue or cell type that shows the disease phenotype, perhaps making it impossible to amplify a critical portion of the gene in question. When this situation arises, or as a matter of preference of the investigator, mutation analysis can proceed via PCR amplification of individual exons of the gene under analysis. However, in most cases, even with genes that have been cloned and characterized, the genomic structure is not known.

Ideally the complete genomic sequence of the entire critical region will someday be known, enabling identification of intron–exon borders and design of primers for exon-specific amplification. For the present, however, analysis at this level will require rapid characterization of the genomic structure for each of the candidate transcripts.

PCR-based methods to determine the exon–intron borders include amplification across introns via primers within the adjacent exonic sequence, amplification across the region between an intronic Alu repeat to exonic sequence, inverse PCR, and bubble-linker PCR. Each of these requires a genomic clone for the gene of interest, such as a YAC or BAC or cosmid clone. In addition, each requires a number of specific primers from within the coding sequence for one or more of the primers. In many cases these primers will be available, having been synthesized from both strands during the sequencing of the full-length transcript.

The simplest PCR approach to this problem is to attempt to amplify across introns by taking a series of forward-facing primers in the coding sequence and using them in tandem with a set of reverse-facing primers from the coding sequence. To increase the chances of getting across large introns, a long-range thermal cycling protocol or kit can be used. Any primer pair that amplifies a fragment longer than the length expected by the cDNA sequence crosses an intron. Such a fragment can be excised from the gel and sequenced. This approach can identify a number of the borders, and this is a good place to begin, especially given a set of presynthesized primers from both strands of the cDNA sequence. This approach may not be capable of amplification across very large introns. These border sequences can be determined using a gene-specific primer in tandem with an Alu repeat primer to amplify from the coding region to a repetitive element within in the adjacent intron.

Inverse PCR and bubble-linker PCR can also be used. In the case of inverse PCR, DNA is cut and circularized similarly to previously described dimensions for YAC end rescue, except that in this case the one of several four-base-pair recognition enzymes having 3′ or 5′ overhangs can be chosen (Groden et al., 1991). The primers chosen are close to each other but extend in opposite directions. Bubble-linker PCR is performed in a manner quite similar to that described for YAC end rescue (Tumer et al., 1995). In this case, however, the bubble primer is used in conjunction with transcript-specific primers to amplify unique fragments from the genomic clone. Amplified fragments will extend from a point within an exon to the first restriction endonuclease site within an adjacent intron.

In general, it is technically demanding to obtain quality sequence directly from cosmids and lambda phage clones, although direct cosmid and lambda sequencing is sometimes performed in limited situations, such as sequencing the ends of genomic clones during contig building. For larger clones such as PACs, BACs, or YACs, direct sequencing is usually not an option. Alternatives involve subcloning a genomic clone containing the gene into plasmid or phagemid, followed by isolation of subclones by hybridization with the cDNA probe. These can be sequenced by means of both vector arm and cDNA primers to obtain intron–exon border sequence.

Comparison of the cDNA and genomic sequence will reveal the positions of the intron–exon borders as well as intronic sequence within which to design the primers. To amplify each exon, ideally one would like primers 10–20 base pairs beyond the borders and within intron sequence. This arrangement allows identification of mutations in the coding sequence adjacent to the exon border, as well as many intronic mutations that might lead to splicing defects. Usually, it is best to have at least 100 base pairs of intron sequence bordering each exon to provide a suitable amount of sequence for primer design. Otherwise, runs of repetitive sequence such as pyrimidine-rich stretches within the intron may prevent the design of workable primers. A number of computer programs are available for primer design, and any of these are better than the eye at choosing suitable primers for PCR. Examples include PRIMER 3, accessible through the Whitehead Institute/MIT Genome Center Web page (*http://www.genome.wi.mit.edu*), and OLIGO (National Biosciences).

Scanning for Mutations

In principle, once the individual exons or portions of the transcript have been amplified from the patient cohort, the most straightforward approach to mutation identification is to directly sequence each gene from all the patients. As the number of possible candidate genes becomes greater, however, as well as the number of affected individuals, it becomes necessary to prescreen or scan each sample for base pair changes. When a region of a gene or transcript from a patient is identified as harboring a base pair change, that sample is sequenced to characterize the change. This strategy reduced the total number of sequencing reactions that must be performed and analyzed. There are a number of techniques to scan the amplified fragments for sequence differences. Perhaps no area of positional cloning has as many different protocols to choose from and opinions on which is the most efficient and cost-effective. At some level this diversity represents the personal biases of investigators toward procedures with which they are familiar. This section briefly discusses a number of the most popular procedures and modifications thereof, especially with regard to their strengths and weaknesses. A protocol that has been fine-tuned until it is running well in the laboratory will be a powerful tool for mutation analysis. Thus it is imperative to choose a method that can be easily adopted in the investigator's laboratory.

These protocols are classified according to the underlying principle behind the technique. The first three, heteroduplex analysis, single-strand conformation polymorphism analysis (SSCP), and denaturing gradient gel electrophoresis (DGGE), are based on aberrant migration of mutant molecules through a gel matrix. Once the DNA has been amplified via PCR, very little additional preparative work is required. The protocols are distinguished primarily by the type of gel electrophoresis that follows amplification. Other techniques require additional steps which follow DNA amplification. In dideoxy fingerprinting, a single dideoxy-DNA sequencing reaction is run as an SSCP assay. In other procedures, the heteroduplexes are cleaved at the site of the mismatched bases. Finally, with a limited number of samples, direct DNA sequencing can in fact be an effective tool, if gel-loading is changed to emphasize rapid visual inspection for different patterns, rather than actual base-calling.

Heteroduplex analysis is a method of identifying mismatches in base pairing of double-stranded DNA (White et al., 1992). It is especially useful in the examination of individuals who might be heterozygous for the mutation in question. This technique is based on the fact that duplexes of DNA with even a single mismatched base pair may exhibit retarded migration on gels. The DNA is amplified and in the final step slowly cooled to allow heteroduplex formation. Heteroduplexes occur when one strand derives from the normal sequence, but the complementary strand derives from the mutant sequence. These can be separated on polyacrylamide or HydroLink MDE gels under conditions that are only slightly denaturing. The heteroduplexes will migrate more slowly than homoduplexes and are visualized as bands migrating above the normal sequence.

Not all base pair changes produce an alternative migration pattern on the gel, however, and this is the most serious drawback of this procedure. In general, deletions or insertions of a single base (or more) will usually produce visible shifts. Many but not all base pair changes will produce altered mobility. The relevant variables are not all known, but fragment size and position of the mutation within the sequence are implicated. Fragments should be no more than 500 base pairs for effective screening.

Single-strand conformation polymorphism analysis, or SSCP (Orita, 1989a, 1989b), is a method of identifying a conformational change that results in a mobility shift due to sequence changes in single-stranded DNA. Amplified DNA is denatured, but reannealing conditions are adjusted to favor intramolecular events. Each single-stranded DNA molecule adopts one or more particular conformations as a result of intramolecular base pairing, and each DNA molecule will migrate in a unique position on the gel. Samples are run on a nondenaturing polyacrylamide gel. Those with base pair changes will show an additional band relative to the normal pattern of bands. This band can be excised, reamplified, and sequenced to identify the base pair change.

SSCP, like heteroduplex analysis, is technically quite simple and thus capable of the rapid screening of a large number of samples. Fragments in the range of 200–300 bp can be screened, and SSCP seems to be capable of identifying nearly 100% of point mutations if two or three different gel running conditions are used with fragments less than 300 bp (Glavac and Dean, 1993; Ravnik-Glavac et al., 1994). Some potential difficulties include the need to run the gels at a constant temperature, usually 4°C or room temperature. Since the factors that result in different conformations are unknown, beginning conditions for running temperature, exact gel composition, running time, and pattern to expect are largely empirical. Because of its ease of operation, however, SSCP is often the method of choice of most investigators for surveying a large number of genes from a large patient cohort.

The variation of SSCP known as dideoxy fingerprinting is essentially a combination of the SSCP and dideoxy-DNA sequencing procedures (Sarkar et al., 1992; Liu and Sommer, 1994). It is performed by electrophoresis through a nondenaturing gel of a single dideoxy termination reaction. Mutations can be detected as a nested series of ddNTP termination products visible as aberrant SSCP bands beginning from the point of the mutation within the fragment. These termination reactions increase the number of fragment lengths and sequences within the gel lane, with each one having the potential to migrate aberrantly. In control experiments this procedure is 100% effective at identifying a large number of known mutations.

Denaturing gradient gel electrophoresis is a powerful if more technically demanding procedure for identifying mutations based on the differences in melting temperatures of duplex DNA (Myers et al., 1985b). Melting behavior is dependent on the sequence of the fragment and occurs within discrete domains of the fragment. Because of this a GC clamp is often incorporated into one of the primers for amplification (Sheffield et al., 1989). This addition of a 30–40 base sequence at the end of one primer creates the highest temperature melting domain within the clamp itself, which can force the other domains within the amplified fragment to act as a single domain. The amplified DNA is run on a gel containing a gradient of a chemical denaturant. The mobility of the fragment is influenced by strand separation once a domain has begun to melt, such that the fragment is retarded. Differences in the sequence will influence the point at which fragment retardation occurs.

High-throughput mutation screening of a given fragment using DGGE requires some knowledge of its melting profile. This can be accomplished via computer algorithms or with perpendicular DGGE over a broad range of denaturant. In perpendicular DGGE, the sample is run through the gel perpendicular to the concentration gradient of the denaturant. Melting domains are visible as separations of the normal and mutant homoduplexes within each domain at a particular position (denaturant concentration) in the gel. When suitable conditions have been chosen investigators can then select one of two variations for the screening stage: parallel DGGE or constant denaturant gel electrophoresis. With parallel DGGE, multiple samples are run through a concentration gradient that is parallel to the direction of electrophoresis. With constant denaturing electrophoresis, a constant concentration of denaturant is chosen for the gel, one for each particular melting domain (Hovig et al., 1991). Unfortunately, DGGE requires special equipment that is not generally used in a standard molecular biology laboratory. Nonetheless, the efficiency of this technique approaches 100% when properly done, so once up and running in a laboratory, it is a powerful approach to mutation detection.

Neither heteroduplex analysis, SSCP, nor DGGE gives information about the position of the mutations in the amplified fragment. Other methods are based on cleavage of DNA or RNA hybrids at the site of cleavage. These have the advantage of localizing the position of the mutation in the amplified fragment as well as in general being able to work on fragments in the size range of 1 kb or more. Mismatches in heteroduplex DNA (or DNA/RNA hybrids) create cleavage sites that can be identified by means of a variety of procedures. Chemicals can be used to recognize the mismatched base and modify it so that it can be cleaved in a subsequent reaction. Other procedures are based on enzymatic recognition of the mismatch and subsequent cleavage.

Heteroduplex DNA can be cleaved at the site of the mismatched base via chemical modification and cleavage (Cotton et al., 1988; Grompe et al., 1989). In the chemical cleavage method, mismatched cytidine and thymidine bases in heteroduplex DNA become more reactive with hydroxylamine and osmium tetroxide, respectively. After treatment with piperidine, cleavage occurs at the site of the modified base. Upon analysis of both strands, all possible base pair mutations can be identified. Large fragments (≤ 1 kb) can be analyzed, and there is high sensitivity, approaching 100% of mutations. However the procedure requires the use and disposal of hazardous chemicals.

Recently developed enzymatic cleavage systems are based on bacteriophage resolvases, such as T4 endonuclease VII and T7 endonuclease I (Babon et al., 1995; Mashal et al., 1995; Youil et al., 1995). These enzymes recognize and cleave branched DNA and can be used to cleave heteroduplex DNA at the mismatch. Fragments of up to 1 kb or larger can be cleaved and the sensitivity can approach 100% (Youil et a., 1996). Improvements in this technique, including a commercially available source of the resolvases, should be forthcoming, placing this technique very near the top of the list for efficiency and cost-effectiveness. Small deletions and insertions are obvious as superimposed frame shifts of the normal pattern.

For RNase A cleavage (Myers et al., 1985a; Rosenzweig et al., 1991), a PCR product from patient DNA is hybridized to a radiolabeled RNA probe synthesized from normal sequence, resulting in a DNA/RNA hybrid. Of the 10 possible base pair mismatches that can occur, 8 can be completely cleaved by RNase A and the other 2 are partially cleaved. Thus, with riboprobes designed for both strands of a given fragment, this methodology can identify greater than 80% of mismatches. Fragments should be on the order of 1 kb or less in size. This procedure is dependent on the synthesis of RNA probes for the region using bacteriophage (T3, T7, or SP6) promoters incorporated into PCR amplification primers. Working with RNA requires special handling to accommodate the ubiquitous presence of RNases. Synthesis of quality full-length RNA probe tends to be the limiting factor in this procedure.

The time and effort required to perform, read, and analyze DNA sequencing gels means that direct DNA sequencing is not often the first choice for mutation scanning. However, lane loading can be changed such that each sample need not be read, but rather can be inspected for a different band pattern in the lane. This is accomplished by loading in adjacent wells all the dideoxy lanes for the patient cohort: dd-adenosine, dd-cytidine, dd-guanosine, and dd-thymidine. New bands that appear in a given lane will readily stand out as aberrations from the normal pattern, indicating base pairs substitution at such points in the sequence. Small deletions and insertions are obvious as superimposed frameshifts of the normal pattern.

MUTATIONS VERSUS POLYMORPHISMS

Although premature termination codons and frameshift mutations usually indicate deleterious mutations within the gene, base pair substitutions leading to amino acid changes can indicate only potential mutations. These must be distinguished from polymorphisms within the normal population. Size or charge changes between the normal and substituted amino acid residues offer one indication as to significance that one may attach to these substitutions. However, even the presence of such dramatic changes is not necessarily indicative of a mutation. Another potential indication of causative mutation, especially with members of gene families, is conservation of the analogous residue (or charge) in the other family members.

In all cases, however, a population survey must be done to show that the base pair change does not occur in the normal population. This normal cohort should be matched with regard to racial, ethnic, and geographic background with the disease cohort. By definition, a polymorphism is a base pair change that occurs in 1% or more of the population. In principle then, at least 50 individuals (100 autosomes) should be checked for the sequence change. However, to approach statistical significance that an observed change is not a polymorphism, 150–200 individuals (300–400 chromosomes) or more should be surveyed for the base pair change.

There are four commonly used methods to survey a population cohort for a specific sequence nucleotide changes. The first of these is the easiest to set up but requires that the putative mutation create or destroy a restriction endonuclease site within the amplified product. This can be determined by comparing the restriction maps of the normal and mutant sequence generated by means of computer programs. Very often such a site can be identified. DNA is amplified from the cohort and digested with the enzyme. It should be borne in mind that whereas the creation of a restriction site usually requires the specific base change from the normal sequence, the destruction of a site can occur at any base in the recognition sequence. Thus in either case, but particularly when the proposed mutation destroys a site, any potential positive within the population cohort should be sequenced to confirm that it is the identical nucleotide change as the purported mutation.

A second method of screening a population cohort is by allele-specific oligonucleotide hybridization (Saiki et al., 1986). In this technique two specific oligonucleotides are synthesized. These nucleotides are the same size (13–19 nucleotides in length, depending on the sequence) and differ only at the site of the proposed mutation, which is centrally located within the oligonucleotides. The melting temperature for both will be quite similar for the cognate sequence, but because of the central location of the mismatch, will be very different for the noncognate sequence. DNA template is amplified via primers such as exon-specific primers or others that will amplify fragments from both the wild-type and mutant alleles. These are run on an agarose gel, blotted to nylon membrane, and probed sequentially with radiolabeled normal and mutant oligonucleotide probes. Whereas all samples should hybridize to the normal sequence, only samples with the identical mutation will hybridize to the mutant oligonucleotide.

Another method of screening the patient cohort is based entirely on PCR amplification and is called the amplification–refractory mutation system or ARMS (Newton et al., 1989; Ferrie et al., 1992). This technique starts with the design of two sets of PCR primers that share one primer in common; the other primer differs at the site of the putative mutation. In this case the nucleotide that differs is placed at the 3′ end of each PCR primer. This base is the most critical for primer extension and should in principle allow for allele-specific extension from the normal and mutant primers. However, even with this difference, it is often difficult to obtain allele-specific amplification using the wild-type and mutant primers. Thus the penultimate base (next to the 3′ end) is often made to be a mismatch in both wild-type and mutant primers. In practice, primers must be synthesized and control reactions run to determine conditions under which allele-specific amplification occurs for the wild type and mutant pair. This procedure is powerful but often requires a number of attempts at primer design and establishment of reaction conditions to obtain sequence-specific amplification from the normal and mutant primer sets.

A final method, conceptually similar to ARMS, is the use of the ligase chain reaction (Nickerson et al., 1990; Barany et al., 1991). In this procedure primers are designed that lie on the same strand and directly adjoin each other at the site of the purported mutation. Another set is designed at the same position but on the complementary strand. If the primers perfectly match the complementary sequence, these can be covalently linked by a thermostable ligase. The reaction occurs in a similar fashion to PCR, where ligation products from earlier rounds serve as template for later rounds and exponential amplification occurs.

CONCLUSIONS

The progression from the establishment of genetic linkage to the eventual identification of causative mutations in a gene requires a number of intermediate goals. These include the construction of a physical map of the critical region, the construction of a transcript map of the region, mutation analysis of candidate genes within the region, and last, a population survey of the purported mutations. Each of these stages involves a number of different but sometimes related procedures. Most of these have been streamlined by the incorporation of PCR amplification as part of the procedure, allowing for more rapid and efficient analysis.

With the advances in physical mapping due to the Human Genome Project, for most regions a complete physical map can be accessed through electronic databases. Progress toward the second goal, to create a transcript map of the region, has been much slower. Nonetheless, many regions of the genome have a number of genes and expressed sequenced tags already mapped. These must be supplemented in the laboratory using protocols to identify additional transcripts. Each of these transcripts must be completely, or at least partially, sequenced to permit the assignment of putative functions based on sequence homology with known genes. A preliminary expression profile especially surveying tissues relevant to the disease will also help rank the candidate genes as to their relevance in the pathophysiology of the disease.

In the distant future, the physical and transcript maps will be a reality for the entire human genome. At first this will be possible because of the availability of complete physical maps for the entire genome, with most if not all of the known transcripts placed on these maps. Beyond this, at some point the complete sequence of the entire genome will become available, allowing instant access to the sequence of all the genes in a critical region. Even when this information is available, however, the following stages must still be performed to identify causative mutations for complex disease.

The strongest candidate genes for the disease must be sequenced from a patient cohort. If these genes can be amplified from mRNA from a tissue easily obtained from the patient cohort, such as peripheral blood lymphocytes, mutation analysis can proceed at the cDNA level. If not, a minimal amount of intron–exon border sequence must be obtained for each candidate and exon-specific primers designed. Mutation analysis can then proceed using the genomic DNA from the patient cohort.

To increase the efficiency of the mutation analysis, a number of mutation scanning methods are available. Each of these takes advantage of some property of the DNA fragment harboring the mutant sequence to create an aberrant migration pattern during gel electrophoresis. Samples showing aberrant patterns on the scanning gels are analyzed further by direct DNA sequencing to identify base pair changes.

All base pair changes are only potential mutations until it has been demonstrated that they are not polymorphic substitutions occurring within the normal population. Once this has been shown on a population of suitable size and composition, one can be reasonably certain that the gene(s) for a given phenotype have been found. Nonetheless, the ultimate proof becomes the creation of in vitro and in vivo models of the disease phenotype, using the mutant genes uncovered by means of these processes.

REFERENCES

Altschul SF, Gish W, Miller W, Myers EW, Lipman DJ (1990): Basic local alignment search tool. *J Mol Biol* 215:403–410.

Altschul SF, Boguski MS, Gish W, Wootton JC (1994): Issues in searching molecular sequence databases. *Nat Genet* 6:119–129.

Andreadis A, Nisson PE, Kosik KS, Watkins PC (1993): The exon trapping assay partly discriminates against alternatively spliced exons. *Nucleic Acids Res* 21:2217–2221.

Arratia R, Lander ES, Tavare S, Waterman MS (1991): Genomic mapping by anchoring random clones: A mathematical analysis. *Genomics* 11:806–827.

Babon JJ, Youil R, Cotton RG (1995): Improved strategy for mutation detection—A modification to the enzyme mismatch cleavage method. *Nucleic Acids Res* 23:5082–5084.

Barany F (1991): Genetic disease detection and DNA amplification using cloned thermostable ligase. *Proc Nat Acad Sci USA* 88:189–193.

Bellane-Chantelot C, Lacroix B, Ougen P, Billault A, Beaufils S, Bertrand S, Georges I, Gilbert F, Gros I, Lucotte G, Susini L, Codani J-J, Gesnouin P, Pook, S, Vaysseix G, Lu-Kuo J, Ried T, Ward D, Chumakov I, Paslier DL, Barillot E, Cohen D (1992): Mapping the whole human genome by fingerprinting yeast artificial chromosomes. *Cell* 70:1059–1068.

Breukel C, Wijnen J, Tops C, Ven der Klift H, Dauwerse H, Khan PM (1990): Vector-Alu PCR: A rapid step in mapping cosmids and YACs. *Nucleic Acids Res* 18:3097.

Borodovsky M, McIninch JD (1993): GENEMARK: Parallel gene recognition for both DNA strands. *Comput and Chem* 17:123–133.

Brody LC, Abel KJ, Castilla LH, Couch FJ, McKinley DR, Yin G, Ho PP, Merajver S, Chandrasekharappa SC, Xu J, Cole JL, Struewing JP, Valdes JM, Collins FS, Weber BL (1995): Construction of a transcription map surrounding the *BRCA1* locus of human chromosome 17. *Genomics* 25:238–247.

Buckler AJ, Chang DD, Graw SL, Brook JD, Haber DA, Sharp PA, Housman DE (1991): Exon amplication: A strategy to isolate mammalian genes based on RNA splicing. *Proc Natl Acad Sci USA* 88:4005–4009.

Burke DT, Carle GF, Olson MV (1987): Cloning of large segments of exogenous DNA into yeast by means of artificial chromosome vectors. *Science* 236:806–812.

Burn TC, Connors TD, Klinger KW, Landes GM (1995): Increased exon-trapping efficiency through modifications to the pSPL3 splicing vector. *Gene* 161:183–187.

Chumakov I, Rigault P, Guillou S, Ougen P, Billaut A, Guasconi G, Gervy P, LeGall I, Soularue P, Grinas L, Bougueleret L, Bellanne-Chantelot C, Lacroix B, Barillot E, Gesnouin P, Pook S, Vaysseix G, Frelat G, Schmitz A, Sambucy J-L, Bosch A, Estivill X, Weissenbach J, Vignal A, Riethman H, Cox D, Patterson D, Gardiner K, Hattori M, Sakaki Y, Ichikawa H, Ohki M, Paslier DL, Heilig R, Antonarakis S, Cohen D (1992): Continuum of overlapping clones spanning the entire human chromosome 21q. *Nature* 359:380–387.

Church DM, Stotler CJ, Rutter JL, Murrell JR, Trofatter JA, Buckler AJ (1994): Isolation of genes from complex sources of mammalian DNA using exon amplification. *Nat Genet* 6:98–105.

Cotton RGH, Rodrigues NR, Campbell RD (1988): Reactivity of cytosine and thymine in single-base-pair mismatches with hydroxylamine and osmium tetroxide and its application to the study of mutations. *Proc Natl Acad Sci USA* 85:4397–4401.

Dracopoli NC, Haines JL, Korf BR, Moir DT, Morton CC, Seidman CE, Seidman CE, Seidman JG, Smith DR (1994): *Current Protocols in Human Genetics.* New York: Wiley.

Duyk GM, Kim SW, Myers RM, Cox DR (1990): Exon trapping: A genetic screen to identify candidate transcribed sequences in cloned mammalian genomic DNA. *Proc Natl Acad Sci USA* 87:8995–8999.

Ferrie RM, Schwarz MJ, Robertson NH, Vaudin S, Super M, Malone G, Little S (1992): Development, multiplexing and application of ARMS tests for common mutations in the *CFTR* gene. *Am J Hum Genet* 51:251–262.

Friedman LS, Ostermeyer EA, Lynch ED, Welcsh P, Szabo CI, Meza JE, Anderson LA, Dowd P, Lee MK, Rowell SE, Ellison J, Boyd J, King M-C (1995): 22 Genes from chromosome 17q21: Cloning, sequencing, and characterization of mutations in breast cancer families and tumors. *Genomics* 25:256–263.

Frohman MA, Dush MK, Martin GR (1988): Rapid production of full-length cDNAs from rare transcripts: Amplication using a single gene-specific oligonucleotide primer. *Proc Natl Acad Sci USA* 85:8998–9002.

Gibbons IR, Asai DJ, Ching NS, Dolecki GH, Mocz G, Phillipson CA, Ren H, Tang W-JY, Gibbons BH (1991): A PCR procedure to determine the sequence of large polypeptides by rapid walking through a cDNA library. *Proc Natl Acad Sci USA* 88:8563–8567.

Glavac D, Dean M (1993): Optimization of the single-strand conformation polymorphism (SSCP) technique for detection of point mutations. *Hum Mutat* 2:404–414.

Green ED, Olson MV (1990): Systematic screening of yeast artificial-chromosome libraries by use of the polymerase chain reaction. *Proc Natl Acad Sci USA* 87:1213–1217.

Green P (1994): Ancient conserved regions in gene sequences. *Curr Opin Struc Biol.* 4:404–412.

Groden J, Thliveris A, Samowitz W, Carlson M, Gelbert L, Albertsen H, Joslyn G, Stevens J, Spirio L, Robertson M, Sargent L, Krapcho K, Wolff E, Burt R, Hughes JP, Warrington J, McPhersen J, Wasmuth J, Le Paslier D, Abderrahim H, Cohen D, Leppert M, White R (1991): Identification and characterization of the familial adenomatous *coli* gene. *Cell* 66:589–600.

Gyapay G, Schmitt K, Fizames C, Jones H, Vega-Czarny N, Spillet D, Muselet D, Prud'Homme J-F, Dib C, Auffray C, Morissette J, Weissenbach J, Goodfellow PN (1996): A radiation hybrid map of the human genome. *Hum Mol Genet* 5:339–346.

Grompe M, Muzny DM, Caskey CT (1989): Scanning detection of mutations in human ornithine transcarboxylase (OTC) by chemical mismatch cleavage. *Proc Natl Acad Sci USA* 86:5888–5892.

Harris NL (1997): Genotator: a workbench for sequence annotation. *Genome Res* 7:754–762.

Hermanson GG, Hoekstra MF, McElligott DL, Evans GA (1991): Rescue of end fragments of yeast artificial chromosomes by homologous recombination in yeast. *Nucleic Acids Res* 19:4943–4948.

Hochgeschwender U, Sutcliffe JG, Brennan MB (1989): Construction and screening of a genomic library specific for mouse chromosome 16. *Proc Natl Acad Sci USA* 86:8482–8486.

Hovig E, Smith-Sorensen B, Brogger A, Borresen A-L (1991): Constant denaturant gel electrophoresis, a modification of denaturing gradient gel electrophoresis, in mutation detection. *Mutant Res* 262:63–71.

Hudson TJ, Engelstein M, Lee MK, Ho EC, Rubenfield MJ, Adams CP, Housman DE, Dracopoli NC (1992): Isolation and chromosomal assignment of 100 highly informative human simple sequence repeat polymorphisms. *Genomics* 13:622–629.

Krizman DB, Berget SM (1993): Efficient selection of 3′-terminal exons from vertebrate DNA. *Nucleic Acids Res* 21:5198–5202.

Krizman DB, Hofmann TA, DeSilva U, Green ED, Meltzer PS, Trent JM (1995): Identification of 3′-terminal exons from yeast artificial chromosomes. *PCR Methods Appl* 4:322–326.

Ioannou PA, Amemiya CT, Garnes J, Kroisel PM, Shizuya H, Chen C, Batzer MA, de Jong PJ (1994): A new bacteriophage P1-derived vector for the propagation of large human DNA fragments. *Nat Genet* 6:84–89.

Kulp D, Haussler D, Reese MG, Eeckman FH (1996): A generalized hidden Markov model for the recognition of human genes in DNA. In *Proceedings of the Conference on Intelligent Systems in Molecular Biology*. AAAI/MIT Press, St. Louis MO.

Larin Z, Monaco AP, Lehrach H (1991): Yeast artificial chromosome libraries containing large inserts from mouse and human DNA. *Proc Natl Acad Sci USA* 88:4123–4127.

Liu Q, Sommer SS (1994): Parameters affecting the sensitivities of dideoxy fingerprinting and SSCP. *PCR Methods Appl* 4:97–108.

Lovett M, Kere J, Hinton LM (1991): Direct selection: A method for the isolation of cDNAs encoded by large genomic regions. *Proc Natl Acad Sci USA* 88:9628–9632.

Marchuk DA, Tavakkol R, Wallace MR, Brownstein BH, Taillon-Miller P, Fong C-T, Legius E, Andersen LB, Glover TW, Collins FS (1992): A yeast artificial chromosome contig encompassing the type 1 neurofibromatosis gene. *Genomics* 13:672–680.

Marshal RD, Koontz J, Sklar J (1995): Detection of mutations by cleavage of DNA heteroduplexes with bacteriophage resolvases. *Nat Genet* 9:177–183.

Morgan JG, Dolganov GM, Robbins SE, Hinton LM, Lovett M (1992): The selective isolation of novel cDNAs encoded by the regions surrounding the human interleukin 4 and 5 genes. *Nucleic Acids Res* 20:5173–5179.

Myers RM, Larin Z, Maniatis T (1985a): Detection of single base substitutions by ribonuclease cleavage at mismatches in RNA:DNA duplexes. *Science* 230:1242–1246.

Myers RM, Lumelsky N, Lerman LS, Maniatis T (1985b): Detection of single base substitutions in total genomic DNA. *Nature* 313:495–498.

Nelson DL, Ledbetter SA, Corbo L, Victoria MF, Ramirez-Solis R, Webster TD, Ledbetter DH, Caskey CT (1989): Alu polymerase chain reaction: A method for rapid isolation of human-specific sequences for complex DNA sources. *Proc Natl Acad Sci USA* 86:6686–6690.

Nelson DL, Ballabio A, Victoria MF, Pieretti M, Bies RD, Gibbs RA, Maley JA, Chinault AC, Webster TD, Caskey CT (1991): Alu-primed polymerase chain reaction for regional assignment of 110 yeast artificial chromosome clones from the human X chromosome: Identification of clones associated with a disease locus. *Proc Natl Acad Sci USA* 88:6157–6161.

Newton CR, Graham A, Heptinstall LE, Powell SJ, Summers C, Kalsheker N, Smith J, Markham AF (1989): Analysis of any point mutation in DNA: The amplification refractory mutation system (ARMS): *Nucleic Acid Res* 17:2503–2516.

Nickerson DA, Kaiser R, Lappin S, Stewart J, Hood L, Landegren U (1990): Automated DNA diagnostics using an ELISA-based oligonucleotide ligation assay. *Proc Natl Acad Sci USA* 87:8923–8927.

Nisson PE, Rashtchian A, Watkins PC (1991): Rapid and efficient cloning of Alu-PCR products using uracil DNA glycosylase. *PCR Methods Appl* 1:120–123.

Ochman H, Gerber AS, Hartl D (1988): Genetic applications of an inverse polymerase chain reaction. *Genetics* 120:621–623.

Orita M, Iwahana H, Kanazawa H, Hayashi K, Sekiya T (1989a): Detection of polymorphisms of human DNA by gel electrophoresis as single-strand conformation polymorphisms. *Proc Natl Acad Sci USA* 86:2766–2770.

Orita M, Suzuki Y, Sekiya T, Hayashi K (1989b): A rapid and sensitive detection of point mutations and DNA polymorphisms using the polymerase chain reaction. *Genomics* 5:874–879.

Osborne-Lawrence S, Welcsh PL, Spillman M, Chandrasekharappa SC, Gallardo TD, Lovett M, Bowcock (1995): Direct selection of expressed sequences within a 1-MB region flanking *BRCA1* on human chromosome 17q21. *Genomics* 25:248–255.

Ostrander EO, Jong PM, Rine J, Duyk G (1992): Construction of small-insert genomic DNA libraries highly enriched for microsatellite repeat sequences. *Proc Natl Acad Sci USA* 89:3419–3423.

Parimoo S, Patanjali SR, Shukla H, Chaplin DD, Weissman SM (1991): cDNA selection: Efficient PCR approach for the selection of cDNAs encoded in large chromosomal DNA fragments. *Proc Natl Acad Sci USA* 88:9623–9627.

Ravnik-Glavac M, Glavac D, Dean M (1994): Sensitivity of single-strand conformation polymorphism and heteroduplex method for mutation detection in the cystic fibrosis gene. *Hum Mol Genet* 3:801–807.

Riley J, Butler R, Ogilvie D, Finniear R, Jenner D, Powell S, Anand R, Smith JC, Markham AF (1990): A novel, rapid method for the isolation of terminal sequences from yeast artificial chromosome (YAC) clones. *Nucleic Acids Res* 18:2887–2890.

Rommens JM, Iannuzzi MC, Kerem B-S, Drumm ML, Melmer G, Dean M, Rozmahel R, Cole JL, Kennedy D, Hidaka N, Zsiga M, Buchwald M, Riordan JR, Tsui L-C, Collins FS (1989): Identification of the cystic fibrosis gene: Chromosome walking and jumping. *Science* 245:1059–1065.

Rosen DR, Siddique T, Patterson D, Figlewciz DA, Sapp P, Hentati A, Donaldson D, Gotto J, O'Regan JP, Deng HG, Rahmani Z, Krizus A, McKenna-Yasek D, Cayabyab A, Gaston S, Tanzi R, Halperin JJ, Herzfeldt B, Van den Berg R, Hung W, Bird T, Deng G, Mulder DW, Smyth C, Laing GL, Soriano E, Pericak-Vance M, Haines JL, Rouleau GA, Gusella J, Horvitz HR, Brown RH (1993): Mutations in the gene encoding Cu/Zn superoxide dismutase are associated with familial amyotrophic lateral sclerosis. *Nature* 362:59–62.

Rosenzweig A, Watkins H, Hwang D-S, Miri M, McKenna WJ, Traill T, Seidman JG, Seidman CE (1991): Preclinical diagnosis of familial hypertrophic cardiomyopathy by genetic analysis of blood lymphocytes. *New Engl J Med* 325:1753–1760.

Saiki RK, Bugawan TL, Horn GT, Mullis KB, Erlich HA (1986): Analysis of enzymatically amplified beta-globin and HLA-DQ alpha DNA with allele-specific oligonucleotide probes. *Nature* 324:163–166.

Sarkar G, Sommer S (1989): Access to a messenger RNA sequence or its protein product is not limited by tissue or species specificity. *Science* 244:331–334.

Sarkar G, Yoon H, Sommer SS (1992): Dideoxy fingerprinting (ddF): A rapid and efficient screen for the presence of mutations. *Genomics* 13:441–443.

Schuler GD et al. (1996): A gene map of the human genome. *Science* 274:540–546.

Sheffield VC, Cox DR, Lerman LS, Myers RM (1989): Attachment of a 40-base-pair G+C rich sequence (GC-clamp) to genomic DNA fragments by the polymerase chain reaction results in improved detection of single-base changes. *Proc Natl Acad Sci USA* 86:232–236.

Shizuya H, Birren B, Kim UJ, Mancino V, Slepak T, Tachiiri Y, Simon M (1992): Cloning and stable maintenance of 300-kilobase-pair fragments of human DNA in *Escherichia coli* using an F-factor-based vector. *Proc Natl Acad Sci USA* 89:8794–8797.

Silverman GA (1993): Isolating vector–insert junctions from yeast artificial chromosomes. *PCR Methods Appl* 3:141–150.

Sternberg N (1994): The P1 cloning system: Past and future. *Mamm Genome* 5:397–404.

Sternberg N, Ruether J, deRiel K (1990): Generation of a 50,000-member human DNA library with an average DNA insert size of 75–100 kbp in a bacteriophage P1 cloning vector. *New Biol* 2:151–162.

Stewart EA, McKusick KB, Aggarwal A, Bajorek E, Brady S, Chu A, Fang N, Hadley D, Harris M, Hussain S, Lee R, Maratukulam A, O'Connor K, Perkins S, Piercy M, Qin F, Reif T, Sanders C, She X, Sun W-L, Tabar P, Voyticky S, Cowles S, Fan J-B, Mader C, Quackenbush J, Myers RM, Cox DR (1997): An STS-based radiation hybrid map of the human genome. *Genome Res* 7:422–433.

Tagle DA, Collins FS (1992): An optimized Alu-PCR primer pair for human-specific amplification of YACs and somatic cell hybrids. *Hum Mol Genet* 1:121–122.

Thomas A, Skolnick MH (1994): A probabilistic model for detecting coding regions in DNA sequences. *IMA J Math Appl Med Biol* 11:149–160.

Tumer Z, Vural B, Tonnesen T, Chelly J, Monaco AP, Horn N (1995): Characterization of the exon structure of the Menkes disease gene using vectorette PCR. *Genomics* 26:437–442.

White MB, Carvalho M, Derse D, O'Brien SJ, Dean M (1992): Detecting single base substitutions as heteroduplexes polymorphisms. *Genomics* 12:301–306.

Xu Y, Mural R, Shah M, Uberbacher E (1994): Recognizing exons in genomic sequence using GRAIL II. In: Setlow JK, ed. *Genetic Engineering, Principles and Methods,* vol. 15. New York: Plenum Press, pp. 241–253.

Yaspo ML, Gellen L, Mott R, Korn B, Nizetic D, Poustka AM, Lehrach H (1995): Model for a transcript map of human chromosome 21: Isolation of new coding sequences from exon and enriched cDNA libraries. *Hum Mol Genet* 4:1291–1304.

Youil R, Kemper BW, Cotton RG (1995): Screening for mutation by enzyme mismatch cleavage with T4 endonuclease VII. *Proc Natl Acad Sci USA* 92:87–91.

Youil R, Kemper B, Cotton RG (1996): Detection of 81 of 81 known mouse beta-globin promoter mutations with T4 endonuclease VII--the EMC method. *Genomics* 32:431–435.

18

Examining Complex Genetic Interactions

Joellen M. Schildkraut
Community Family Medicine
Duke University Comprehensive Cancer Center
Durham, North Carolina

INTRODUCTION

In the study of the etiology of complex traits, there is a need for analytic strategies that can determine the cause of the discordance between genetic susceptibility and gene expression. Genetic models of complex traits need to account for genetic heterogeneity and interactions with other genes and the environment. Traditional linkage methods alone may not be powerful enough for studying the genetics for many complex diseases. The combination of genetic and epidemiologic methods in the study of complex genetic interactions has the potential for melding complementary approaches for investigating the multifactorial nature of many common diseases.

There are several ways in which genetic susceptibility may influence disease. Genetic susceptibility can influence the risk of disease by itself; it may exacerbate the expression of an environmental risk factor; or the risk factor may exacerbate the genetic effect (Ottman, 1990). Through the use of epidemiologic research designs, interactions between genotype or family history and environmental influences on disease can be evaluated, thereby enhancing the ability to uncover genetic influences on disease (Ottman, 1990). Since exposure to some environmental factors that influence genetic risk are modifiable, the discovery of such relationships has important public health implications that can be assessed through epidemiologic

Approaches to Gene Mapping in Complex Human Diseases, Edited by Jonathan L. Haines and Margaret A. Pericak-Vance. ISBN 0-471-17195-6

measures. Employing epidemiologic principles in the study of complex genetic disease also introduces strategies for minimizing bias and increasing the generalizability of the results. This chapter reviews genetic epidemiologic methods applied to investigate the relationship between genetic susceptibility and other factors. The integration of genetic and epidemiologic approaches is also discussed.

EVIDENCE FOR COMPLEX GENETIC INTERACTIONS

Lack of one-to-one correspondence between a given genotype and a given phenotype is characteristically observed in complex traits with inheritance patterns that deviate from Mendelian models. Genetic models of complex diseases relate genetic heterogeneity, modifying genes, and environmental effects to the concept of reduced penetrance among genetically susceptible individuals (Ottman, 1990).

Genetic Heterogeneity

Genetic heterogeneity exists when several genes are associated with the same disease (Chapter 3). Genetic heterogeneity can also be due to variations in the same genes that are associated with differences in disease risk (allelic heterogeneity). Genetic heterogeneity may affect the ability to identify genetic and nongenetic factors associated with the risk of disease, since etiologically distinct subgroups may not be distinguished by phenotype alone. Ignoring the possibility of different genetic pathways could result in the improper reduction of the magnitude of epidemiologic measures of association between important risk factors and the disease (Hodge, 1994; Newman et al., 1995; Slattery et al., 1995). Thus, failure to correct for genetic heterogeneity of disease often leads to inaccurate report of weak to moderate odds ratios depicting the associations between the disease and marker alleles (Hodge, 1994).

Alzheimer disease (AD) is an example of a complex trait related to multiple genes. There is significant evidence that several genes as well as the environment influence the risk of this disease. Age at onset is one factor that distinguishes subgroups of AD. Genes for rare early-onset AD (i.e., onset before the age of 60 years) are found on chromosomes 1, 14, and 21 (St. George-Hyslop et al., 1990; Goate et al., 1991; Mullan et al., 1992; Van Broeckhoven et al., 1992; Levy-Lehad et al., 1995; Rogaev et al., 1995; Sherrington et al., 1995). Late onset of a more common form of AD has been associated with the apolipoprotein E (APOE) locus on chromosome 19 (Pericak-Vance et al., 1991; Corder et al., 1993; Saunders et al., 1993). Allelic variation in the *APOE* gene has been shown to be associated with the risk of AD, with increased risk at a decreased age at onset associated with the *APOE*E4* allele and a decreased risk of AD associated with the *APOE*E2* allele (Pericak-Vance and Haines, 1995).

Gene–Gene Interaction (Epistasis)

Complex genetic traits may be influenced by other modifying genes that are not linked to the primary (or major) gene for that trait. Polygenic traits, which require the simultaneous presence of variations in multiple genes, are also an important type of genetic interaction to be considered in the analysis of complex traits (Lander and Schork, 1994; Cordell and Todd, 1995). The degree and type of genetic interaction strongly influences the chance of detecting genes through linkage analysis (Risch, 1990a, 1990b). Polygenic inheritance is difficult to demonstrate in humans and complicates genetic mapping because no single locus is solely responsible for producing the trait (Lander and Schork, 1994).

Examples of gene–gene interaction illustrate the need for a combination of analytic approaches to understand the complexity of disease susceptibility. Linkage data for diabetes mellitus (IDDM) (Thomson, 1994; Cordell and Todd, 1995) using sib pair analyses have demonstrated an association between human leukocyte antigen (HLA) genes and IDDM. Case–control association data have indicated that non-HLA genetic factors also contribute to IDDM and additional linkage data have revealed genetic heterogeneity as well as evidence for gene–gene interaction at the *IDDM1* and *IDDM2* loci on chromosomes 6p and 11p, respectively (Cordell and Todd, 1995). A possible gene–gene interaction has been suggested between the *NAT1* and *NAT2* enzyme genetic polymorphisms, important in N- or O-acetylation, and colorectal cancer risk (Bell et al., 1995). Although the *NAT*10* allele is associated with a 1.9-fold [95% confidence interval (CI) = 1.2–3.2; $P = 0.009$] increased risk of colorectal cancer, the association with this *NAT1* variant is found to be stronger among *NAT2* rapid acetylators (OR = 2.8; 95% CI; 1.4–5.7; $P = 0.003$).

Gene–Environment Interaction

In addition to genetic heterogeneity and gene–gene interaction, it is important to consider the influence of environmental factors in the study of complex genetic traits. Once a major genetic effect has been identified, gene–environmental interactions can be explored. The examination of candidate genes in the study of disease–exposure associations, or gene–environment interactions, can reveal effects of an environmental factor on the risk of disease that would be concealed if genetic susceptibility were ignored (Hwang et al., 1994). Three plausible models of gene–environment interaction have been described by Khoury et al. (1988):

1. Disease risk is increased only in the presence of the susceptibility genotype and the environmental risk factor.
2. Disease risk is increased by the environmental risk factor alone but not by the genotype alone.
3. Disease risk is increased by the genotype in the absence of the environmental risk factor but not by the risk factor alone.

There are several examples of gene–environment interaction in the literature. For example, both genetic and environmental factors are felt to be important in the susceptibility of IDDM (Thomson 1994; Cordell and Todd, 1995). Although twin studies suggest that environmental risk factors may be important in IDDM risk, given a 36% concordance rate between monozygotic twins, little is known about the environmental determinants of this disease (Dahlquist, 1994). Models for carcinogenesis are also illustrative of complex traits with multiple risk factors and etiologic pathways. It is thought that interactions between environmental factors and inherited polymorphisms of genes are important in the multistage process of carcinogenesis. For example, interactions between environmental chemical exposures and N- and O-acetylation phenotypes encoded by two loci, *NAT1* and *NAT2,* have been associated with an increase in the risk of bladder cancer (Sim et al., 1995). Prior reports relating head trauma as a causative factor in AD were clarified with the discovery of a synergistic effect between head trauma and the *APOE*E4* allele (Mayeux et al., 1995; Pericak-Vance and Haines, 1995).

ANALYTIC APPROACHES FOR STUDYING COMPLEX GENETIC INTERACTIONS

It may be necessary to combine several analytic approaches to explore causes of reduced penetrance due to genetic heterogeneity and interactions between other genes and the environment. The integration of genetic and epidemiologic methods in the study of complex genetic interactions has the potential of providing a complementary approach for investigating the multifactorial nature of many common diseases (Weeks and Lathrop, 1995).

Genetic Approaches

Complex genetic diseases challenge the capability of genetic analytic approaches to detect a major gene effect. Although at times computationally complex, statistical methods, including sib pair and relative pair analyses, and methods that combine genetic analytic approaches, have been shown to be effective in the estimation of genetic parameters of complex traits. Modeling joint effects of several loci and environmental factors may aid in the elucidation of important biological information about the interaction between these disease determinants. Genetic approaches in the determination of complex genetic interactions are discussed below. For more complete information on a specific genetic analyses, consult other chapters in the text.

Linkage and Allele Sharing Analyses. Applying linkage analysis of complex traits can be problematic, since the genetic model to explain the inheritance pattern may be difficult to determine (Lander and Schork, 1994). However, by narrowing the definition of the disease or by restricting the patient population, the chances for unraveling the genetic etiology of a complex disease may be improved. For example, focusing on early-age-at-onset cases, such as with Alzheimer disease (St. George Hyslop et al., 1990; Mullan et al., 1992; Van Broeckhoven et al., 1992) and breast cancer (Hall et al., 1990), has proved to be advantageous and has allowed the identification of several causative genes (Goate et al., 1991; Futreal et al., 1994; Miki et al., 1994; Levy-Lehad et al., 1995; Rogaev et al., 1995; Sherrington et al., 1995).

For complex traits, where the mode of inheritance is unclear, nonparametric allele sharing methods, such as affect sib pair analyses, are recommended (Thomson, 1994; Cordell and Todd, 1995; Weeks and Lathrop, 1995). These methods do not require that the specific mode of inheritance be assumed, and they appear to be more robust than traditional linkage analysis. With technological advances, large-scale, rapid genotyping is becoming more feasible. Consideration is recommended for genomic screening in data sets large enough to examine the joint action of several loci simultaneously. Nonparametric methods, which can account for the joint effects of several underlying genetic components, are more successful in detecting disease loci than single locus methods (Cordell and Todd, 1995). The affected sib pair method may also be used to provide evidence for polygenic inheritance (Weeks and Lathrop, 1995).

The affected sib pair method may serve, as well, for determining gene–gene interaction. Genes may interact in several ways, either in a multiplicative or additive manner. In the multiplicative model, where N loci may be involved, the risk ratio λ_i depends on the increase in risk attributable to the ith locus and the overall risk ratio λ_S is the product of the locus-specific risk ratios (Risch, 1990a). Evidence for multiplicative genetic interaction is seen when the risk ratios decrease exponentially across different degrees of relation (Risch, 1990a). In the additive model, the locus-specific risk ratios are added together to determine the joint effect. Evidence for linkage can be assessed by a maximum lod score (MLS) statistic T, where a lower threshold for linkage is applied (T = 1.0, rather than 3.0) to allow for the distinction between convincing evidence for linkage and regions of interest to help ensure that loci with weaker effects are not missed (Weeks and Lathrop, 1995).

Once candidate chromosomal regions have been identified, more thorough genotyping can be pursued and additional sib pairs can be examined as a test data set to obtain more convincing evidence for linkage. The affected sib pair method has been applied successfully to numerous diseases including IDDM, where 18 different chromosome regions have been implicated in the genetic etiology of this disease (Faden and Beaucharp, 1986; Wertz et al., 1994; Weeks and Lathrop, 1995).

Another method of detecting polygenic inheritance is the extended affected sib pair method, a complementary likelihood-based approach to confirm linkage (Hyer et al., 1991). This method allows for genetic heterogeneity and includes marker information on parents and unaffected siblings as well as the use of multiple loci simultaneously. It has the advantage of handling partially informative data and may be more powerful than the nonparametric MLS approach. It has also been suggested that new methods for detecting linkage in extended pedigrees without any assumptions concerning the mode of transmission may yield more information for detection of linkage than sib pairs containing the same number of affected subjects (Curtis et al., 1995).

Finally, the affected relative pair (ARP) analysis takes advantage of extended pedigree structures by including affected relatives of arbitrary relationship (Weeks and Lange, 1988; Weeks and Lathrop, 1995; Davis et al., 1996; Kruglyak et al., 1996). This method also does not require assumptions about the disease model and tests for an excess of sharing of alleles between pairs of relatives or all relatives simultaneously. This method does not require genotypic information on unaffected individuals, including parents, and is therefore useful for diseases of late onset. While the older affected-pedigree-member (APM) method is particularly sensitive to misspecification of marker allele frequency and has a high false positive rate (on the order of 20%), the newer SimIBD approach is more robust to allele frequency specification and has a lower false positive rate (see Chapter 14).

Segregation Analysis. Although segregation analysis is limited to the investigation of modes of inheritance of complex traits, the development of statistical methods using mixed genetic models has been successful in estimating genetic parameters (Weeks and Lathrop, 1995). Regressive models incorporate a major gene effect as well as other sources of familial correlation (Bonney 1984; Demenais et al., 1992). Regressive models, which condition the phenotype of each individual on the phenotype of preceding relatives, have been shown to be useful in the analysis of both qualitative (categorical) and quantitative traits (Bonney et al., 1989; Demenais and Lathrop, 1993; Shields et al., 1994). The use of regressive models also has been effective in exploring interaction between loci (Weeks and Lathrop, 1995). Segregation analysis for determining major gene effects, using the regressive model by Bonney et al. (1984), incorporates covariates such as age, sex, and specific genetic polymorphisms to assess the effects of complex genetic interactions. However, like all segregation analyses, this approach is very sensitive to sample size and to ascertainment bias; moreover, it generally requires substantial computer resources (see Chapter 5).

Combining segregation and linkage analyses has been shown to be a powerful modeling technique (Bonney et al., 1988; Shields et al., 1994). Together, these analytic approaches have been successful for determining genetic interactions and improving lod scores when genotype-dependent effects of covariates are incorporated into the linkage analysis (Bonney et al., 1988; Shields et al., 1994; Weeks and Lathrop, 1995). This approach makes it possible to detect a major gene when the effects of genetic modifiers are accounted for in the model. The development of highly polymorphic markers has also contributed to the success of this strategy.

An example of this approach is seen in a study by Thein et al. (1994), which detected a major gene for "heterocellular hereditary persistence of fetal hemoglobin" (HPFH) by means of a regressive model approach and linkage analysis. Evidence of the presence of a major gene was found through segregation analysis, with residual familial correlations when the effect of genetic modifiers, β-thalassemia, and the Xmn I-$^{G}\gamma$ polymorphism, were taken into account. Linkage analysis determined that the major gene was located outside the β-globin cluster, with the exact position yet to be determined. This example illustrates how regressing on genetic cofactors can enhance analysis of complex genetic traits and lead to the eventual identification of the genes involved.

Although the practice of combining segregation and linkage analyses has shown some successes, certain drawbacks limit the usage of this approach. These include the possibility of ascertainment bias, thus limiting generalizability, computational constraints that occur when it is necessary to maximize variables, and the nature of the method itself, which is better adapted for the analysis of quantitative traits.

Epidemiologic Approaches

The application of epidemiologic methods to the study of genetics in complex diseases has been increasing. Epidemiologic research designs can be used to test models for the relationship between genetic susceptibility and environmental risk factors. In particular, the case–control method has emerged as a powerful approach for addressing the role of genetic factors and their interaction with environmental factors (Khoury and Beaty, 1994). Using a case–control approach, subjects are selected according to their disease status. Cases are defined as those affected with the particular disease of interest. The controls, although at risk for developing the disease (i.e., old enough to develop a disease with a characteristically late onset, or the appropriate sex for developing a disease such as ovarian cancer), are unaffected with the disease. The case and control subjects are then compared with respect to the proportion having the exposure or characteristic of interest. In genetic epidemiologic studies, the cases and controls are classified according to the presence or absence of the genotype (or family history) as well as the presence or absence of an additional risk factor, either an environmental factor or a modifying gene.

To achieve valid case–control comparisons, controls should be selected so that they represent the same source population as the cases. Thus the selection of controls should include the same exclusion or restriction criteria applied to the cases. Another important issue in selecting controls is the source of the subjects. The source could be the hospital from which the cases were identified, the general population, or friends, neighbors, or relatives of cases. Each set offers advantages and disadvantages that must be considered in the context of the source of the case selection, the nature and goals of the study, and the type of information obtained (Hennekens and Buring, 1987). Control subjects must be at risk for developing the disease under study, and to avoid introducing secular influences of disease risk that differ in the cases versus controls, it is preferable to choose control subjects during the same time period as the cases.

Association studies that use the case–control method provide a complementary strategy for detection of disease susceptibility genes and are used for identifying such genes once the region of the chromosome has been determined by linkage (Khoury and Beaty, 1994). In the candidate gene approach, the frequency of the disease susceptibility gene or genetic marker of interest in a group of affected individuals or cases is compared to the frequency in a group of controls or noncases (Khoury and Beaty, 1994). This strategy can be expanded to permit examination of complex genetic interactions by analyzing the joint effects of genes on disease susceptibility. As with all case–control studies, selection of an appropriate control group, such as matching on ethnicity and other factors, is important to avoid spurious associations due to variation in allelic frequencies in different populations. Spurious associations can also result from confounding due to recent admixture and selection or drift between unlinked loci (Khoury et al., 1994; Lander and Schork, 1994; Weeks and Lathrop, 1995).

A second epidemiologic design is the cohort study. Rather than selecting study subjects on the basis of their disease status and looking backward to a possible cause, study subjects are selected on the basis of their exposure status, are disease free at the start of the follow-up period, and are followed forward in time to onset of disease. In the context of genetic epidemiology, the exposure can be defined as family history of disease or a susceptibility genotype. Some of the advantages of a cohort design are the ability to establish a temporal relationship between exposure and disease, the suitability for the study of rare exposures, and the ability to study multiple effects of a single exposure. Cohort studies are more costly and time-consuming than case–control studies and are not efficient for the investigation of relatively rare disease outcomes such as cancer or diseases with late onset such as Alzheimer disease. However, existing family history data, derived from a case–control study, have been applied in retrospective cohort analyses of the relatives of cases and controls, avoiding some of these drawbacks (Schildkraut et al., 1989a; Claus et al., 1990).

Measurement error, such as misclassification of the susceptibility genotype or of an environmental exposure, can occur regardless of the study design. In studies that use family history in place of an unknown genetic factor, it is possible that genetic susceptibility will be misclassified. Such misclassification, if nondifferential with respect to an environmental risk factor, will cause a dilution in the estimate of association between the genetic susceptibility factor and the disease (Ottman, 1990). The attenuation in the magnitude of the association make it more difficult to detect an association that actually exists. However, nondifferential misclassification does not induce a spurious association between the risk factor and disease. Therefore, if an association is detected in the presence of nondifferential misclassification, the true value for the measure is at least as great as what is observed.

Another important aspect to consider when evaluating an epidemiologic study is the generalizability of the results or how findings in the study population are applicable to other populations. Generalizability depends on how the study population is selected or defined. Replication of the results in other populations is one way to address this question directly.

Case–control analyses using logistic regression and survival analysis of cohort data (described below) have aided investigators in the determination of familial risk and in the examination of other covariates. These analyses have included the affected status of the relatives of the affected versus unaffected probands and age at onset.

Epidemiologic Measures. There are several measures that are often applied in epidemiologic research to quantity the magnitude of the association between the risk factor and the disease. These include the relative risk (RR), the odds ratio (OR), and the attributable risk (AR), which quantifies the importance or impact of that factor in the occurrence of the disease relative to other risk factors for that disease (see Table 18.1).

The RR is the ratio of the incidence of disease among the "exposed" group, which can be defined as individuals with the susceptibility genotype or with a positive family history, compared to the incidence of disease in the "nonexposed" group, or individuals without the susceptibility alleles. Figure 18.1 is a 2 × 2 table representing the joint distribution of the presence or absence of the genotype and disease in the study population. If the disease represents the occurrence of "newly"

Table 18.1 Summary of Epidemiologic Measures

Measure	Definition
Relative risk (RR)	Measure of association between genetic exposure and disease, the ratio of the incidence of the disease in the "exposed" and "unexposed" groups using prospective/cohort data: $RR = I_e/I_o = \frac{a(a+b)}{c(c+d)}$ (see Fig. 18.1).
Odds ratio (OR)	Measure of association between genetic exposure and disease, the ratio of the odds of exposure (genetic susceptibility) among affected subject or cases to the odds or exposure (genetic susceptibility) among unaffected individuals or controls calculated using data from case–control studies: $OR = ad/bc$ (see Fig 18.1).
Attributable risk (AR)	Measure of impact or the excess risk of disease among genetically susceptible individuals compared to those who are not. The genetic AR is dependent on the proportion of disease due to the susceptibility gene and the magnitude of the relative risk among gene carriers and noncarriers: $AR = \frac{P_{Aa}[1 - 2q(1 - P_{aa})]}{P_{Aa}(1 - 2q)}$ (Claus et al., 1996).

diagnosed cases of the disease over a defined period of time, the incidence of disease D among the exposed (i.e., those with a mutation in a susceptibility gene or family history of disease D) is I_e and the incidence of disease D in the unexposed (i.e., absence of a genetic mutation or a family history of disease D) is I_o.

A RR can be determined from a risk ratio or a rate ratio. A risk ratio is the ratio of two proportions and determines the number of times by which the risk in exposed persons exceeds that for nonexposed persons. A rate ratio is the ratio of incidence rates in the exposed versus nonexposed groups and is computed with follow-up time or the total person-time units in the denominator of the incidence rate, rather than the number of individuals in each group. Over a short period of time the risk

	Disease		
Genetic Test	Yes	No	**Total**
Positive	a	b	a + b
Negative	c	d	c + d
Total	a + c	b + d	a + b + c + d

Figure 18.1. *2 × 2 contingency table for case–control analysis.*

and rate ratios are approximately equal (Rothman, 1986a). If the individual person-time units are known (the length of follow-up or observation time of each study subject), survival analysis methods are used to calculate cumulative incidence from follow-up data. These methods, such as the product–limit method (Kaplan and Meier, 1958) and the Cox proportional hazards model (Cox, 1972), can account for varying lengths of observation by calculating the risk for developing disease at each point in time when there is a change in the occurrence of disease and at each withdrawal (or death) of an individual who has not developed disease.

If the individual follow-up times are unknown, person-time is multiplied by the number of affected and unaffected individuals, respectively, with the assumption that disease onset occurred by the midpoint of the follow-up period. Both relative risk measures can be used to determine the association between the genetic marker or family history and the disease.

From Figure 18.1, the risk ratio = $I_e/I_o = a(a + b)/c(c + d)$. To calculate a rate ratio for a 5 year follow-up period, the midpoint of 2.5 is multiplied by the number of cases in the exposed and unexposed groups a and c. The number of noncases, b and d, would be multiplied by 5, the length of the follow-up period. If the length of follow-up for each individual is known, the denominator for the incidence rates in the exposed and nonexposed is the total person-time, or the sum of the length of follow-up for each individual in the study within the respective exposure groups.

An illustrative example comes from the Framingham Study cohort, through which the relationship between risk of early-onset coronary artery disease (CAD) and parental history of early CAD death was investigated (Schildkraut et al., 1989b). Members of the cohort were followed biennially from 1948 to exam 16, or the 16th biennial exam in 1980. The length of follow-up for each participant without a CAD end point was equal to the number of days between the date of exam 2 and whichever came first, date of examination 16 or the date of death. For those who developed CAD between exam 2 and exam 16, the length of follow-up was calculated from the date of exam 2 to the date of the CAD event with a total length of follow-up of 28 years. Relative risks were calculated to determine the association between family history of CAD, defined as CAD death before the age of 65, and the risk of developing early-onset CAD by comparing the incidence of CAD among those with a family history (I_e) to those without a family history of CAD (I_o). The denominators were the sum of the days of follow-up of each individual in the exposed (those with a family history) and unexposed (those without a family history) groups, respectively. A RR of 1.6 was derived from the ratio of the incidence rate of CAD in each group using proportional hazards analysis. Therefore, early age at death due to CAD in a parent was associated with a 60% increase in risk of early-onset CAD in the offspring.

The absolute risk for an individual may vary according to the specific characteristics of that individual such as age, sex, race, and other characteristics associated with risk of the disease. In addition, the absolute risk of the exposed population may be small even when the relative risk is high. This is illustrated in the case of a rare disease, where I_e = 10 per 100,000 per year and I_o = 1 per 100,000 per year and the relative risk is 10.0, despite the very low risk of disease.

Incidence rates and relative risks are derived from cohort study data (prospective studies), where the exposure is defined "before" disease occurs. Study subjects who are disease free at baseline are followed over time and observed for subsequent disease onset. However, as is often the case in genetic linkage studies, cohort data are not available or are impractical to obtain, especially for late-onset and rare disease. Case–control data (retrospective studies), where disease status is established first and exposure subsequently determined, are often utilized to estimate the relative risk between a study factor and disease. If the data in Figure 18.1 were obtained from a case–control study, the marginal totals for the exposed ($a + b$) and nonexposed ($c + d$) groups of the population would not be known, and the incidence rate could not be calculated directly. In this scenario, an OR may be calculated to estimate the RR. The OR is defined as the ratio of the odds of exposure among the cases, the subjects diagnosed with disease D, to that among the noncases or controls. Based on the 2 × 2 table in Figure 18.1, the odds ratio is calculated by applying the following formula:

$$\mathrm{OR} = \frac{a/c}{b/d} = \frac{ad}{bc}$$

Confidence intervals (CI), which specify the range of the true values for the OR, can be derived by estimating the variance and are calculated from the following equation:

$$\mathrm{CI} = (\mathrm{OR})\exp z[\pm(\mathrm{var}(\ln \mathrm{OR})]^{1/2}$$

where, for a 95% confidence limit or a significance level of 0.05, $z = 1.96$. The confidence interval for the odds ratio can be approximated using a Taylor series expansion as follows (Hennekens and Buring, 1987):

$$\mathrm{CI} = (ad/bc)\exp[\pm z(1/a + 1/b + 1/c + 1/d)^{1/2}]$$

If the confidence interval does not include 1.0 (the null value for a measure of association), the association is statistically significant at the specified level.

As shown in Figure 18.2, using data from a study conducted by Mayeux et al. (1995a), the crude odds ratio of 2.3 (95% CI = 1.3–3.9) for the association between *APOE*E4* and Alzheimer disease is statistically significant.

In most case–control studies, in the absence of bias, the odds ratio provides a valid estimate of the relative risk (Hennekens and Buring, 1987). The validity of the relative risk estimate and the inferences based on this estimate depend on the absence of measurement error, the representativeness of the study population (absence of selection bias), and the absence of an imbalance of other factors related to the disease and the exposure in the exposed and nonexposed groups (confounding bias).

The AR represents the excess risk of disease in those exposed compared with those unexposed to the factor of interest. The AR depends not only on the magni-

	Disease Status	
APOE *E4	AD Present	AD Absent
Yes	55	36
No	58	87

$$\mathrm{OR} = \frac{ac}{bd} = \frac{(55)(87)}{(58)(36)} = 2.3$$

$$95\%\ \mathrm{CI} = \frac{ad}{bc}\ \exp\left[\pm\, z\left(\frac{1}{a} + \frac{1}{b} + \frac{1}{c} + \frac{1}{d}\right)^{1/2}\right]$$

$$= (2.3)\ \exp\left[\pm\, 1.96\left(\frac{1}{55} + \frac{1}{36} + \frac{1}{58} + \frac{1}{87}\right)^{1/2}\right] = (2.3)e^{\pm 0.53}$$

or

Lower bound: $= (2.3)e^{-0.53} = (2.3)\ (0.59) = 1.3$

Upper bound: $= (2.3)e^{+0.53} = (2.3)\ (1.69) = 3.9$

Figure 18.2. *Crude odds ratio (OR) and 95% confidence intervals (CI) for the association between APOE*E4 and Alzheimer disease (AD). (From Mayeux et al., 1995.)*

tude of the relative risk, but on the prevalence of the exposure (the proportion of individuals in a population that are exposed at a specific instant in time such as the proportion carrying a susceptibility allele) in the study population. The AR reflects the "impact" of the exposure on the occurrence of the disease or how much disease would be "prevented" if that factor were somehow eliminated. The AR can be expressed as a percentage and is calculated (Rothman, 1986a) by incorporating the relative risk (or OR) using the following formula:

$$\mathrm{AR}\% = \frac{\mathrm{RR} - 1}{\mathrm{RR}} \times 100$$

with RR = 4.0, the AR% = (4.0 – 1)/4.0 × 100 = 75%, which suggests that 75% of D may be attributed to a family history of D in at least one first-degree relative. If the familial cause of the disease could be prevented or treated, 75% would be eliminated. The following equation, for the population attributable risk percent (PAR%), is used to estimate the excess rate of disease in the total study population of exposed and nonexposed individuals that is attributable to the exposure:

$$\mathrm{PAR}\% = \frac{(P_e)(\mathrm{RR} - 1)}{(P_e)(\mathrm{RR} - 1) + 1} \times 100$$

where P_e is the prevalence of exposure in the population, which can either be estimated in the control group or obtained from another source (Hennekens and Buring, 1987). The AR may also be expressed as the "risk difference" or $I_e - I_o$.

Attributable risk for genetic susceptibility is related to allele frequency and perhaps a second factor such as age (Claus et al., 1996). The genetic attributable risk can be calculated by incorporating the risk (R_e) for developing a disease and the allele frequency of the abnormal allele (q). The proportion of disease due to the susceptibility allele is calculated by comparing the risk of disease among those likely to be carriers of the susceptibility allele (e.g., heterozygotes, given a susceptibility allele, with genotype $e = Aa$) to the risk of disease among those who are unlikely to be carriers of the susceptibility allele (homozygotes with genotype $e = aa$). The following formula estimates the proportion of disease (P_e) likely to occur in those who carry a rare autosomal dominant susceptibility allele (genotype $e = Aa$, where the homozygous genotype is ignored):

$$P_e = \frac{2pq \times (2R_{Aa} - R_{aa})}{2pq \times (2R_{Aa} - R_{aa}) + p^2 \times R_{aa}}$$

The risk attributable to a susceptibility allele is calculated (Claus et al., 1996) by noting that the following relationship exists between P_{Aa} and the genetic PAR%:

$$\text{PAR\%} = P_{Aa}\frac{1 - 2q(1 - P_{aa})}{P_{Aa}(1 - 2q)} \times 100$$

Claus et al. (1996) applied this formula to estimate age-specific $AR_{e,i}$ values of ovarian cancer that are due to the susceptibility allele for breast/ovarian cancer using data from a case–control study of breast cancer. Age-specific genetic attributable risks can be calculated by incorporating age-specific risks ($R_{e,i}$) for developing a disease within age-specific categories (I). Carrier probabilities, or the probability of genotype e ($e = Aa$ or aa) were calculated via likelihood equations conditional on the breast cancer status of the mother and sisters as well as the ages at onset of any affected relatives. Age-specific risks ($R_{e,i}$) for ovarian cancer in mothers and sisters of women who developed breast cancer were generated using estimates derived from the product–limit method (Kaplan and Meier, 1958). The allele frequency for the abnormal gene was determined from earlier analyses using maximum likelihood methods. The proportions of ovarian cases due to the susceptibility gene ($P_{e,i}$) were determined by comparing the risks in first-degree relatives of breast cancer patients who were likely to be carriers (genotype $e = Aa$) with the risk in first-degree relatives of breast cancer patients who are unlikely to be carriers (genotype $e = aa$) within age categories. Results demonstrated that the proportion of ovarian cancers predicted to be due to the susceptibility allele ranged from 14% among patients diagnosed in their thirties to 7% among those diagnosed in their fifties.

For gene-environment interactions, Khoury and Wagener (1995) defined the AR as the population-attributable fraction (PAF), a value associated with the joint effect of genetic and environmental factors,

$$\text{PAF(ge)} = \frac{\text{ge}(R'_{\text{ge}} - 1)}{1 + \text{ge}(R'_{\text{ge}} - 1)}$$

where ge is the prevalence of those exposed to both the genotype and the environmental factor and R'_{ge} is the apparent relative risk of disease for exposed individuals, those with both the environmental risk factor and the susceptibility genotype, compared to everyone else.

For a given R'_{ge}, the PAF(ge) may vary according to the type of interaction (see above: Gene–Environment Interaction). Suppose, for example, that disease risk increases only in the presence of the susceptibility genotype and the risk factor, such as for phenylketonuria, which is dependent on both the presence of phenylalanine hydroxylase enzyme deficiency and phenylalanine in the diet (Scriver et al., 1989). In this case the PAF(ge) will be less than or equal to the PAF when disease risk is increased by genotype in the absence of the risk factor but not by the risk factor alone. The synergistic relationship between head trauma and *APOE*E4* allele for Alzheimer disease (AD) is illustrative. In this example, the risk of AD is increased in the presence of *APOE*E4* alone but not in the presence of head trauma alone (Mayeux et al., 1995). The PAF also increases as the prevalence of the susceptibility genotype increases (Khoury and Wagener, 1995).

Epidemiologic Analyses of Complex Genetic Interactions

Defining complex genetic interactions in the context of a case–control study is an important strategy for determining whether a genetic factor confers disease susceptibility. In contrast to the family studies, this approach does not incorporate the inheritance pattern of the trait. The analytic strategy is the 2 × 2 contingency table approach and logistic regression models, which allow for simultaneous consideration of other risk factors (see below). In the context of genetic epidemiology of complex traits, the genetic risk factor under study can be family history, a putative susceptibility allele, or a biological marker for genetic susceptibility. Associations, determined by the odds ratio (OR), can be calculated for disease risk and family history alone, for the risk factor alone, and for both family history and the risk factor, thus determining the relationships between susceptibility allele, environmental factors (or other alleles), and disease risk (Ottman, 1990). Differences in these relationships may be related to different models of interaction between genes and the environment and can help discriminate between various etiologic pathways (Ottman, 1990). Such models of genetic interaction provide the ability to discriminate between models where there is an independent effect of the risk factor on disease, regardless of the presence of other susceptibility factors. Not considering the relationship of a genetic factor of interest in the presence of other genes or an environmental risk factor could lead to considerable dilution of the odds ratio (Khoury et al., 1988; Hodge, 1994; Khoury and Beaty, 1994; Newman et al., 1995; Slattery et al., 1995).

Assessing Confounding Bias. Confounding occurs when an observed association is due to the mixing of effects between the exposure (e.g., family history), the disease, and a third factor (or confounder) that is associated with the exposure and independently affects the risk of developing the disease (Hennekens and Buring, 1987).

Confounding bias in epidemiologic studies can be addressed through various statistical analytic strategies to control for such bias. For case–control studies, stratification and logistic regression are the most common methods for assessing an association between the disease and a risk factor while controlling for confounding factors. Survival analyses are often used to analyze data in cohort studies.

Stratified analyses can be useful to control for confounding of either genes or environmental factors. Using this approach, the genetic susceptibility–disease association can be estimated within various levels of the confounding variable. For example, in a study of Native Americans and type 2 diabetes mellitus, a strong negative association between the Gm haplotype $Gm^{3;5,13,14}$ from the Gm system of human immunoglobin was found to be confounded by the effect of Caucasian admixture (Knowler et al., 1988). The crude OR for the association between $Gm^{3;5,13,14}$ and diabetes in a sample of 4920 Native Americans was 0.21 (95% CI = 0.14–0.32) and was statistically significant, suggesting that the absence of this haplotype (or absence of a closely linked gene) is a causal factor in this disease. However, $Gm^{3;5,13,14}$ is also a marker for Caucasian admixture, and therefore the degree of Indian heritage is a potential confounder for this association. Stratum-specific odds ratios, according to the degree of Indian heritage, either none, half, or full, revealed that there was no significant relationship between $Gm^{3;5,13,14}$ and diabetes mellitus. The stratum-specific ORs and 95% CI for full, half, and no Indian heritage were 0.63 (95% CI = 0.34–1.18), 0.69 (95% CI = 0.24–1.96), and 0.75 (95% CI = 0.11–5.32), respectively (see Fig. 18.3). Further adjustments for confounding by age resulted in an even weaker association between the marker and disease within these strata. Therefore, the relationship between $Gm^{3;5,13,14}$ and diabetes mellitus was explained by the inverse relationship between the genetic marker and Caucasian admixture (Knowler et al., 1988).

As strata become numerous, and in the presence of a need to control for more than a few variables, stratified analyses become cumbersome, sample sizes decrease, and the results are difficult to describe and summarize. In addition, using a stratified analytic approach that involves the categorization of a continuous variable can introduce a degree of arbitrariness and perhaps result in loss of information. A multivariate approach allows for the simultaneous control of several factors to determine which ones are independent effects in predicting disease.

Multivariate analysis involves construction of a mathematical model to describe the association between a genetic exposure and disease and can incorporate variables, either continuous or categorical, that confound or modify the genetic effect. In many genetic epidemiologic studies where the outcome of interest is a binary variable, such as diseased versus nondiseased, the most appropriate analytic approach is logistic regression analysis (Hennekens and Buring, 1987). In the logistic model, the risk of developing a disease is expressed as a function of independent predictor variables such as genotype, as well as other known risk factors, either genetic or environmental, which could potentially confound or modify the disease–gene association. The dependent disease status variable is defined as the

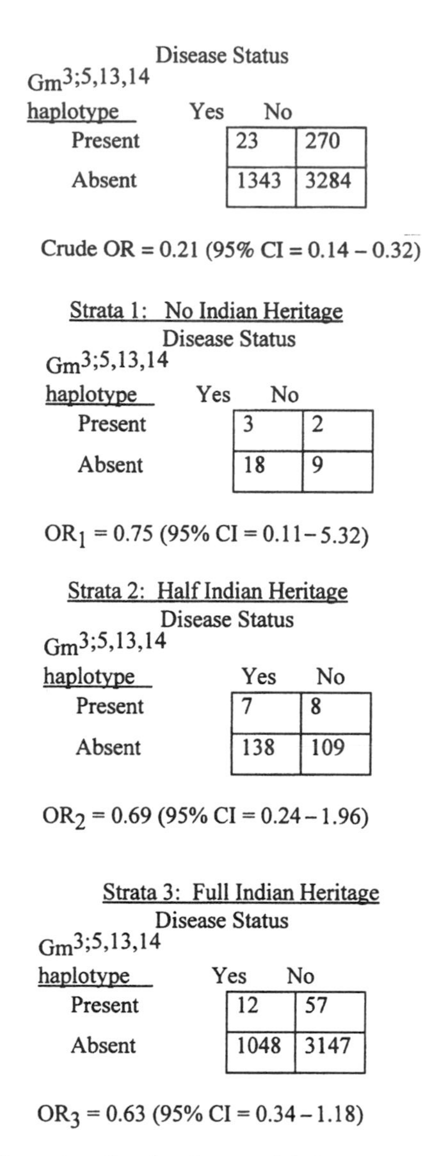

$Gm^{3;5,13,14}$ haplotype	Disease Status: Yes	No
Present	23	270
Absent	1343	3284

Crude OR = 0.21 (95% CI = 0.14 – 0.32)

Strata 1: No Indian Heritage

$Gm^{3;5,13,14}$ haplotype	Disease Status: Yes	No
Present	3	2
Absent	18	9

OR_1 = 0.75 (95% CI = 0.11 – 5.32)

Strata 2: Half Indian Heritage

$Gm^{3;5,13,14}$ haplotype	Disease Status: Yes	No
Present	7	8
Absent	138	109

OR_2 = 0.69 (95% CI = 0.24 – 1.96)

Strata 3: Full Indian Heritage

$Gm^{3;5,13,14}$ haplotype	Disease Status: Yes	No
Present	12	57
Absent	1048	3147

OR_3 = 0.63 (95% CI = 0.34 – 1.18)

Figure 18.3. *Example of confounding by degree of Indian heritage: stratified analysis of $Gm^{3;5,13,14}$ haplotype and diabetes mellitus. (From Knowler et al., 1988.)*

natural logarithm (ln) of the odds of disease, or logit. If Y is the probability of disease, then $Y/(1 - Y)$ is the odds of developing the disease and log odds of disease is $\ln[Y/(1 - Y)]$. The log odds of disease is expressed as a simple linear function of the independent predictor variables using the following formula:

$$\ln \frac{Y}{1-Y} = a + b_1x_1 + b_2x_2 + \cdots + b_n\mathrm{x}_\mathrm{n}$$

The coefficients (b_i) obtained through logistic regression are the magnitude of the increase or decrease of the log odds produced by one unit change in the value of the independent variable, which is indicative of the effect of an individual factor on the log odds of the disease outcome with all remaining variables held constant. By taking the antilogarithm, one can directly convert the logistic regression coefficients to an odds ratio that is an estimate of the relative risk adjusted for confounding (Hennekens and Buring, 1987):

$$\mathrm{OR}(x_i) = e^{bi}$$

Confidence limits can be calculated from the coefficient and its standard error (SE):

$$95\%\ \mathrm{CI} = \exp[bi \pm 1.96\mathrm{SE}(bi)]$$

The use of logistic regression in genetic epidemiology is illustrated in an example of an analysis of the association between family history and the risk of ovarian cancer (Schildkraut and Thompson, 1988). In this case–control analysis the crude odds ratio (the odds ratio not adjusted for confounders) was 3.6 (95% CI = 1.8–7.1). Adjustment for additional risk factors in the probands and the relatives in a logistic regression model can determine whether there is excess disease risk associated with family history (or genetic susceptibility) among cases compared to controls. In this example, family size and total months pregnant (an important protective factor in the development of ovarian cancer) were incorporated into a logistic regression analysis. The logistic model, adjusting for these factors, is depicted by the following equation:

$$\ln\left(\frac{Y}{1-Y}\right) = a + b_1X_1 + b_2X_2 + b_3X_3 + \cdots + b_nX_\mathrm{n}$$

where Y = probability of ovarian cancer

X_1 = family history of ovarian cancer in a first-degree relative (1 = at least one relative, 0 = none)

X_2 = number of first-degree relatives at risk for ovarian cancer

X_3 = total months pregnant

X_n = other independent variables (e.g., age)

$b_1, \ldots, b_n$ = respective coefficients for the dependent variables

This model assumes that there is no effect modification. The coefficients, standard errors, and corresponding ORs for the variables of interest in this model are given in Table 18.2. The coefficient associated with family history, controlling for the simultaneous effects of these variables, was 1.2809, with a standard error (SE) of 0.3537.

Table 18.2 Logistic Regression Analysis of the Association Between Family History and Ovarian Cancer

Variable	Coefficient	Standard Error	OR (95% CI)
Primary family history of ovarian cancer (X_1)	1.2809	0.3537	3.59 (1.79–7.23)
Number of first-degree relatives (X_2)	0.0365	0.0316	1.04 (0.97–1.10)
Months pregnant (6 months) (X_3)	−1.3344	0.0771	0.26 (0.19–0.35)

Thus the relative risk estimate for the family history and ovarian cancer controlling for confounding is $e^{1.2809}$, or 3.6 (1.8–7.2). The lack of a difference between the crude and adjusted odds ratios thus provides no evidence for confounding by other variables. Therefore, women with a family history were found to be at 3.6 times the risk of developing ovarian cancer compared to women without a family history.

Other multivariate analytic methods, including survival analysis techniques, may also be employed to assess familial aggregation. In the study of diseases with variable age of onset, it is also important to account for the age of those at risk for developing the disease. Survival analyses, applied in the analysis of cohort data, account for varying lengths of observation of study subjects. These techniques are particularly appropriate for late-onset disease, where censoring of data (when a family member dies as a result of a competing cause or is lost to follow-up before the disease is diagnosed) is important, since these methods weight the contribution of older family members more than that of younger members.

Age-of-onset distributions of relatives of cases and controls are often examined to help determine recurrence risk patterns and the influence of age and age-specific risk of disease (Schildkraut et al., 1989a, 1989b; Claus et al., 1991; Khoury and Beaty, 1994). The Cox proportional hazards model (1972) is one survival method that takes into account the unequal lengths of time or variable age at onset in family members at risk for developing the disease of interest. The proportional hazards model describes the relation of the independent variable (disease status) to the natural logarithm of the incidence rate of disease rather than to the odds of disease (Hennekens and Buring, 1987). The following equation describe the proportional hazards model:

$$\ln[\text{incidence rate } (t)] = a(t) + b_1X_1 + \cdots + b_nX_n$$

where $a(t)$ = baseline incidence rate expressed as a function of age (or time)

$X_1, \ldots, X_n$ = independent variables, such as family history, status of the probands (affected vs. unaffected) genotype, or other risk factors for the disease under study

$b_1, \ldots, b_n$ = respective coefficients for each independent variable

Proportional hazards analysis, as logistic regression, allows for simultaneous inclusion of other factors in the model to assess the independent effects and control for confounding and test for interaction. The coefficients from the proportional hazards model are estimates of the incidence density ratio and are analogous to the coefficients from the logistic regression model.

One example of the use of survival analyses is in a study investigating the genetic relationship between ovarian, breast, and endometrial cancers (Schildkraut et al., 1989a). The data analysis was based on information about cancer in relatives of cancer cases and controls from large population-based, case–control study. By employing the proportional hazards model, age-adjusted relative risks (RRs) and 95% confidence intervals for breast, ovarian, and endometrial cancers among mothers and sisters of women diagnosed with cancer at these three sites (or cases) were calculated. The reference group for all comparisons consisted of the mothers and sisters of population controls.

Using the proportional hazards model, the age-specific incidence rates for relatives of cases and controls were compared under the assumption that the hazard ratio for case and control relatives is constant throughout the risk period. In this example, $a(t)$ is the incidence rate expressed as a function of age or age at death (for those without cancer)/age at diagnosis of cancer (for diagnosed with cancer) in mothers and sisters. At each point in time, women who developed cancer of the breast, ovary, or endometrium were compared to a risk set comprising all women who survived at least as long and did not develop cancer. The variable (X_1) included in this model was the case–control status of the proband. The result demonstrated there were significantly elevated age-adjusted RRs for ovarian cancer (RR = 2.8; 95% CI = 1.6–4.9) and breast cancer (RR = 1.6; 95% CI = 1.1–2.1) among relatives of ovarian cancer probands and for breast cancer (RR = 2.1, 95% CI = 1.7–2.5) and ovarian cancer (RR = 1.7; 95% CI = 1.0–2.0) among relatives of breast cancer probands. Relatives of endometrial cancer probands had an elevated RR for endometrial cancer only (RR = 2.7; 95% CI = 1.6–4.8). The relative risks that were derived from proportional hazards models were then used in an analysis in which a multivariate polygenic threshold model was chosen to estimate heritability and to determine genetic relationships by estimating the genetic correlation among traits. The heritability estimates and genetic correlations derived from the multivariate polygenic threshold model were consistent with the results of the proportional hazards analysis.

Assessing Interaction. For complex genetic traits, one or more additional factors, M, may be found to modify the effect of the genetic determinant of a specific trait. Effect modification is referred to statistically as an interaction. Case–control methods, such as stratification and logistic regression, are used to assess possible interaction. Examining strata-specific odds ratios allows for a preliminary investigation of possible interaction between a genetic or environmental exposure and the exposure–disease association under study.

One indication that a factor is an effect modifier would be notable differences in the association between the genetic exposure and the risk of the disease in the presence versus the absence of the effect modifier or when stratum-specific estimates of the relative risk vary sufficiently. For example, if M modifies the effect of genotype

AA in the risk of developing *D,* an example of effect modification would be an OR of 15.5 observed among those with the *AA* genotype who had been exposed to *M* while an OR of 1.5 appeared among those with the *AA* genotype who had not been exposed to *M.* These stratum-specific ORs are indicative of multiplicativity of effects. However, if the strata-specific ORs are the same across all levels of *M* (which is what is anticipated if the joint effects are multiplicative), a different causal mechanism, possibly including additivity of effects, cannot be ruled out (Thompson, 1994). To assess interaction statistically, one must decide on a definition of interaction, either as multiplicative or additive. It has been argued that additivity is a better definition of independent effects (Rothman, 1986b). Both can be assessed via logistic regression (Thompson, 1991, 1994).

Examples of gene–environment interaction in case–control studies demonstrate the importance of taking the assessment of interactions into account in the design of a case–control study. The synergistic relationship between the apolipoprotein E4 allele, head trauma, and the risk of Alzheimer disease serves as an example of gene–environment interaction. It is hypothesized that head injury may contribute to the pathogenesis of AD by increasing the deposition of β-amyloid in the cortex (Mayeux et al., 1995). Stratum-specific estimates of the relative risk for AD are illustrative of a possible synergistic relationship between head injury and *APOE*E4* and the risk of AD (Mayeux et al., 1995). Using persons with neither head injury nor an *APOE*E4* Allele as the reference group, the odds ratio for a history of head injury alone was 1.0 while the odds ratio for at least one *APOE*E4* allele alone was 2.0. However, the joint effect of these two factors produced an OR of 10.5 (see Table 18.3), which exceeds what would be expected from independent effects of both risk factors based on the multiplicative scale (i.e., $1.0 \times 2.0 = 2.0$) (Mayeux et al., 1995).

Logistic regression analysis can also be used to assess interaction. Because stratified analyses suggested possible effect modification, the data from the familial ovarian cancer case–control study described above were analyzed further. A product term, X_1X_4, representing a combination of the genetic exposure (family history) and the potential effect modifier (total months of pregnancy) was added into the logistic regression model. The coefficient for the product term allows for the assessment of

Table 18.3 Case–Control Data of the Effect of the Synergistic Effect Head Injury and *APOE*E4* on the Risk of Alzheimer Disease

*APOE*E4* Allele	History of Head Injury	Number of Cases	Number of Controls	OR
Absent	No	52	78	1.0 (referent)
Absent	Yes	6	9	1.0
Absent	No	52	78	1.0 (referent)
Present	No	48	35	2.0
Absent	No	52	78	1.0 (referent)
Present	Yes	7	1	10.5

Source: Mayeux et al. (1995).

whether the magnitude of the contribution of family history of the disease varies according to the total months of pregnancy. The result of $e^{0.1823}$ or 1.2 (95% CI = 0.9–1.6) is interpreted as a ratio of the odds ratios for the effect of family history in subgroups that differed by one pregnancy. That is, the magnitude of the odds ratios for family history of ovarian cancer was estimated to increase by 20% with each pregnancy that a woman had had. However, this effect was not statistically significant.

Implication of Disease Susceptibility/Alleles for Complex Traits in the Population

With a final goal of prevention, advances in molecular genetics will make it possible to target intervention to individuals with risk factors who carry disease susceptibility alleles (Khoury and Wagener, 1995; Mayeux and Schupf, 1995). For complex traits, genetic factors that have been found to be associated with a disease may be part of a causal pathway and a necessary but not sufficient cause of the disease (Khoury and Wagener, 1995; Mayeux and Schupf, 1995). Sensitivity, specificity, and predictive value of a positive genetic test are dependent on the proportion of risk that is attributable to the allele being tested (Mayeux and Schupf, 1995). Therefore, for complex diseases, assessment of the ability for a genetic test to predict disease risk is not straightforward.

An epidemiologic approach can be taken to assess the impact of genetic testing (Khoury and Wagener, 1995). If the genetic test is positive, what is the chance that the susceptibility allele is truly present, and what are the chances of disease onset? If the test result is negative, what is the chance there is truly no susceptibility allele present and what is the chance that the disease will develop through an alternate etiologic pathway? For genetic susceptibility screening, the validity of a screening test is measured by its ability to correctly classify persons who have the susceptibility allele. In addition, further consideration is necessary to determine the chance of developing the disease given a positive genetic test as well as the chance of disease when the genetic test is negative. For example, not everyone with *APOE-4* allele develop AD, while some without the *APOE-4* allele do (Mayeux and Schupf, 1995). Similarly for breast cancer, women who test positive for a mutation in *BRCA1* are estimated to have an 85% risk of developing breast cancer in their lifetime (Easton et al., 1995). However, women who do not test positive are still at risk for developing breast cancer, although to a lesser degree, with a lifetime risk of 1 in 8, or 12%.

The value of a genetic marker for predicting a specific disease entity can be characterized by measures used in screening for undiagnosed illness in asymptomatic subjects. *Sensitivity* and *specificity* are two measures that address the validity of screening tests. In the genetic context, sensitivity is defined as the probability of testing positive for the genetic test if the disease is truly present. As the sensitivity of the test increases, the number of persons with the disease who are missed by being classified as test-negative (false negative) decreases. Specificity is defined as the probability of testing negative for the genetic test if the disease is truly absent. A highly specific test will rarely be positive in the absence of the disease and will therefore result in a lower proportion of persons without the disease who are incorrectly classified as test-positive (false positive).

In the evaluation of a screening test it is necessary to answer questions concerning the test's ability to accurately predict disease. Two measures that directly address the estimation of probability of disease are the *positive predictive value* (PV+) and the *negative predictive value* (PV–). The PV+ is an estimate of the accuracy of the test in predicting the presence of disease, and the PV– is an estimate of the accuracy of the test in predicting the absence of disease. The PV+ is a useful measure because it can be used for both clinical and preventive risk communication and counseling (Khoury and Wagener, 1995). The predictive values are a function of the sensitivity, specificity, and prevalence of disease. For complex genetic traits, the predictive values are also strongly influenced by the age-specific penetrance of the gene. The *accuracy* of a test is a measure of the percent of all results that are true results, whether positive or negative or the total correct test results. Figure 18.4 summarizes the relationship between results of a genetic screening test and the presence of a susceptibility allele.

The population in which the genetic test is applied influences the utility of the test results. Although the sensitivity and specificity of the test remain constant, the PV+ and PV– will vary according to the prevalence of the gene in the population being screened and the age distribution of that population. For a rare gene, the greater the prevalence of the gene in the population being screened, the greater the

	Truth (Presence of Disease)	
Genetic Test	**Disease Present**	**Disease Absent**
Positive	TP	FP
Negative	FN	TN

Figure 18.4. *Evaluation of genetic susceptibility screening:*

Sensitivity = TP(TP + FN) × 100

Specificity = TN(TN Apl FP) × 100

PV^+ (over a lifetime, assuming 100% penetrance) = TP(TP + FP) × 100

PV^- (over a lifetime, assuming 100% penetrance) = TN(TN + FN) × 100

Accuracy = TPTN/(TP + TN + FP + FN) × 100

where TP (true positive) = the number of subjects who develop the disease and who are correctly classified by the genetic test

FP (false negative) = number of disease-free subjects who are incorrectly predicted to develop the disease by the genetic test

TN (true negative) = number of disease-free subjects who are correctly classified by the genetic test

FN (false negative) = number of subjects who develop the disease who are incorrectly classified as disease-free by the genetic test

PV+. When one is screening for a rare gene, a high PV– is expected because most of the individuals screened will not have the mutation. This underscores the importance of screening appropriate at-risk populations for a particular disease, especially since the clinical implications of an abnormal test result are not benign.

For complex genetic diseases, the impact of genetic interaction can be taken into account when the predictive value of genetic testing is calculated. The following equation (Khoury and Wagener, 1995) for the PV+ includes the effect of a gene–environment interaction:

$$\text{PV+} = \frac{dR'_{ge}}{1 + \text{ge}(R'_{ge} - 1)}$$

where d is the lifetime risk in the population and ge and R'_{ge} are as defined for the PAF (ge). The PV+ decreases with increasing prevalence of the susceptibility genotype and increases with increasing gene–environment interaction and increasing RR associated with the environmental risk factor (Khoury and Wagener, 1995). The prevalence of the disease also impacts the PV+. For example, when the disease is very rare, 1/1000, the PV+ for the risk factor–genotype interaction is generally low ($\leqslant 1\%$) compared to some cancers with a lifetime risk of 10%, where there is a much higher PV+, for the same values of the relative risk (Khoury and Wagener, 1995).

With the ever expanding discovery of susceptibility alleles, many different questions will have to be addressed, including social, ethical, and legal concerns (Lancaster et al., 1996; see also Chapter 6). Because prevention and treatment options among susceptible individuals may not be available or straightforward, many questions have been raised concerning the value of tests based on susceptibility alleles. Issues concerning prenatal diagnosis of diseases with an age at onset occurring between 30 and 50 years are very challenging, and universal guidelines may be difficult to formulate (Lancaster et al., 1996; see also Chapter 6).

Integrating Genetic and Epidemiologic Analytic Methods

Research in familial breast cancer and Alzheimer disease provides examples of how the application of both epidemiologic analyses and family studies led to the discovery of the genetic susceptibility alleles. The subsections that follow illustrate how our understanding of the genetic epidemiology of these heterogeneous diseases has evolved.

Breast Cancer Data from case–control studies have shown a two- to threefold increase in the risk of breast cancer associated with a family history of breast cancer in a mother or sister (Ottman et al., 1986; Claus et al., 1990; Kelsey et al., 1993). Patterns of recurrence risk in mothers and sisters, observed in proportional hazards analyses, demonstrated that the estimated age-specific rates for breast cancer in relatives of cases and controls were consistent with a genetic model (Claus et al., 1990, 1991).

Applying proportional hazards analyses to the distribution of age at diagnosis of breast cancer in first-degree relatives of breast cancer cases and controls led to the detection of age- and genotype-specific penetrances and demonstrated the effect of genotype to be a function of age (Claus et al., 1991). Formal genetic analyses, incorporating the age-associated risks, provided strong evidence for the association of one or more rare autosomal dominant inherited genes with an increased susceptibility to breast cancer (Newman et al., 1988; Claus et al., 1991). Using segregation analysis, Claus et al. (1991) found that the genetic model that best fits the breast cancer recurrence rates in first-degree relatives of the cases and controls specified a rare (allele frequency = 0.0033) allele that is dominantly inherited along with a high frequency of phenocopies or nonfamilial cases. This result was in agreement with other studies (Bishop et al., 1988; Newman et al., 1988).

In 1990 the *BRCA1* gene was localized to chromosome 17q21 by means of genetic linkage analysis in families with a high incidence of early-onset breast cancer (Hall et al., 1990). The same locus was shown to be linked in families with breast and ovarian cancer (Narod et al., 1991). The *BRCA1* gene for breast cancer (Futreal et al., 1994; Miki et al., 1994) was isolated by positional cloning in 1994. However, evidence presented from a polygenic model suggested that there may be more than one locus underlying breast cancer and that shared allele(s) and unique allele(s) may exist for breast and ovarian cancer (Schildkraut et al., 1989a). In addition, linkage studies determined that not all breast cancer families were linked to *BRCA1* (Hall et al., 1990). Subsequently, a second susceptibility gene, *BRCA2,* was found to be linked to chromosome 13q (Wooster et al., 1994); shortly thereafter, it was identified (Wooster et al., 1995). Familial breast cancer is therefore genetically heterogeneous, and the possibility of a third susceptibility gene appears likely (Phelan et al., 1996a).

The cumulative lifetime cancer risks of *BRCA1* mutation carriers, the proportion of gene carriers that become newly diagnosed with the disease by age 70, are approximately 85% for breast cancer and 63% for ovarian cancer, with an attributable risk of up to 5% for all breast and ovarian cancers (Easton et al., 1995; Shattuck-Eidens et al., 1995). To better understand factors effecting penetrance of *BRCA1,* further investigation of gene–gene and gene–environment interactions are under way. Proportional hazards analysis from a recent study of ovarian cancer risk in BRCA1 carriers suggests that the *HRAS1* variable number of tandem repeats (VNTR) polymorphism is a possible genetic modifier of cancer penetrance (Phelan et al., 1996b). Individuals with rare alleles of this VNTR appear to have an increased risk for developing ovarian cancer. A second study examining environmental risk modifiers in carriers of BRCA1 mutations suggested that ovarian cancer risk increases significantly with increasing parity, with a 40% increase in relative risk for up to five births (Narod et al., 1995).

Our understanding of familial breast cancer is rapidly evolving, with the possibility of the identification of additional susceptibility alleles on the horizon. This necessitates that population-based case–control studies, cohort studies, and familial cancer registries be undertaken to provide appropriate data sets for more fully understanding this complex genetic disease.

Future genetic epidemiologic research in this area will, in part, be directed toward understanding factors that modify the penetrance of *BRCA1* and *BRCA2,* with the possibility that specific types of mutation, other genes, and environmental factors may influence gene expression. Further delineation of the constellation of cancer sites associated with *BRCA1* and *BRCA2* will be addressed. Although genetic susceptibility does not appear to be important in late onset breast cancer, the possibility that genetics plays a role cannot be ruled out.

With the discovery of susceptibility genes such as *BRCA1* and *BRCA2,* new challenges in cancer screening arise with the need for appropriate screening recommendations in high risk subgroups. Not only are those who inherit a mutation for cancer susceptibility at higher risk for developing cancer, but often the age at onset among such individuals is younger than the age at onset in the general population. The susceptibility alleles for breast cancer illustrate a perplexing situation inasmuch as recommendations for screening with mammography do not address the early age at onset of this disease. In addition, Lancaster et al. (1996) recently addressed the topic of prenatal testing for mutations in *BRCA1* among gene carriers and discussed the need for the individual clinician and the patient to work together in formulating decisions. This example underscores the difficult issues associated with genetic screening.

Alzheimer Disease Similar to breast cancer, familial aggregation of AD, observed in epidemiologic studies, has been explained by genetic factors. Alzheimer disease is also genetically heterogeneous, with mutations in several different genes resulting in the same clinical and pathological phenotype (Pericak-Vance and Haines, 1995). Epidemiologic studies have suggested that various environmental factors may also play a role in the etiology of AD including head injury, smoking, exposure to heavy metals, and education level (Pericak-Vance and Haines, 1995).

The incidence of AD increases with age leading to a prevalence of about 10% after the age of 80 (Schoenberg et al., 1987). Several case–control studies have provided evidence that family history of dementia is significantly associated with a higher risk of developing AD (Breteler et al., 1992). A reanalysis of case–control study data demonstrated an association between family history of dementia and both early-onset and late-onset AD (van Duijn et al., 1991), with an overall RR of 3.5 (95% CI = 2.6–4.6). Similar to breast cancer, the RR was greater for early-onset AD and decreased with increasing age at onset. In contrast to breast cancer, the increased risk was still evident among cases diagnosed after the age of 80 years (van Duijn et al., 1991).

As with breast cancer, our understanding of the genetic etiology of AD has been aided through the recognition of the importance of age at onset, and survival analysis techniques have been useful in studies estimating the risk of AD among relatives. This is evident in one study that demonstrated differences in the distribution of risk of AD in first-degree relatives with early- versus late-onset disease, suggesting different causes that were possibly genetic (Breitner et al., 1988). From results obtained using the Kaplan–Meier life table method, it became apparent that there was a mixture of two distributions of age-specific risk. In contrast to breast cancer, the role of genetic susceptibility was apparent in both early- and late-onset AD.

Consistent with the findings of epidemiologic studies, genetic linkage analyses have detected differences in the genetic etiology of AD related to age at onset. Although one form of early-onset AD was initially linked to chromosome 21 (St. George-Hyslop et al., 1987), subsequent studies failed to confirm this finding, suggesting genetic heterogeneity in AD (Pericak-Vance and Haines, 1995). This suggestion was also supported by statistical evidence for heterogeneity (St. George Hyslop et al., 1990). A subset of early-onset families were later shown to have a mutation in the amyloid protein (*APP*) gene on chromosome 21 (Goate et al., 1991). Given the evidence for nonallelic heterogeneity, the search for additional loci was begun. Using a genomic screen, a locus on chromosome 14 was found to be linked to the majority of early-onset families not segregating APP mutations (Schellenberg et al., 1992; St. George-Hyslop et al., 1992). Although the chromosome 14 linked families are the same pathologically and clinically as to the *APP* families, they tend to have an earlier age at onset. Still, some early-onset AD families were not found to be linked to either chromosome 21 or 14, providing evidence for additional susceptibility genes for early-onset AD (Pericak-Vance and Haines, 1995). In fact, mutations in the presenilin II gene on chromosome 1 have been found in some Volga German families (Levy-Lehad et al., 1995; Rogaev et al., 1995).

The task of distilling the genetics of late-onset AD, the most common form of the disease, was complicated by various factors, including competing causes of death and diagnostic phenocopies (Pericak-Vance and Haines, 1995). However, with the ascertainment of a number of multiplex families, a genomic screen revealed evidence for linkage of AD to a region on chromosome 19 (Pericak-Vance et al., 1991). Both classical linkage methods (Chapter 12) and affected pedigree member (APM) analysis (Chapter 14) (where it is not necessary to assume the mode of inheritance) were utilized in the localization of the chromosomal region. Subsequently, by combining the results from the linkage studies with what was known biologically, the apolipoprotein E gene (*APOE*), which encodes a plasma lipoprotein localized on chromosome 19, was identified as harboring a susceptibility allele for late-onset AD (Saunders et al., 1993; Strittmatter et al., 1993). Specifically, the *APOE*E4* allele was found to be associated with an increased risk of both familial late-onset and sporadic AD (Saunders et al., 1993; Strittmatter et al., 1993). Additionally, the *APOE*E2* allele has been found to have a protective effect (Chartier-Harlin et al., 1994; Corder et al., 1994).

Further insight into the multifactorial etiology of late-onset AD was revealed in studies examining possible gene–environment interaction. As described earlier in this chapter (and in Table 18.3), a study of head trauma and AD has demonstrated evidence for a synergistic relationship between head trauma and the *APOE*E4* allele (Mayeux et al., 1995). Using a case–control design, Mayeux et al. (1995) determined the genotype of *APOE* in 113 AD patients and 123 healthy controls. Logistic regression analysis revealed a 10-fold increase in the risk of AD associated with both *APOE*E4* and a history of head trauma, compared to a twofold increase in risk with *APOE*E4* alone. Head injury alone (e.g., in the absence of *APOE*E4*) was not found to be associated with an increased risk of AD. The authors hypothesized that head injury may contribute to an increased risk of AD by causing an increase in cerebral amyloid protein (βAPP) deposition, leading to neuritic plaques and neuronal death, and exacerbating the effect the *APOE*E4* allele, which is thought to be related to cerebral βAPP deposition.

As with other complex genetic traits there is an incomplete correlation between the *APOE*E4* genotype and AD phenotype. *APOE*E4* accounts for approximately 30–50% of AD. Although *APOE*E4* is clearly a risk factor for AD, the combination of other genetic and environmental factors modify its effect on AD susceptibility. Therefore, while *APOE*E4* may contribute to the risk of developing AD, risk prediction is not straightforward. As with other complex genetic traits, genetic testing for *APOE*E4* and susceptibility for AD raises scientific, social, legal, and ethical questions (Mayeux et al., 1995; see also Chapter 6).

Genetic research in AD has provided evidence for genetic heterogeneity in the risk of AD at both the locus level and the gene level. In comparison, although there are clearly multiple susceptibility genes for breast cancer, variations in risk at the gene level have yet to be explored thoroughly. Unlike the association between *APOE* and the risk of AD, where some understanding of the biological function of the susceptibility gene, in part, led to its discovery, knowledge of the function of the *BRCA1* and *BRCA2* genes was not a factor in their identification. To date, the biological functions of *BRCA1* and *BRCA2* remain unknown. Unlike AD, a relationship with sporadic breast cancer and *BRCA1* and *BRCA2* mutations has not been uncovered.

CONCLUSION

Combining epidemiologic and genetic approaches can unravel the complexities of gene–gene and gene–environment interactions. Results of epidemiologic studies have helped define syndromes and have aided in the identification of phenotypes associated with specific genetic syndromes.

Genetic epidemiologic research in breast cancer and Alzheimer disease illustrates that it is important that research to complex multifactorial traits address the characterization of how genes and the environment interact. Discovery of major genetic effects allows exploration of these avenues of research and can further our understanding of complex genetic traits. Advances in molecular technology promise to provide powerful tools for identifying genetic and environmental factors in the etiology of complex diseases. Finally, further study is needed to establish efficacious screening protocols for those who are genetically predisposed.

REFERENCES

Bell DA, Stephens EA, Castranio T, et al (1995): Polyadenylation–polymorphism in the acetyl transferase 1 gene (*NAT1*) increases risk of colorectal cancer. *Cancer Res* 55:3537–3542.

Bishop DT, Cannon-Albright L, McLellan T, et al (1988): Segregation and linkage analysis of nine Utah breast cancer pedigrees. *Genet Epidemiol* 5:151–169.

Bonney GE (1984): On the statistical determination of major gene mechanisms in continuous human traits: Regressive models. *Am J Med Genet* 18:731–749.

Bonney GE, Lathrop GM, Lalouel JM (1988): Combined linkage and segregation analysis using regressive models. *Am J Hum Genet* 43:29–37.

Bonney GE, Dunston GM, Wilson J (1989): Regressive logistic model for ordered and unordered polychotomous traits: Application to affective disorders. *Genet Epidemiol* 6:211–215.

Breitner CS, Murphy EA, Silverman JM, et al (1988): Age-dependent expression of familial risk in Alzheimer's disease. *Am J Epidemiol* 128:536–548.

Breteler MMB, Claus JJ, van Duijn CM, et al (1992): Epidemiology of Alzheimer's disease. *Epidemiol Rev* 14:59–82.

Chartier-Harlin MC, Parfitt M, Legrain S, et al (1994): Apolipoprotein E epsilon 4 allele as a major risk factor for sporadic early and late-onset forms of Alzheimer's disease: Analysis of 19q13 chromosomal region. *Hum Mol Genet* 3:569–574.

Claus EB, Risch N, Thompson WD (1990): Age of onset as an indicator of familial risk of breast cancer. *Am J Epidemiol* 131:961–972.

Claus EB, Risch N, Thompson WD (1991): Genetic analysis of breast cancer in the Cancer and Steroid Hormone study. *Am J Hum Genet* 48:232–242.

Claus EB, Schildkraut JM, Thompson WD, Risch NJ (1996): The genetic attributable risk of breast and ovarian cancer. *Cancer* 77:2318–2324.

Cordell HJ, Todd JA (1995): Multifactorial inheritance in type 1 diabetes. *Trends Genet* 11:499–504.

Corder EH, Saunders AM, Strittmatter WJ, Schmechel DE, Gaskell PC, Small GW, Roses AD, Haines JL, Pericak-Vance MA (1993): Apolipoprotein E4 gene dose and the risk of Alzheimer disease in late onset families. *Science* 261:921–923.

Corder EH, Saunders AM, Risch NJ, et al (1994): Protective effect of apolipoprotein E type 2 allele for late onset Alzheimer's disease. *Nat Genet* 7:190–184.

Cox DR (1972): Regression models and lifetables (with discussion). *J R Stat Soc* B105:488–495.

Curtis D, Sham PC, Vallada HP (1995): Genetic analysis of complex disease. *Nat Genet* 9:13.

Dahlquist G (1994): Non-genetic risk determinants of type I diabetes (review). *Diabetes Metab* 20:251–257.

Davis S, Schroeder M, Goldin LR, Weeks DE (1996): Nonparametric simulation-based statistics for detecting linkage in general pedigrees. *Am J Hum Genet* 58:867–880.

Demenais F, Lathrop M (1993): Use of the regressive models in linkage analysis of quantitative traits. *Genet Epidemiol* 10:587–592.

Demenais FM, Laing AE, Bonney GE (1992): Numerical comparisons of two formulations of logistic regressive models with the mixed model in segregation analysis of discrete traits. *Genet Epidemiol* 9:419–435.

Easton DF, Ford D, Bishop DT, et al (1995): Breast and ovarian cancer incidence in *BRCA1*-mutation carriers. *Am J Hum Genet* 56:265–271.

Faden R, Beaucharp T (1986): *A History and Theory of Informed Consent.* Oxford: Oxford University Press.

Futreal PA, Liu Q, Shattuck-Eidens D, et al (1994): *BRCA1* mutations in primary breast and ovarian carcinomas. *Science* 266:120–122.

Goate AM, Chartier-Harlin MC, Mullan M, et al (1991): Segregation analysis of a missense mutation in the amyloid precursor protein gene with familial Alzheimer's disease. *Nature* 349:704–706.

Hall JM, Lee MK, Newman B, et al (1990): Linkage of early-onset familial breast cancer chromosome 17q12. *Science* 250:1684–1689.

Hennekens CH, Buring JE (1987): Measures of disease frequency and association. In: Mayrent SL, ed. *Epidemiology in Medicine.* Toronto: Little, Brown, pp. 54–98.

Hodge SE (1994): What association analysis can and cannot tell us about the genetics of complex disease. *Am J Med Genet* 54:318–323.

Hwang SJ, Beaty TH, Liang K (1994): Minimum sample size estimation to detect gene–environment interaction in case–control designs. *Am J Epidemiol* 140:1029–1037.

Hyer RN, Julier C, Buckley JD, et al (1991): High-resolution linkage mapping for susceptibility genen in human polygenic disease: Insulin-dependent diabetes mellitus and chromosome 11q. *Am J Hum Genet* 48:243–257.

Kaplan EL, Meier P (1958): Nonparametric estimation from incomplete observations. *J Am Stat Assoc* 53:457–481.

Kelsey JL, Gammon MD, John EM (1993): Reproductive factors and breast cancer. *Epidemiol Rev* 15:36–47.

Khoury MJ, Beaty TH (1994): Applications of the case–control method in genetic epidemiology. *Epidemiol Rev* 16(1):134–150.

Khoury MJ, Wagener DK (1995); Epidemiological evaluation of the use of genetics to improve the predictive value of disease risk factors. *Am J Hum Genet* 56:835–844.

Khoury MJ, Beaty TH, Liang K (1988): Can familial aggregation of disease be explained by familial aggregation of environmental risk factors? *Am J Epidemiol* 127:674–683.

Knowler WC, William RC, Pettitt DJ, Steinberg A (1988): $GM^{3;5,13,14}$ and type 2 diabetes mellitus: An association in American Indians by genetic admixture. *Am J Hum Genet* 43:520–526.

Kruglyak L, Daly MJ, Reeve-Daly M, Lander ES (1996): Parametric and nonparametric linkage analysis: A unified multipoint approach. *Am J Hum Genet* 58:1347–1363.

Lancaster JM, Wiseman RW, Berchuck A (1996): An inevitable dilemma: Prenatal testing for mutations in the *BRCA1* breast–ovarian cancer susceptibility gene. *Obstet Gynecol* 87:306–309.

Lander ES, Schork NJ (1994): Genetic dissection of complex traits. *Science* 265:2037–2048.

Levy-Lehad E, Wasco W, Poorkaj P, et al (1995): Candidate gene for the chromosome 1 familial Alzheimer's disease locus. *Science* 269:970–973.

Mayeux R, Schupf N (1995): Apolipoprotein E and Alzheimer's disease: The implications of progress in molecular medicine. *Am J Public Health* 85(9):1280–1284.

Mayeux R, Ottman R, Maestre G, et al (1995): Synergistic effects of traumatic head injury and apolipoprotein-4 in patients with Alzheimer's disease. *Neurology* 45:555–557.

Miki Y, Swenson J, Shattuck-Eidens DJ, et al (1994): A strong candidate for the breast and ovarian cancer susceptibility gene *BRCA1*. *Science* 266:66–71.

Mullan M, Houlden H, Windelspecht M, et al (1992): A locus for familial early-onset

Alzheimer's disease on the long arm of chromosome 14, proximal to the alpha-1-antichymotrypsin gene. *Nat Genet* 2:340–342.

Narod SA, Feunteun J, Lynch HT, et al (1991): Familial breast-ovarian cancer locus on chromosome 17q21-q23. *Lancet* 338:82–83.

Narod SA, Goldgar D, Cannon-Albright L, et al (1995): Risk modifiers in carriers of *BRCA1* mutations. *Int J Cancer (Predict Oncol)* 64:394–398.

Newman B, Austin MA, lee MK, et al (1988): Inheritance of human breast cancer: Evidence of autosomal dominant transmission of high-risk families. *Proc Natl Acad Sci USA* 85:3044–3048.

Newman B, Moorman PG, Millikan R, et al (1995): The Carolina Breast Cancer Study: Integrating population-based epidemiology and molecular biology. *Breast Cancer Res Treat* 35:51–60

Ottman R (1990): An epidemiologic approach to gene–environment interaction. *Genet Epidemiol* 7:177–185.

Ottman R, Pike MC, King MC, et al (1986): Familial breast cancer in a population-based series. *Am J Epidemiol* 123:15–21.

Pericak-Vance MA, Haines JL (1995): Genetic susceptibility to Alzheimer disease. *Trends Genet* 11:504–508.

Pericak-Vance MA, Bebout JL, Gaskell PC, et al (1991): Linkage studies in familial Alzheimer disease: Evidence for chromosome 19 linkage. *Am J Hum Genet* 48:1034–1050.

Phelan CM, Lancaster JM, Tonin P, et al (1996a): Mutation analysis of the *BRCA2* gene in 49 site-specific breast cancer families. *Nat Genet* 13:120–122.

Phelan CM, Weber BL, Ruttledge MH, et al (1996b); Ovarian cancer risk in *BRCA1* carriers is modified by the *HRAS1* variable number of tandem repeat (VNTR) locus. *Nat Genet* 12:309–311.

Risch N (1990a): Linkage strategies for genetically complex traits. I. Multilocus models. *Am J Hum Genet* 46:222–228.

Risch N (1990b): Linkage strategies for genetically complex traits. II. The power of affected relative pairs. *Am J Hum Genet* 46:229–241.

Rogaev EI, Sherrington R, Rogaeva EA, et al (1995): Familial Alzheimer's disease in kindreds with missense mutations in a gene on chromosome 1 related to the Alzheimer's disease type 3 gene. *Nature* 376:775–778.

Rothman KJ (1986a): Measures of effect. In: *Modern Epidemiology.* Boston: Little, Brown, pp. 23–32.

Rothman KJ (1986b): Interactions between causes. In: *Modern Epidemiology.* Boston: Little, Brown, pp. 321–326.

Saunders AM, Strittmatter WJ, Schmechel D, et al (1993): Association of apolipoprotein E allele epsilon 4 with late-onset familial and sporadic Alzheimer's disease. *Neurology* 43:1467–1472.

Schellenberg GD, Bird ID, Wijsman M, et al (1992): Genetic linkage evidence for familial Alzheimer's disease locus on chromosome 14. *Science* 258:668–671.

Schildkraut JM, Thompson WD (1988): Familial ovarian cancer: A population-based case–control study. *Am J Epidemiol* 128:456–466.

Schildkraut JM, Risch N, Thompson WD (1989a): Evaluating genetic association among ovarian, breast, and endometrial cancer: Evidence for a breast/ovarian relationship. *Am J Hum Genet* 45:521–529.

Schildkraut JM, Myers RH, Cupples AL, Keily DK, Kannel WB (1989b): Coronary risk associated with age and sex of parental heart disease in the Framingham Study. *Am J Cardiol* 64:555–559.

Schoenberg BS, Kokmen E, Okazaki H (1987): Alzheimer's disease and other dementing illnesses in a defined United States population: Incidence rates and clinical features. *Ann Neurol* 22:724–729.

Scriver CR, Kaufman S, Woo SLC (1989): The hyperphenylalaninemias. In: Scrier CR, Beaudet AL, Sly WS, Valle D, eds. *The Metabolic Basis for Inherited Disease,* 6th ed, vol 1. New York: McGraw-Hill Information Services, pp. 495–546.

Shattuck-Eidens D, McClure M, Simard J, et al (1995): A collaborative survey of 80 mutations in the *BRCA1* breast and ovarian cancer susceptibility gene. *JAMA* 273:7:535–541.

Sherrington R, Rogaev EI, Liang Y, Rogaeva EA, Levesque G, Ikeda M, Chi H, Li G, Holman K, Tsuda T, Mar L, Foncin J-F, Bruni AC, Montesi MP, Sorbi S, Rainero I, Pinessi L, Nee L, Chumakov I, Pollen D, Brookes A, Sanseau P, Polinsky RJ, Wasco W, Da Silva HAR, Haines JL, Pericak-Vance MA, Tanzi RE, Roses AD, Fraser PE, Rommens JM, St George-Hyslop PH (1995): Cloning of a gene bearing missense mutations in early-onset familial Alzheimer's disease. *Nature* 375:755–760.

Shields DC, Ratanachalyavong S, McGregor AM, et al (1994): Combined segregation and linkage analysis of graves disease with a thyroid autoantibody diathesis. *Am J Hum Genet* 55:540–554.

Sun F, Stanely LA, Risch A, et al (1995): Xenogenetics in multifactorial disease susceptibility. *Trends Genet* 11:509–512.

Slattery ML, O'Brien E, Mori M (1995): Disease heterogeneity: Does it impact our ability to detect dietary associations with breast cancer? *Nutr Cancer* 24:213–220.

St George-Hyslop PH, Tanzi RE, Polinsky RJ, et al (1987) The genetic defect causing familial Alzheimer's disease maps chromosome 21. *Science* 235:885–890.

St George-Hyslop PH, Haines JL, Farrer LA, et al (1990): Genetic linkage studies suggest Alzheimer's disease is not a single homogeneous disorder. *Nature* 347:194–196.

St George-Hyslop PH, Haines JL, Rogaev EI, Mortilla M, Vaula G, Pericak-Vance M, Foncin JF, Montesi MP, Bruni AC, Sorbi S, Rainero I, Pinessi L, Pollen D, Polinsky RJ, Nee L, Kennedy JL, Macciardi F, Rogaeva EA, Liang Y, Alexandrova N, Lukiw WJ, Schlumpf K, Tanzi R, Tsuda T, Farrer LA, Cantu JM, Duara R, Amaducci L, Bergamini L, Gusella JF, Roses AD, Crapper-Maclachlan DR (1992): Genetic evidence for a novel familial Alzheimer's disease locus on chromosome 14. *Nat Genet* 2:330–334.

Strittmatter WJ, Saunder AM, Schmechel D, et al (1993): Apolipoprotein E: High-avidity binding to beta-amyloid and increased frequency of type 4 allele in late onset familial Alzheimer's disease. *Proc Natl Acad Sci USA* 90:1977–1981.

Thein SL, Sampietro M, Rohde K, et al (1994): Detection of a major gene for heterocellular hereditary persistence of fetal hemoglobin after accounting for genetic modifiers. *Am J Hum Genet* 54:214–228.

Thomson G (1994): Identifying complex disease genes: Progress and paradigms. *Nat Genet* 8:108–110.

Thompson WD (1991): Effect modification and the limits of biological inference from epidemiologic data. *J Clin Epidemiol* 42:221–232.

Thompson WD (1994): Statistical analysis of case–control studies. *Epidemiol Rev* 16(1):33–50.

Van Broeckhoven C, Backhovers H, Cruts M, et al (1992): Mapping of a gene predisposing to early-onset Alzheimer's disease to chromosome 14q24.3. *Nat Genet* 2:335–339.

van Duijn CM, Stijnen T, Hofman A, eds. (1991): Risk factors for Alzheimer's disease: A collaborative re-analysis of case–controls studies. *Int J Epidemiol* 20(suppl 2):S1–73.

Weeks DE, Lange K (1988): The affected-pedigree-member method of linkage analysis. *Am J Hum Genet* 42:315–326.

Weeks DE, Lathrop GM (1995): Polygenic disease: Methods for mapping complex disease traits. *Trends Genet* 11:513–519.

Wertz D, Fanos JH, Reilly PH (1994): Genetic testing for children and adolescents. Who decides? *JAMA* 272:875–881.

Wooster R, Neuhausen S, Morgan J, et al (1994): Localization of breast cancer susceptibility gene, *BRCA1*, on chromosome 13q. *Science* 265:2088–2090.

Wooster R, Bignall G, Lancaster J, et al (1995): Identification for the breast cancer susceptibility gene *BRCA2*. *Nature* 378:789–762.

Glossary

Additive genetic effects The effects of alleles at two different loci are additive when their combined effect is equal to the sum of their individual effects. Additive effects are most easily understood in the context of continuous (quantitative) traits. Consider a disease in which two loci, locus 1 with alleles *A* and *a*, and locus 2 with alleles *B* and *b*, contribute to the phenotype. If each allele represented by a capital letter contributes a score of 2 to the phenotype, and each allele represented by a lowercase letter contributes a score of 3 to the phenotype, then if the alleles at loci 1 and 2 are additive, the resultant phenotypes for the possible genotypes are as follows: *AABB*, 8; *AABb*, 9; *Aabb*, 11; *AaBB*, 9; *AaBb*, 10; *Aabb*, 11; *aaBB*, 10; *aaBb*, 9; *aabb*, 12.

Affected relative pair A general term describing a set of individuals related by blood, each of whom is affected with the trait in question. The most common types of affected relative pair are affected sibling pairs, affected cousin pairs, and affected avuncular pairs.

Allele Alternative forms of a gene or marker due to changes at the level of DNA. For instance, at the ABO blood group locus, there are three alleles *A, B,* and *O.*

Allelic association See Linkage disequilibrium.

Alpha (α) See Type I error.

Anticipation The phenomenon whereby disease severity increases with each passing generation. Since disease severity is often difficult to measure, anticipation is frequently measured in terms of patient-reported age of onset of the disorder. For example, in a disease showing anticipation, a grandchild may have earlier onset than the affected parent, who has earlier onset than the affected grandparent.

Ascertainment The scheme by which individuals are selected, identified, and recruited for participation in a research study.

Association In human genetic linkage studies, association studies frequently involve the comparison of allele frequencies for a marker locus between a disease population and a control population. When statistically significant differences in the frequency of an allele(s) are found between a disease and control population, the disease and allele(s) are said to be in association.

Avuncular relationship The nieces and nephews (or aunts and uncles) of individuals are related in an avuncular fashion.

Bandwidth The range of frequencies on a channel (or connection) between two sites gives the bandwidth. A larger bandwidth correlates to more information being transmitted in a given amount of time.

Beta (β) See Type II error.

CentiMorgan (cM) A measure of genetic distance equivalent to 1/100 of a Morgan. On a global level, a centiMorgan covers roughly one million base pairs of DNA. However this can vary by orders of magnitudes either way. It is usually equivalent to approximately 1% recombination.

Centiray (cR) A measure of physical distance based on the rate of chromosomal breakage induced by radiation exposure in hybrid cell lines. The conversion between length in base pairs and cR is dependent upon the strength of the radiation exposure.

CEPH pedigree Pedigrees collected by the Centre D'Étude du Polymorphisme Humain that are appropriate for reference genetic mapping. These pedigrees are characterized by the availability of a large number of offspring (average 8.5) and usually both sets of paternal and maternal grandparents. The structure of these pedigrees renders them "linkage phase known."

Chromosome A linear structure found in all nucleated cells consisting of both DNA and proteins. Genes are organized on the strands of DNA. The normal human complement of chromosomes is 46 per cell.

Complex trait A trait (most often disease) having a genetic component that is not strictly Mendelian (dominant, recessive, or sex-linked). Complex traits are characterized by the risk to relatives of an affected individual that is greater than the incidence of the trait in the population. Complex traits may involve the interaction of two or more genes to produce a phenotype, or may involve the interaction of genes with environmental factors.

Consanguineous mating A mating is consanguineous when the members of the mating couple are related to each other through a common ancester (e.g., "by blood"). A consanguineous mating is one that occurs between related individuals (e.g., cousins, siblings).

Crossover The physical process that results in the exchange of genetic material between two paired chromosomes during a recombination event.

Database Data organized in a structure designed for easy retrieval, updating, and deleting.

Database engine The proprietary database management system software that provides tools for creating the schema (database design), storing data in the database, and submitting queries.

Database query A statement, generally written in SQL (Structured Query Language), which requests data from the database.

Epistasis Two or more genes interacting with one another in a multiplicative fashion.

Ethernet A data transmission format that runs over existing building wiring (e.g., twisted-pair phone cable). Computers that have an Ethernet interface card installed and the appropriate software (e.g., TCP/IP) can then share data and connect to the Internet.

Exon The part of gene's DNA sequence that codes for proteins.

Expression How a disease gene manifests itself. If a disease gene carrier shows signs of the disease gene, then that gene is "expressed." Expression is often qualified as "variable." For example, the spectrum of clinical signs and symptoms for myotonic dystrophy includes myotonia, frontal balding, characteristic cataracts, narcolepsy, male infertility, and ptosis. One individual who carries the myotonic dystrophy gene may express only cataracts, while another may have only myotonia, narcolepsy, and ptosis. Note that for the term "expression" to be applicable, an individual must carry the disease gene and be penetrant for it.

FTP (file transfer protocol) A standard that allows digital files to be transferred between computers of different architectures and on different networks.

Gamma (γ) The genotype-specific relative risk. Measures the increase in the risk for a trait to an individual of a particular genotype.

Gene An individual unit of heredity. It is a specific instruction that encodes a protein or RNA product. Each gene is located on a specific place (locus) on a specific chromosome.

Genetic model The overall specification of how genes act to influence a given trait. For parametric (model-dependent) linkage analysis, the genetic model must be specified for the analysis. Components of the genetic model include information on whether the trait is autosomal or X-linked, dominant or recessive, as well as the frequency and penetrance of the disease allele, the frequency of phenocopies, the mutation rate, and any marker allele frequency for complex traits. The genetic model also demonstrates how multiple genes interact with each other.

Genotype The observed alleles at a genetic locus for an individual. For autosomal loci, a genotype is composed of two alleles: one is paternally transmitted and the other is maternally transmitted. For X-linked loci, a genotype for a female includes two alleles while a genotype for a male, since he is hemizygous, includes only one allele. For example, the genotype for an individual whose blood type is (phenotype)

O is designated as OO; the genotype for an individual whose blood type is (phenotype) A is either AA or AO.

Haplotype The linear, ordered arrangement of alleles on a chromosome. Haplotype analysis is useful in identifying recombination events.

Hemizygous In a diploid organism, a hemizygous individual has only one copy of a particular gene. For example, all cytogenetically normal male humans are hemizygous for all X and Y loci.

Heritability Can be defined either narrowly or broadly. In the narrow sense, heritability is defined as the proportion of the total phenotypic variance in a trait that is due to the additive effects of genes, as opposed to dominance or environmental effects. In the broad sense, heritability is the proportion of the total phenotypic variance of a trait that is due to all genetic effects, including additive and dominance effects.

Heterogeneity Different causes for the same disease phenotype. Four different types of heterogeneity exist. Clinical heterogeneity is present when a disorder expresses itself differently in different individuals. For example, Charcot–Marie–Tooth disease has one form that involves decreased nerve conduction velocities (NCVs) and another form that has normal NCVs. Genetic heterogeneity is present when a disorder has different types of inheritance patterns. Limb–girdle muscular dystrophy, for example, has both dominant and recessive forms. Locus heterogeneity occurs when different genes lead to the same clinical phenotype in different families. In familial spastic paraplegia, at least three different genes have been localized (including those on chromosomes 2, 14, and 15). Allelic heterogeneity occurs when different alleles (or mutations) at the same locus are present and result in the same phenotype. Cystic fibrosis is a classic example of allelic heterogeneity in which the most frequent mutation is ΔF508; however over 500 other unique mutations within the *CFTR* gene also have been identified.

Heterozygosity value A measure of the informativeness of a genetic marker. The heterozygosity measure defines the probability that a random individual has distinct alleles at a locus.

Heterozygous The alleles at a genetic locus are different from one another. An individual with blood type AB is heterozygous at the ABO blood group locus.

Homozygous The alleles at a genetic locus are identical. An individual who has blood type O (genotype *OO*) is homozygous for the *O* allele.

Identity by descent Two alleles are identical by descent (IBD) when it can be determined with certainty that they have been inherited from a common ancestor. For instance, a mother with blood type O and father with blood type AB have two children, each with blood type A. Since the genotypes of the children are *AO*, the children share one allele IBD, the *A'* allele. Whether the maternally inherited *O* allele is IBD in the children is unclear, since the mother is homozygous for the *O* allele. Alleles that are identical by descent are always identical by state.

Identity by state Two alleles are identical by state (IBS) when they share the same state. For example, two unrelated individuals each with blood group AB share two alleles IBS. Two unrelated individuals one with alleles 115 and 117, and the other with alleles 113 and 117 would share one allele IBS. Alleles that are IBS are not always identical by descent.

Imprinting The phenomenon in which the phenotype of the disease depends on which parent passed on the disease gene. For instance, both Prader–Willi syndrome and Angelman syndrome are inherited when the same part of chromosome 15 is missing. When the father's complement of 15 is missing, then the child has Prader–Willi, but when the mother's complement of 15 is missing, the child has Angelman syndrome.

Informed consent The process by which an individual willingly and voluntarily agrees to participate in an activity or research study after arriving at an understanding of the risks and benefits of participation versus nonparticipation. In a genetic study, potential participants should be appraised of the study goals, risks, benefits, alternatives to participation, disclosure policies, and financial and time commitments involved. The informed consent process should be documented, typically with a signed consent form approved by an institutional review board. Special considerations apply to vulnerable populations (e.g., minors, mentally handicapped individuals).

Interference The phenomenon whereby the presence of one crossover in a region affects the chances that another crossover will occur nearby. Positive interference is the same as interference and indicates that the probability of an additional crossover forming is inhibited by the first crossover event. Negative interference indicates that the probability of a second crossover is increased by the presence of the first crossover event.

Intron The noncoding regions of genes. The introns are spliced out of the messenger RNA following transcription.

IP address The Internet Protocol address is a 32-bit number unique across all computers in the world that completely identifies a single computer. It is aliased to a text-based host name for ease of use.

LAN (local area network) A network usually bounded in a single building.

Lambda (λ) The ratio of the recurrence risk of a disorder in a specified degree of relative to the prevalence of the disorder in the general population. In general, a high value for λ suggests that there is a genetic component to the disorder. The higher the λ value, the greater potential genetic contribution to the disorder. These values should be interpreted with caution, because λ values can also be due to common environmental factors or other nongenetic causes.

Linkage Two loci that are physically connected on the same chromosome at a distance that is measured at less than 50% recombination are said to be linked. Two traits are linked when they fail to be transmitted to offspring independently from one another. The more closely linked two loci are, the greater the chance that both loci will be transmitted together to offspring.

Linkage disequilibrium Linkage disequilibrium is often termed "allelic association." When alleles at two distinctive loci occur in gametes more frequently than expected given the known allele frequencies and recombination fraction between the two loci, the alleles are said to be in linkage disequilibrium. Evidence for linkage disequilibrium can be helpful in mapping disease genes because it suggests that the two may be very close to one another.

Locus (plural: loci) The physical location of a gene.

Location score The base *e* (natural) logarithm of the likelihood of the odds ratio for linkage. To convert a lod score to a location score, multiply the lod score by 4.6.

Lod score The base_{10} logarithm of the likelihood of the odds ratio for linkage. Traditionally, lod scores have been used to investigate genetic disorders where the mode of inheritance is well defined. Lod scores of 3.0 or more provide evidence in favor of linkage; lod scores of –2.0 or less provide evidence against linkage; and lod scores between –2.0 and 3.0 indicate a requirement for additional data before a definite conclusion can be reached. Lod scores are always reported in association with a value of the recombination fraction θ.

Map distances Maps of the human genome are generally of two types, genetic and physical. Genetic maps measure the amount of recombination between two loci, and genetic map units are quantified as either percentage recombination (θ) or centiMorgans between two loci. Physical maps aim to quantify the actual amount of DNA, usually in base pairs, between two loci. There is a rough correspondence between 1% recombination and 10^6 base pairs of DNA, but this can vary by orders of magnitude over short distances.

Meiotic drive A deviation from the expected 50:50 ratio of segregation of two chromosomes or alleles in an individual. Also called segregation distortion.

MLS score Defined originally by N. Risch (*Am J Hum Genet* 46:222–228, 1990), the MLS score is a model-free analogue of the lod score (i.e., the log 10 likelihood) dependent on the availability of a data set of a single type of concordant affected (or discordant) relative pairs. The MLS statistic increases, thereby suggesting evidence for linkage, when the proportion of relative pairs sharing more than 1 allele identical by descent is greater than that expected based on their degree of relationship. Like lod scores, MLS scores are additive over pedigrees and are interpreted using the traditional criteria for parametric linkage analysis. In contrast, MLS scores provide no estimate of the recombination fraction.

Model-free analysis Any type of linkage analysis that requires no user specification of the components of the underlying genetic model. The study of sib pair data in which both parents are available and genotyped (e.g., there is no inference of missing data, which inherently relies on allele frequency data) is an example of model-free analysis.

Multifactorial A trait is considered to be multifactorial in origin when two or more genes, together with an environmental effect, work together to lead to a phenotype.

Multiplicative effect The effects of alleles at multiple loci that together contribute to a phenotype are considered multiplicative when their combined effect is not equal to the sum of the individual contributions of each allele, but rather by multiplying their individual effects.

Multipoint location score With respect to the mapping of a trait, a multipoint lod score is the base *e* (natural) logarithm of the ratio of the likelihood of the pedigree data, assuming a known map order, and distances for the trait gene and marker loci to the likelihood of the pedigree data, assuming that the trait gene is unlinked to the established genetic map for the marker loci.

Multipoint lod score The base_{10} logarithm of the same ratio of likelihoods described in connection with multipoint location score. Note that there is some confusion in the literature about the distinction between the multipoint lod and location scores.

Mutation A change in the DNA. A mutation occurring in a germ cell is a heritable change in that it can be transmitted from generation to generation. Mutations may also occur in somatic cells and are not heritable in the traditional sense of the word, but are transmitted to all daughter cells.

Nonparametric analysis A type of linkage analysis that relies on some specification of components of the genetic model, but usually not penetrance or the degree of dominance. For instance, affected relative member methods of linkage analysis require the specification of allele frequencies and MLS sib pair models of inheritance require the specification of a λ value. This is contrasted with parametric analysis, where all components of the genetic model are specified.

Odds ratio The odds ratio (OR) may be calculated from either a prospective or retrospective (case–control) study. In a case–control study, the OR is usually interpreted as the ratio of the odds of exposure in cases versus the odds of exposure in controls. In a prospective study or a case–control study where the exposure is known to precede the disease (as in most genetic studies), it is interpreted as the ratio of odds of disease in exposed people versus odds of disease in unexposed people.

Oligogenic A trait is considered to be oligogenic when two or more genes work together to produce the phenotype. Oligogenicity, which implies that "*few*" genes are involved, should be contrasted with polygenicity, which implies that "*many*" genes are involved in phenotype expression.

Parametric A statistical test is considered to be parametric when it requires the specification of an underlying model. For instance, the traditional lod score approach to linkage analysis is a parametric test because it requires the specification of all components of the genetic model.

Penetrance The probability of expressing a phenotype given a genotype. Penetrance is described as either "complete" or "incomplete." For example, individuals who carry the gene for tuberous sclerosis have an 80% chance of expressing the disorder. Penetrance may also be dependent on a susceptible individual's current age. For example, 20% of all gene carriers for myotonic dystrophy express the gene to some degree by age 15, while 80% of all gene carriers express it by age 60.

Phenocopy A trait that appears to be identical to a genetic trait but is caused by nongenetic factors.

Phenotype The outward appearance of an individual with a disorder. The phenotype of a trait may be expressed physically, biochemically, or physiologically.

PIC value The polymorphism information content (PIC) is similar to the heterozygosity value except that it excludes the probability that a set of parents who are identically heterozygous at a marker locus have a heterozygous offspring. The PIC value is always less than or equal to the heterozygosity value.

Pleiotropy One gene loading to many different phenotypic expressions. An excellent example of a gene with pleiotropic effects is the gene for myotonic dystrophy. Affected individuals can have one or more of a range of signs and symptoms including characteristic Christmas-tree-like cataracts, myotonia, narcolepsy, testicular atrophy, frontal balding, mental retardation, and cardiac abnormalities.

Polygenic A trait is considered to be polygenic in origin when it is caused by the combined effects of three or more loci.

Polymorphism A piece of DNA that has more than one form (allele), each of which occurs with at least 1% frequency is said to be polymorphic ("poly," many; "morph," forms). Polymorphisms are a normal part of genetic variability. Polymorphisms of the same gene may or may not have different functions.

Power The power of a study is the probability of correctly rejecting the null hypothesis. In the case of linkage analysis, the power is the probability of correctly identifying a true linkage.

Proband The individual in a pedigree who causes the pedigree to come to the attention of medical or research personnel.

***P* value** The probability that an observation occurred by chance alone. See also Type I error.

Recombination The process during meiosis (cell division resulting in egg or sperm) by which homologous chromosomes exchange material.

Recombination fraction (θ) The frequency of crossing over between two loci. Estimates of the recombination fraction between two loci of less than 0.50 are consistent with the loci being linked to one another; estimates of the recombination fraction equal to 0.50 are consistent with the loci being unlinked to one another. In theory, the recombination fraction between two loci should not be greater than 0.50;

however, estimates exceeding 0.50 may suggest a highly recombinogenic area or data error. When $\theta = 0$, then the two loci are at the same location.

Relational database Composed of tables, the relational database associates records from different tables through a common field. In contrast, a nonrelational database has records associated through devices such as sets.

Relative risk A relative risk (RR) quantifies how many times more or less likely the disease is in "exposed" people compared to "unexposed" people. Traditionally, exposure has been considered in terms of environmental agents; but in genetic studies, exposure can represent the underlying genotype or allele. A null value of 1.0 indicates that the disease is equally likely in exposed and unexposed people; a value greater than 1.0 indicates that the disease is more likely in the exposed people; and a value less than 1.0 suggests that the disease is more likely in the unexposed people. The relative risk is calculated from prospective data only.

Schema A schema is a description of a database that defines the flow of information (e.g., how each table relates to other tables in the database). It also specifics the data structure, primary keys, candidate keys, foreign keys, and integrity constraints in each table. The schema describes the structure of the relational database.

Segregation distortion See Meiotic drive.

Sporadic case An individual who is the only member within a pedigree possessing the trait of interest.

Synteny Two genes that occur on the same chromosome are syntenic; however, syntenic genes may or may not be "linked."

TCP/IP TCP is a common data transmission protocol that runs on computer systems of all types. TCP/IP runs over Ethernet connections (among others) and allows the transfer of data often via the Internet between computers using different operating systems.

Telnet A protocol that allows for connecting a remote computer to a server.

Theta (θ) See Recombination fraction.

Type I error (α) The incorrect rejection of the null hypothesis. In linkage analysis, type I errors are expressed in terms of the probability of declaring evidence in favor of linkage when in truth there is no linkage. A type I error is a false positive result.

Type II error (β) The incorrect failure to reject the null hypothesis. In linkage analysis, type II errors are expressed in terms of the probability of failing to declare linkage when linkage is present. A type II error is a false negative result.

Appendix

Useful Web Sites for Genetic Analysis

(For additional Web sites of chromosome-specific databases, see text, Table 16.1.)

Alliance of Genetic Support Groups
http://medhlp.netusa.net/www/agsg/agsgscp.htm

American Society of Human Genetics
http://www.faseb.org/genetics/ashg

Center for the Study of Inherited and Neurological Disorders (Duke University Medical Center)
http://www2.mc.duke.edu/depts/medicine/medgen/index.html

Centre D'Étude du Polymorphisme Humain (CEPH)
http://www.cephb.fr

Cooperative Human Linkage Center (CHLC)
http://www.chlc.org/

Coriell Institute of Medical Research
http://arginine.umdnj.edu/

Eccles Institute of Human Genetics (Utah)
http://www.genetics.utah.edu/home.html

Ethical, Legal, and Social Implications of Biotechnology
http://www.ncgr.org/gpi/

Généthon
http://www.genethon.fr/genethon_en.html

GenLink (Washington University at St. Louis)
http://www.genlink.wustl.edu

Genome Database
http://gdbwww.gdb.org/

Genome Sequence Data Base
http://www.ncgr.org/gsdb

Helix: A Directory of Medical Genetics Laboratories*
http://www.hslib.washington.edu/helix

Human Gene Mutation Database
http://www.uwcm.ac.uk/uwcm/mg/hgmd0.html

Human Genetic Analysis Resource (HGAR) at Case Western University Medical Center (site for SAGE information)
http://darwin.mhmc.cwru.edu

Human Genome Project: From Maps to Medicine
http://www.nhgri.nih.gov/Policy_and_public_affairs /Communications/ Publications/

HUM-MOLGEN (Largest Internet mailing list in human molecular genetics)
http://www.informatik.uni-rostock.de /HUM-MOLGEN/

Information for Genetics Professionals
http://www.kumc.edu/GEC/prof/geneprof.html

Jackson Laboratory/Mouse Genome Informatics
http://www.informatics.jax.org/

Linkage Analysis at Rockefeller University (Jurg Ott)
http://linkage.rockefeller.edu

LDB (the Genetic Location Database)
http://cedar.genetics.soton.ac.uk/public_html

*Helix is password-protected; information on access is limited to medical professionals.

Marshfield Medical Research Foundation/Center for Medical Genetics
http://www.marshmed.org/genetics/

National Center for Biotechnology Information (GenBank, Medline, OMIM)
http://www.ncbi.nlm.nih.gov

National Human Genome Research Institute
http://www.nhgri.nih.gov/home.html

Online Mendelian Inheritance in Man (OMIM)
http://www3.ncbi.nlm.nih.gov/omim/

Science Human Gene Maps 1996: Human Transcript Map
http://www.ncbi.nlm.nih.gov/SCIENCE96

Stanford Human Genome Center
http://shgc-www.stanford.edu

Whitehead Institute for Biomedical Research
http://www-genome.wi.mit.edu

Index